Progress in Mathematics
Volume 256

Series Editors
Hyman Bass
Joseph Oesterlé
Alan Weinstein

David Borthwick

Spectral Theory of Infinite-Area Hyperbolic Surfaces

Birkhäuser
Boston • Basel • Berlin

David Borthwick
Department of Mathematics and Computer Science
Emory University
Atlanta, GA 30322
U.S.A.
davidb@mathcs.emory.edu

Mathematics Subject Classification (2000): 58J50, 47A40, 11F72, 30F35

Library of Congress Control Number: 2007932363

ISBN-13: 978-0-8176-4524-3   e-ISBN-13: 978-0-8176-4653-0

Printed on acid-free paper.

©2007 Birkhäuser Boston

9 8 7 6 5 4 3 2 1

www.birkhauser.com   (TXQ/SB)

For Sarah, Julia, and Benjamin

Preface

I first encountered the spectral theory of hyperbolic surfaces as an undergraduate physics student, through the intriguing expository article of Balazs–Voros [11] on relations between the Selberg theory of automorphic forms and quantum chaos. At the time I was quite impressed at the range of topics represented, including quantum physics, discrete groups, differential geometry, number theory, complex analysis, and spectral theory. In my previous experience these were completely separate realms, but here they were all mixed together in the same setting.

Twenty years later, these topics do not seem so far apart to me. However, I am no less amazed by the rich cross-fertilization of ideas in this subject area. The primary motivation for this book is the conviction that this sort of mathematics that bridges the divides between fields ought to be made accessible to as broad an audience as possible—to graduate students especially, for whom regular coursework often exaggerates the impression of boundaries between disciplines.

The spectral theory of compact and finite-area Riemann surfaces is a classical subject with a history going back to the pioneering work of Atle Selberg, who brought techniques from spectral theory and harmonic analysis into the study of automorphic forms. These cases have been thoroughly covered in various expository sources. In particular, Buser [37] develops the spectral theory for compact Riemann surfaces with a concrete approach based on hyperbolic geometry and cutting and pasting. Most treatments of the finite-area case, for example Venkov [210], emphasize arithmetic surfaces and connections to number theory.

For infinite-area hyperbolic surfaces, a good understanding of the spectral theory has emerged only recently. The assumption of infinite area changes the character of the theory. The resolvent of the Laplacian takes on a predominant role and the emphasis shifts from discrete eigenvalues to scattering theory and resonances. It has only been through dramatic advances in geometric scattering theory that the full development of the infinite-area theory has become possible.

My goal in this book is to present a relatively self-contained account of this recent development. Although many of the results could be stated in greater generality (e.g., higher dimensions), the book is restricted to the hyperbolic surface context for the sake of accessibility. The notes at the end of each chapter include references to more-general results.

The book assumes basic algebra and topology, at the level of a first graduate course. An undergraduate course on curves and surfaces should provide sufficient background in differential geometry. Because spectral theory is the primary topic, the analysis requirements are necessarily somewhat steeper. Beyond the basic real and complex analysis, a student would need basic functional analysis and some introduction to the analysis of linear partial differential equations. The appendix, while not a self-sufficient introduction to these topics, is meant to serve as a guide to readers who need more background information.

I would like to thank my collaborators, Chris Judge and Peter Perry, with whom I learned much of this material, and Edward Taylor, who introduced me to scattering theory on hyperbolic manifolds. Thanks also to Arthur Wightman, who supervised the undergraduate project in which I first learned about the Selberg trace formula, and to Richard Melrose and Rafe Mazzeo, for encouragement when I first undertook to learn some scattering theory. I am very grateful to Colin Guillarmou and Peter Hislop for reading parts of the manuscript and offering corrections and suggestions.

David Borthwick
Atlanta
February 2007

Contents

1

Introduction

A hyperbolic surface X is a surface with geometry modeled on the hyperbolic plane, and spectral theory in this context refers generally to the Laplacian operator Δ_X induced by the hyperbolic structure. Selberg [188] pioneered the study of spectral theory of hyperbolic surfaces in the 1950s, drawing inspiration from earlier work of Maass [120]. Motivated by analogies to the classical zeta and theta functions of number theory, Selberg applied tools and ideas from spectral theory and harmonic analysis to the study of automorphic forms associated to Fuchsian groups. This led in particular to beautiful formulas connecting the geometry of compact hyperbolic surfaces to the spectral theory.

The Selberg theory has since been developed and extended by many others; see, for example, Buser [37], Venkov [210], and Sarnak [186, 187] for background and references. For noncompact but finite-area surfaces, the theory was first interpreted in terms of stationary scattering theory by Faddeev [58] in 1967, a viewpoint developed further by Lax–Phillips [117].

The spectral theory of hyperbolic surfaces of infinite area was first taken up in the early 1970s by Elstrodt [55], Patterson [155], and Fay [60]. Fundamental results about the spectral theory of infinite-volume hyperbolic manifolds were proven in the early 1980s by Lax–Phillips [113, 114]. In the 1990s great progress was made in developing tools for counting resonances, particularly in the work of Melrose, Sjöstrand, Vodev, and Zworski; see [135], [195], [224], [226], and [228] for surveys of these results. The new techniques were applied with spectacular success to the spectral theory of infinite-area hyperbolic surfaces by Guillopé and Zworski [83, 86, 87, 88, 225].

In studying infinite-area hyperbolic surfaces, we will always restrict our attention to geometrically finite surfaces (which is equivalent to requiring finite Euler characteristic). The possible geometries "at infinity" for geometrically finite hyperbolic surfaces are easily categorized—there are only two types of ends, called funnels and cusps. Moreover, we have very simple explicit models for these two geometries (Chapter 2), and these models will prove to be essential tools in our understanding of the spectral theory. (For geometrically infinite surfaces the notion of geometry "at infinity" is ill-defined, and there is virtually nothing we can say about the spectral theory of the Laplacian.)

For any noncompact geometrically finite hyperbolic surface X, the essential spectrum of Δ_X is $[1/4, \infty)$, and this spectrum is absolutely continuous. The discrete spectrum consists of finitely many eigenvalues in the range $(0, 1/4)$. In the finite-area case one may also have embedded eigenvalues in the continuous spectrum, but these do not occur for infinite-area surfaces. These facts were proven by Lax–Phillips [114]. Our development will take a somewhat different route, following Melrose's approach to geometric scattering theory [135]. In particular, we'll first study the meromorphic extension of the resolvent $R_X(s) := (\Delta_X - s(1 - s))^{-1}$ to $s \in \mathbb{C}$ and develop a detailed understanding of the structure of its kernel (Chapter 6). Then we'll apply this knowledge to derive the basic spectral results (Chapter 7).

A pole of the meromorphically continued resolvent $R_X(s)$ is called a resonance. Resonances have a long history in physics, where they describe long-lived but unstable states. For example, the complex frequency of a damped harmonic oscillator is a resonance. In quantum-mechanical scattering theory, resonances arise from states that are unstable due to quantum tunneling effects (see, e.g., Hislop–Sigal [93] for background). In a geometric context, resonances for a noncompact manifold can take on the role played by the discrete spectrum of the Laplacian on a compact manifold, yielding a natural set of physically significant geometric invariants. Resonances also appear in analytic number theory, as zeros of Dirichlet series. For example, the resonances of the Laplacian on the modular surface are the nontrivial zeros of the Riemann zeta function, a result of Selberg [189].

In conjunction with our development of the spectral theory of an infinite-area hyperbolic surface X, we will also present the theory of the associated Selberg zeta function $Z_X(s)$. This can be defined for $\text{Re}\, s > 1$ as a product over the length spectrum of X, another natural set of geometric invariants. We will show that $Z_X(s)$ extends to a meromorphic function, a result of Guillopé [83] at this level of generality. The divisor of $Z_X(s)$ consists of zeros corresponding to resonances plus a sequence of "topological" zeros and poles at negative half-integers (Chapter 10). This description is known from the Selberg trace formula for finite-area surfaces. The structure of the zeta function was established by Patterson–Perry [160] for infinite-volume hyperbolic manifolds without cusps and extended by Borthwick–Judge–Perry [24] to infinite-area hyperbolic surfaces with cusps. The Selberg zeta function links the two very different classes of invariant, the "quantum" resonances and the "classical" length spectrum.

A related connection is given by the Poisson formula, which expresses a regularization of the wave trace as a sum over the resonance set (Chapter 11). This was established by Guillopé–Zworski [87] for hyperbolic surfaces, with arbitrary metric perturbations allowed inside a compact set. Their derivation of the Poisson formula predates the analysis of the zeta function for infinite-area surfaces. We present the zeta function first because for (exactly) hyperbolic surfaces the Poisson formula can be most easily derived from a factorization formula for the zeta function. By this method, Perry extended the Poisson formula to higher-dimensional hyperbolic manifolds without cusps in [169].

Another major topic of the book will be the distribution of resonances. The resonance set $\mathcal{R}_X$ is defined as the set of poles of $R_X(s)$, repeated according to a multiplicity given by the rank of the operator-valued residue at the pole (Chapter 8). In a few "elementary" cases, namely when X is the hyperbolic plane or diffeomorphic to a cylinder, the resonances can be computed exactly.

For a nonelementary, geometrically finite hyperbolic surface of infinite area, the distribution of resonances is pictured in Figure 1.1. There are at most finitely many resonances in the interval $(\frac{1}{2}, 1)$ coming from the discrete spectrum, and infinitely many in the half-plane $\{\mathrm{Re}\, s < \frac{1}{2}\}$. The critical line $\{\mathrm{Re}\, s = \frac{1}{2}\}$ corresponds to the continuous spectrum.

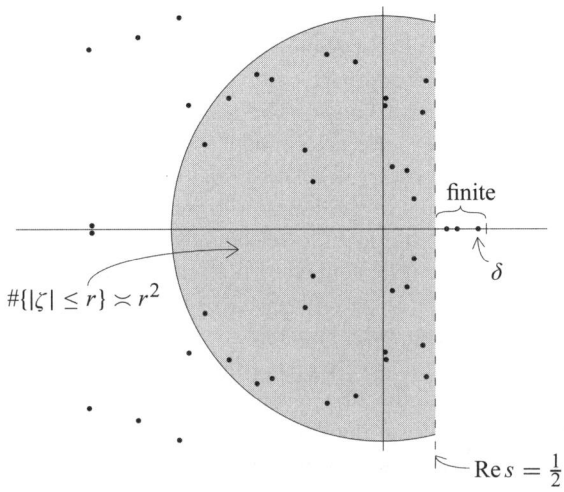

Fig. 1.1. Distribution of resonances.

Here is a summary of the results we will present on the distribution of resonances, when X is a nonelementary geometrically finite hyperbolic surface of infinite area:

1. The resonance counting function satisfies global upper and lower polynomial bounds,

$$\#\{\zeta \in \mathcal{R}_X : |\zeta| \leq r\} \asymp r^2 \qquad (1.1)$$

   (Theorems 9.1 and 12.1).
2. The number of resonances in strips near the continuous spectrum is bounded below. For any $\varepsilon \in (0, \frac{1}{2})$,

$$\#\left\{\zeta \in \mathcal{R}_X : |\zeta| \leq r, \ \mathrm{Re}\, \zeta > \tfrac{1}{2} - \varepsilon^{-1}\right\} \neq O(r^{1-\varepsilon}) \qquad (1.2)$$

   (Theorem 12.3).

3. The first resonance (furthest to the right) is a simple pole at $s = \delta$, where $\delta \in (0, 1)$ is the Hausdorff dimension of the limit set of the Fuchsian group Γ for which $X \cong \Gamma \backslash \mathbb{H}$ (Theorem 14.15). In particular, $\delta < \frac{1}{2}$ if and only if the discrete spectrum is empty.
4. If X has no cusps, then the counting function satisfies a more precise upper bound in vertical strips near the continuous spectrum. For any $M > 0$,

$$\#\left\{ \zeta \in \mathcal{R}_X : |\zeta| \leq r, \ \mathrm{Re}\, \zeta > -M \right\} = O(r^{1+\delta}) \qquad (1.3)$$

(Theorem 15.10).

All of these results could be stated equivalently as theorems on the distribution of zeros of the Selberg zeta function.

The quadratic upper and lower bounds (1.1) are consistent with the Weyl asymptotics of eigenvalues in the compact case, but no such exact asymptotic for the resonance counting function is known. The upper bound was proven by Guillopé–Zworski [86] using the Fredholm determinant method introduced in the context of obstacle scattering by Melrose [133]. The global lower bound was proven in Guillopé–Zworski [87] by means of the Poisson formula for resonances. This result is particularly striking—optimal global lower bounds are known in very few cases for dimension greater than one. The context of both the global upper and lower bounds is the more general setting of hyperbolic surfaces with arbitrary metric perturbations inside some compact set. The lower bound in strips (1.2), derived from the wave trace and Poisson formula by Guillopé–Zworski [88], applies only to hyperbolic surfaces.

The characterization of the first resonance is due to Patterson [156, 158]. It involves the Patterson–Sullivan construction of a special measure on the limit set (Chapter 14). A related result, analogous to the prime number theorem, is the asymptotic behavior of the counting function for the primitive length spectrum $\mathcal{L}_X$,

$$\#\{\ell \in \mathcal{L}_X : \ell \leq t\} \sim \frac{e^{\delta t}}{\delta t},$$

(Theorem 14.20). This was first proven independently by Guillopé [82] and Lalley [111]. Naud [143] extended Patterson's result to show that no other resonances occur in $\mathrm{Re}\, s > \delta - \varepsilon$ for some $\varepsilon > 0$.

The upper bound in vertical strips (1.3), due to Guillopé–Lin–Zworski [84], extends to Schottky hyperbolic manifolds in higher dimensions, and also draws on the Patterson–Sullivan theory. This bound fits neatly with the expectation, originating in the work of Sjöstrand [194] on geometric upper bounds for resonances, that for chaotic systems the number of resonances near the continuous spectrum should satisfy a power law with exponent given by half of the dimension of the trapped set of geodesics. The Patterson–Sullivan theory shows that the dimension of the trapped set is $2 + 2\delta$ for a hyperbolic manifold.

One last topic that we will address is "inverse" spectral theory, which refers to the problem of determining properties of the surface from the data provided by the resonance set. Using the Selberg zeta function we can deduce that the resonance set and length spectrum determine each other (along with the Euler characteristic and number of cusps), up to finitely many possibilities (Theorem 13.1). This is the analogue of Huber's theorem from the compact case and was proved by Borthwick–Judge–Perry [23, 24]. We can exploit this connection to prove that the resonance set determines an infinite-area hyperbolic surface up to finitely possibilities (Theorem 13.10), extending theorems of McKean [130], in the compact case, and Müller [139], in the finite-area case.

2

Hyperbolic Surfaces

For the purposes of this book, a *surface* is a connected, orientable two-dimensional smooth manifold, without boundary unless otherwise specified. Throughout the book we will restrict our attention to surfaces that are *topologically finite*, meaning that the surface is homeomorphic to a compact surface with finitely many points excised. An *end* of the surface is an equivalence class of neighborhoods which are contractible to one of these excised points. Topologically finite surfaces are classified up to diffeomorphism by the genus g and the number of ends n (see, e.g., Figure 2.1). The corresponding value of the Euler characteristic is $\chi = 2 - 2g - n$.

Definition. A *hyperbolic surface* is a smooth surface equipped with a complete Riemannian metric of constant Gaussian curvature -1.

For $\chi \geq 0$ there are only a few special cases of hyperbolic surfaces (the plane and cylinders), but any surface with $\chi < 0$ admits a family of hyperbolic metrics. After a brief introduction to plane hyperbolic geometry, the main point of this chapter will be a classification of hyperbolic surfaces. For the later analysis, we are particularly interested in the structure of the ends.

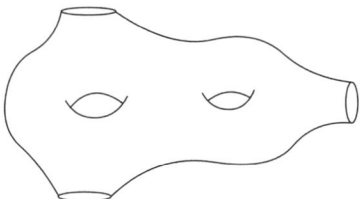

Fig. 2.1. A surface of genus two with three ends.

2.1 The hyperbolic plane

Up to isometry, there is a unique simply connected hyperbolic surface, called the hyperbolic plane, for which there are several standard models. The model we will use most frequently is the upper half-plane,

$$\mathbb{H} := \{z = x + iy \in \mathbb{C} : y > 0\}, \qquad ds^2 = \frac{dx^2 + dy^2}{y^2}. \tag{2.1}$$

The other standard alternative is the unit disk model (or Poincaré disk),

$$\mathbb{B} := \{z \in \mathbb{C} : |z| < 1\}, \qquad ds^2 = 4\frac{dx^2 + dy^2}{(1 - |z|^2)^2}. \tag{2.2}$$

Most calculations are simpler in $\mathbb{H}$, but $\mathbb{B}$ has the advantage that the boundary is treated uniformly.

In either model, the *Möbius transformations* provide a natural set of orientation-preserving maps. Given the matrix,

$$T = \begin{pmatrix} a & b \\ c & d \end{pmatrix}, \tag{2.3}$$

the corresponding Möbius transformation is

$$z \mapsto Tz := \frac{az + b}{cz + d}.$$

Note that T is invertible as a map if and only if $\det T \neq 0$ as a matrix. And rescaling $T \to \lambda T$ does not change the action. Hence Möbius transformations are naturally identified with the matrix group,

$$\mathrm{PSL}(2, \mathbb{C}) := \mathrm{SL}(2, \mathbb{C})/\{\pm I\}.$$

A map $T \in \mathrm{PSL}(2, \mathbb{C})$ will preserve $\mathbb{H}$ if and only if its coefficients are real, so the group of Möbius automorphisms of $\mathbb{H}$ is $\mathrm{PSL}(2, \mathbb{R})$.

Proposition 2.1. *The group of orientation-preserving isometries of $\mathbb{H}$ is the group* $\mathrm{PSL}(2, \mathbb{R})$ *of Möbius transformations preserving the upper half-plane.*

Proof. Because the hyperbolic metric is conformally related to the Euclidean metric, an isometry $\mathbb{H} \to \mathbb{H}$ preserves Euclidean angles in particular and so must be a conformal automorphism of the upper half-plane. The Schwarz lemma implies that the only such automorphisms are Möbius transformations. Thus isometries must be Möbius.

To see the converse, note that in complex coordinates the hyperbolic metric can be written

$$ds^2 = \frac{|dz|^2}{(\operatorname{Im} z)^2}.$$

Suppose that $T \in \mathrm{PSL}(2, \mathbb{R})$ is represented as in (2.3), with $\det T = 1$. We simply compute,

$$T'(z) = \frac{1}{(cz + d)^2}, \qquad \mathrm{Im}\, Tz = \frac{\mathrm{Im}\, z}{|cz + d|^2}, \tag{2.4}$$

where T' denotes the complex derivative. (In the notation we distinguish between the action $z \to Tz$ and the function $T'(z)$.) Using these to compute the pullback of the metric gives

$$T^*(ds^2) = \frac{|T'(z)\, dz|^2}{(\mathrm{Im}\, Tz)^2} = ds^2,$$

which shows that T is an isometry.   □

Any Möbius transformation from the upper half-plane onto the unit disk, for example

$$z \mapsto \frac{z - i}{z + i}, \tag{2.5}$$

will give an isometry $\mathbb{H} \to \mathbb{B}$. From this we can immediately deduce that the (orientation-preserving) isometry group of $\mathbb{B}$ is the group of Möbius transformations preserving the unit disk. This is identified with the matrix group $\mathrm{PSU}(1, 1)$.

We will make frequent use of the topology of the unit sphere metric on the Riemann sphere $\mathbb{C} \cup \{\infty\}$. For $z \in \mathbb{C}$, $w \in \mathbb{C} \cup \{\infty\}$, the unit sphere distance function is given by

$$d_\infty(z, w) := \begin{cases} \dfrac{2|z - w|}{\sqrt{(1 + |z|^2)(1 + |w|^2)}} & w \in \mathbb{C}, \\[3mm] \dfrac{2}{\sqrt{1 + |z|^2}} & w = \infty. \end{cases}$$

For example, we define the boundary of $\mathbb{H}$ with respect to this topology as the one-point compactification of the real line,

$$\partial \mathbb{H} := \mathbb{R} \cup \{\infty\}.$$

For the $\mathbb{B}$ model the Riemann-sphere topology is equivalent to the Euclidean topology, and we simply have $\partial \mathbb{B} := S^1$.

As usual when considering Möbius transformations, it is convenient to define a *circle* in $\mathbb{C}$ in the generalized sense of a circle with respect to d_∞. Any Euclidean circle or straight line in $\mathbb{C}$ is a circle in this sense.

The large isometry group makes it easy to determine the geodesics of the hyperbolic plane, which turn out to be circles of a certain type.

Proposition 2.2. *The geodesics of $\mathbb{H}$ are precisely the arcs of circles intersecting $\partial \mathbb{H}$ orthogonally. Similarly the geodesics of $\mathbb{B}$ are circles intersecting $\partial \mathbb{B}$ orthogonally.*

Proof. First, we claim that the positive y-axis is a geodesic. Let $\eta : [t_1, t_2] \to \mathbb{H}$ be some curve connecting ia to ib, where $a < b$. The hyperbolic length of the curve is given by integrating ds along η, so if we write $\eta(t) = x(t) + iy(t)$, then

$$\ell(\eta) = \int_{t_1}^{t_2} \frac{\sqrt{x'(t)^2 + y'(t)^2}}{y(t)} \, dt$$

$$\geq \int_{t_1}^{t_2} \frac{|y'(t)|}{y(t)} \, dt$$

$$\geq \int_{t_1}^{t_2} (\log y(t))' \, dt$$

$$= \log(b/a).$$

It's clear from this calculation that the minimum is achieved if and only if $y'(t) > 0$ and $x'(t) = 0$ (which implies $x(t) = 0$). Thus the y-axis is a path of shortest distance and hence a geodesic.

Now suppose that $\gamma : \mathbb{R} \to \mathbb{H}$ is an arbitrary geodesic. By a Möbius transformation R we can send $\gamma(0)$ to i and rotate $\gamma'(0)$ to i also. By uniqueness of the geodesic with given starting position and velocity, this implies that $R \circ \gamma$ parametrizes the y-axis. The characterization of γ follows easily because $\mathrm{PSL}(2, \mathbb{R})$ preserves circles as well as angles and fixes $\partial \mathbb{H}$.

Conversely, any arc of a generalized circle intersecting $\partial \mathbb{H}$ orthogonally could be mapped to the y-axis by an isometry and is therefore a geodesic. The same reasoning applies to $\mathbb{B}$.   □

From Proposition 2.2 it follows that there is a unique geodesic arc connecting any two distinct points $z, w \in \mathbb{H} \cup \partial \mathbb{H}$. We will denote this segment by $[z, w]$. When $z, w \in \mathbb{H}$, the hyperbolic distance is defined by

$$d(z, w) := \ell([z, w]).$$

Proposition 2.3. *For $z, z' \in \mathbb{H}$ the hyperbolic distance is given by*

$$\cosh d(z, z') = 1 + \frac{|z - z'|^2}{2yy'}. \tag{2.6}$$

Proof. A simple exercise using the formula for $T'(z)$ from (2.4) shows that

$$|Tz - Tw|^2 = |T'(z)| \, |T'(w)| \, |z - w|^2.$$

The second identity in (2.4) then makes it obvious that the right-hand side of (2.6) is invariant under isometries. Since the distance function is invariant by definition, it suffices to check the formula for two general points on the y-axis. The computation in the proof of Proposition 2.2 shows that $d(ia, ib) = \log(b/a)$, and the formula (2.6) is then easily verified.   □

Elements of $\mathrm{PSL}(2, \mathbb{R})$ can be classified by their fixed points. The solutions of $z = Tz$ are roots of the polynomial $cz^2 + (d - a)z - b$, whose discriminant is $(d - a)^2 + 4bc = (\mathrm{tr}\, T)^2 - 4$. The sign of the discriminant determines how the fixed points are situated within $\mathbb{H}$.

Definition. A transformation $T \in \mathrm{PSL}(2, \mathbb{R})$ is

1. *elliptic* if tr $T < 2$, implying one fixed point within $\mathbb{H}$ (with a matching point in the lower half-plane);
2. *parabolic* if tr $T = 2$ (and $T \neq I$), with a single degenerate fixed point in $\partial \mathbb{H}$;
3. *hyperbolic* if tr $T > 2$, yielding two distinct fixed points in $\partial \mathbb{H}$, one attracting and one repelling.

(The double usage of the term "hyperbolic" here is standard but potentially confusing; note that all three types of transformations could reasonably be called "hyperbolic isometries.") Figure 2.2 shows the fixed points and circles preserved by each type of isometry. Since traces are preserved under conjugation, the same classification by traces applies in $\mathrm{PSU}(1, 1)$ as well.

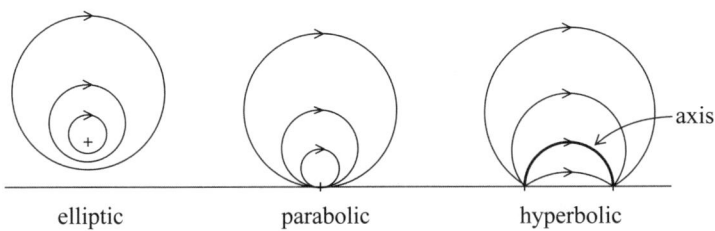

elliptic   parabolic   hyperbolic

Fig. 2.2. Isometries of $\mathbb{H}$.

Consider an elliptic transformation T, with fixed point $z_0 \in \mathbb{H}$. Let Q be a Möbius transformation mapping $\mathbb{H}$ onto $\mathbb{B}$ such that $Q(z_0) = 0$. Then QTQ^{-1} fixes the origin and so must be a rotation of the form $z \mapsto e^{i\theta}z$, by the Schwarz lemma. Hence the conjugacy class of an elliptic transformation is determined by the rotation angle.

A parabolic transformation can be conjugated to a map whose fixed point is ∞. The only such maps are horizontal translations $z \mapsto z + b$ for $b \in \mathbb{R}$. A further conjugation by the dilation $R : z \mapsto |b|^{-1}z$ reduces this translation to $z \mapsto z \pm 1$. Thus, within $\mathrm{PSL}(2, \mathbb{R})$ there are two conjugacy classes of parabolic transformations, corresponding to left or right translations.

The standard form for a hyperbolic transformation is given by conjugating the repelling fixed point to 0 and the attracting fixed point to ∞. The resulting map must be a dilation $z \mapsto e^\ell z$ with $\ell > 0$. The conjugacy classes of hyperbolic elements are indexed by the positive number $\ell = \ell(T)$, called the *displacement length* of T.

There is a unique geodesic $\alpha(T)$ connecting the fixed points of a hyperbolic transformation, which is called the *axis* of T, as shown in Figure 2.2. By conjugating to the standard form as above, we see immediately that the displacement length $\ell(T)$ is the distance by which points on $\alpha(T)$ are translated. Since conjugation preserves traces, we have the formula

$$\mathrm{tr}(T) = 2 \cosh \ell(T)/2.$$

By applying (2.6) to give a simple expression for $\cosh d(z, e^\ell z)$, we can easily see that the displacement length is also given by

$$\ell(T) = \min_{z \in \mathbb{H}} d(z, Tz), \tag{2.7}$$

with the minimum achieved if and only if z lies on $\alpha(T)$.

Other geometric features of $\mathbb{H}$ that will be important to us are the area form,

$$dA(z) = \frac{dx\, dy}{y^2}, \tag{2.8}$$

and the formula for the Laplacian. The (positive) Laplacian on a Riemannian manifold is defined globally by $\Delta = -\operatorname{div}\operatorname{grad}$. In local coordinates x^i, with the metric given by $ds^2 = g_{ij} dx^i dx^j$, this translates to

$$\Delta = -\frac{1}{\sqrt{\det g}} \partial_i \left(g^{ij} \sqrt{\det g}\, \partial_j \right),$$

where g^{ij} denotes the components of the inverse matrix to g_{ij}. For the hyperbolic metric on $\mathbb{H}$, the resulting operator is

$$\Delta_{\mathbb{H}} = -y^2 (\partial_x^2 + \partial_y^2).$$

In addition to the coordinates of the $\mathbb{H}$ and $\mathbb{B}$ models, we will make frequent use of *geodesic coordinates*. These are coordinates (r, t) for which the r-coordinate curves are unit-speed geodesics and the t-coordinate curves are orthogonal to them. This implies a metric of the form

$$ds^2 = dr^2 + \varphi^2 dt^2,$$

for some function $\varphi(r, t)$. In any such coordinate system, the Gaussian curvature is given by the simple formula $K = -(\partial_r^2 \varphi)/\varphi$. For a hyperbolic metric written in geodesic coordinates we must therefore have

$$\partial_r^2 \varphi = \varphi. \tag{2.9}$$

Geodesic *polar* coordinates $(r, \theta) \in \mathbb{R}_+ \times S^1$ are defined through pullback by the exponential map at some point p. In this case $r = 0$ is a coordinate singularity corresponding to the center point p, and φ must satisfy $\varphi \sim r$ as $r \to 0$. Together with (2.9), this implies $\varphi(r, \theta) = \sinh r$. Hence the metric in geodesic polar coordinates for a hyperbolic surface has a unique local form,

$$ds^2 = dr^2 + \sinh^2 r\, d\theta^2. \tag{2.10}$$

Two other obvious solutions of (2.9) will be important for us as well: $\varphi = \cosh r$ and $\varphi = e^{-r}$ are the model metrics for funnel and cusp ends, respectively.

We let $B(w; r)$ denote an open neighborhood with respect to the hyperbolic metric: for $w \in \mathbb{H}$ and $r > 0$,

$$B(w; r) := \{z : d(z, w) < r\}.$$

Using the geodesic polar coordinates centered at w, in which $dA = \sinh r \, dr \, d\theta$, we can see that

$$\text{area}(B(w; r)) = 2\pi \int_0^r \sinh r \, dr = 2\pi (\cosh r - 1). \tag{2.11}$$

2.2 Fuchsian groups

Given the large isometry group of $\mathbb{H}$, a natural way to obtain a hyperbolic surface is as a quotient $\Gamma \backslash \mathbb{H}$, for some subgroup $\Gamma \subset \text{PSL}(2, \mathbb{R})$. Points in the quotient correspond to orbits of Γ, and there is a natural projection $\pi : \mathbb{H} \to \Gamma \backslash \mathbb{H}$ given by $\pi(z) = \Gamma z$. For the quotient to be well-defined as a metric space, the action needs to be *properly discontinuous*, which means that the orbits are locally finite (any compact subset of $\mathbb{H}$ contains only finitely many orbit points). Conveniently, we can characterize the groups that act properly discontinuously on $\mathbb{H}$ by their topology as subsets of $\text{PSL}(2, \mathbb{R})$. On $\text{PSL}(2, \mathbb{R})$ we use the standard matrix topology defined by the norm $\|T\| = (\text{tr } T^* T)^{1/2}$.

Definition. A *Fuchsian group* is a discrete subgroup of $\text{PSL}(2, \mathbb{R})$.

One easy way to obtain examples of Fuchsian groups is to choose an even number of Euclidean disks centered on the real axis, with mutually disjoint closures. Divide the disks up into pairs, and for each pair choose a hyperbolic transformation mapping the exterior of one disk to the interior of the other. These transformations generate a particular kind of Fuchsian group called a *Schottky group*. We'll study this class in more detail in Section 15.1.

Proposition 2.4. *A subgroup* $\Gamma \subset \text{PSL}(2, \mathbb{R})$ *acts properly discontinuously on* $\mathbb{H}$ *if and only if it is Fuchsian.*

Proof. If a subgroup Γ is Fuchsian, then it is easy to see that any orbit Γz is discrete. For any compact $K \subset \mathbb{H}$, the set $\Gamma z \cap K$ is both discrete and compact and therefore finite. Hence Γ acts properly discontinuously.

On the other hand, assume that Γ acts properly discontinuously. We claim that there are points in $\mathbb{H}$ not fixed by any element of Γ except I. Indeed, if $Tw = w$, then for any $z \in \mathbb{H}$, we have

$$d(Tz, z) \leq d(Tz, Tw) + d(Tw, z) = 2d(z, w),$$

by the triangle inequality. Proper discontinuity therefore implies that only finitely many points in any neighborhood of z could be fixed by elements of $\Gamma - \{I\}$.

Hence we can pick a point w not fixed by any element of Γ except I. If Γ is not discrete, then there exists a sequence $\{T_k\} \subset \Gamma$ of distinct elements such that $T_k \to I$. By the choice of w, the sequence $\{T_k w\}$ contains only distinct points, and $T_k w \to w$ contradicts the proper discontinuity of the action. $\square$

For the quotient space to be smooth, we have the additional requirement that Γ act without fixed points. Since only elliptic transformations fix points within $\mathbb{H}$, this is equivalent to the absence of elliptic elements in Γ. If Γ had elliptic elements, then the quotient would be an *orbifold*, with conical singularities corresponding to the elliptic fixed points. Orbifolds are not intractable from a spectral theory point of view, because one can always pass to a finite cover. We omit this case mainly to avoid excessive notational complexity later on.

Hopf's theorem on the classification of manifolds of constant sectional curvature implies, in the two-dimensional case, that all hyperbolic surfaces are associated to Fuchsian groups.

Theorem 2.5 (Hopf). *For any hyperbolic surface X there is a Fuchsian group Γ with no elliptic elements and a Γ-invariant Riemannian covering map $\pi : \mathbb{H} \to X$ realizing the isometry $X \cong \Gamma \backslash \mathbb{H}$.*

Proof (sketch). For $p \in X$ the exponential map $\exp_p : T_p X \to X$ defines geodesic polar coordinates, in which the metric takes the form $ds^2 = dr^2 + \sinh^2 r \, d\theta^2$ by the assumption of Gaussian curvature -1. The lack of singularity in the metric for $r \in (0, \infty)$ implies that $\exp_p : T_p X \to X$ is an immersion. With the geodesic polar coordinates we can identify $T_p X \cong \mathbb{H}$, and $\exp_p$ induces a local isometry $\pi : \mathbb{H} \to X$. A local isometry of complete Riemannian manifolds is automatically a covering map. And since X is a smooth surface, the group of covering transformations must be Fuchsian with no elliptic elements. (The details of these arguments involve some differential geometry that will not be needed for the rest of this book; see, e.g., [118] or [171].) $\square$

A *hyperbolic structure* on a surface is defined by an atlas of coordinate patches identified with open subsets of $\mathbb{H}$, with transition maps given by orientation-preserving isometries. Theorem 2.5 shows that any hyperbolic metric is induced by a hyperbolic structure. Of course, since isometries are Möbius transformations, the hyperbolic structure also induces a complex structure.

A *Riemann surface* is a one-dimensional complex manifold, so that hyperbolic surfaces are a subcategory of Riemann surfaces. One might expect complex structure to be a more general concept than hyperbolic structure, since analytic functions need not be Möbius. But the uniformization theorem for Riemann surfaces says that a smooth Riemann surface is either the Riemann sphere or a quotient of $\mathbb{C}$ or $\mathbb{H}$ by a discrete group of conformal automorphisms (see, e.g., [59]). The Riemann sphere and flat tori are the only compact examples of Riemann surfaces with $\chi \geq 0$. Every Riemann surface with $\chi < 0$ is a hyperbolic surface, so in some sense most of the Riemann surfaces are hyperbolic.

Definition. A *fundamental domain* $\mathcal{F} \subset \mathbb{H}$ for a Fuchsian group Γ is a closed region such that

$$\Gamma\mathcal{F} := \bigcup_{T \in \Gamma} T\mathcal{F} = \mathbb{H},$$

and for each $T \in \Gamma - \{I\}$, the interiors of $\mathcal{F}$ and $T\mathcal{F}$ do not intersect.

A convenient construction of a fundamental domain is given by the *Dirichlet domain* of a point $w \in \mathbb{H}$, defined by

$$\mathcal{D}_w := \{z \in \mathbb{H} : d(z, w) \leq d(z, Tw) \text{ for all } T \in \Gamma\}. \tag{2.12}$$

Convexity in $\mathbb{H}$ is always interpreted in the sense of hyperbolic geodesics, i.e., a set $U \subset \mathbb{H}$ is *convex* if $z, w \in U$ implies that the geodesic arc $[z, w]$ is a subset of U.

Lemma 2.6. *If w is not the fixed point of an elliptic element of Γ, then the Dirichlet domain $\mathcal{D}_w$ is a fundamental domain for Γ. The domain $\mathcal{D}_w$ is convex and bounded by a union of geodesics.*

Proof. Fix such a w with domain $\mathcal{D}_w$. Given $z_0 \in \mathbb{H}$, we can minimize $d(z, w)$ for $z \in \Gamma z_0$ by the discreteness of the orbit. This gives at least one $z \in \Gamma z_0 \cap \mathcal{D}_w$, implying that $z_0 \in \Gamma\mathcal{D}_w$. Hence $\Gamma\mathcal{D}_w = \mathbb{H}$.

Suppose now that both $z \in \mathcal{D}_w$ and $Rz \in \mathcal{D}_w$ for $R \in \Gamma - \{I\}$. Then $z \in \mathcal{D}_w$ implies

$$d(z, w) \leq d(z, Rw),$$

and $Rz \in \mathcal{D}_w$ implies

$$d(Rz, w) \leq d(Rz, Rw) = d(z, w).$$

Hence $d(z, w) = d(z, Rw)$, so z lies on the boundary of $\mathcal{D}_w$. This shows that the interiors of $\mathcal{D}_w$ and $R\mathcal{D}_w$ do not intersect, and thus $\mathcal{D}_w$ is a fundamental domain.

Note that $\mathcal{D}_w$ is an intersection of half-planes of the form

$$H_w(T) := \{z \in \mathbb{H} : d(z, w) \leq d(z, Tw)\},$$

for $T \in \Gamma$. Thus to prove the second statement, it suffices to check that such half-planes have geodesic boundary. By conjugation we can assume $w = i$ and $T : z \mapsto \lambda^2 z$. Then $z \in \partial H_w(T)$ is characterized by $d(z, i) = d(z, i\lambda^2)$. By (2.6) this can easily be reduced to $|z| = \lambda$, which defines a geodesic. $\square$

The action of Γ on a fundamental domain gives a tessellation of $\mathbb{H}$. An example is shown in Figure 2.3. The corresponding quotient surface has genus one with a single end.

Lemma 2.7. *The tessellation $\{T\mathcal{D}_w : T \in \Gamma\}$ is locally finite, meaning that any compact region of $\mathbb{H}$ meets only finitely many copies of $\mathcal{D}_w$, and contains only finitely many vertices and sides of any particular copy.*

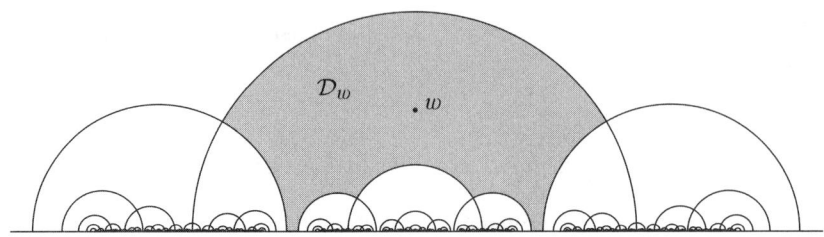

Fig. 2.3. A Dirichlet tessellation of $\mathbb{H}$.

Proof. Suppose that $\overline{B(w;r)}$ contained infinitely many points of the form $z_j = T_j(w_j)$ for $w_j \in \mathcal{D}_w$. Then

$$d(w, T_j w) \leq d(w, z_j) + d(z_j, T_j w)$$
$$= d(w, z_j) + d(w_j, w)$$
$$\leq 2r.$$

Then $\overline{B(w;2r)}$ would contain infinitely many images of w, contradicting the properly discontinuous action of Γ.   □

Definition. For a Fuchsian group Γ, the *limit set* $\Lambda(\Gamma) \subset \partial\mathbb{H}$ is the set of limit points (in the Riemann-sphere topology) of all orbits Γz for $z \in \mathbb{H}$. The complement of the limit set in $\partial\mathbb{H}$ is the set of *ordinary points*.

Lemma 2.8. *If $w \in \mathbb{H}$ is not an elliptic fixed point of Γ, then $\Lambda(\Gamma)$ is the set of limit points of the single orbit Γw. It follows immediately that $\Lambda(\Gamma)$ is closed and invariant under Γ.*

Proof. Let $\mathcal{D}_w$ be the Dirichlet domain centered at w. Suppose $q \in \Lambda(\Gamma)$. Then there is a $z \in \mathcal{D}_w$ and a sequence $\{T_j\} \subset \Gamma$ such that $T_j z \to q$. Applying the triangle inequality for the Riemann-sphere metric d_∞ gives

$$d_\infty(T_j w, q) \leq d_\infty(T_j w, T_j z) + d_\infty(T_j z, q).$$

We claim that the first term on the right converges to zero as $j \to \infty$. The second term does so by assumption, so this would imply $T_j w \to q$, establishing that q is a limit point of Γw.

To prove the claim, suppose that $d_\infty(T_j w, T_j z)$ doesn't converge to zero. Because $\mathbb{H} \cup \partial\mathbb{H}$ is compact in the topology of d_∞, by passing to a subsequence we can assume that $T_j w \to p \in \partial\mathbb{H}$ and $T_j z \to p' \in \partial\mathbb{H}$, where $p \neq p'$ by assumption. Then the geodesic arcs $[T_j w, T_j z]$ accumulate on $[p, p']$, contradicting Lemma 2.7.   □

The standard classification of Fuchsian groups is based on the following characterization of the limit set.

Theorem 2.9 (Poincaré, Fricke–Klein). *The possibilities for the limit set of a Fuchsian group Γ are:*

1. *$\Lambda(\Gamma)$ contains 0, 1, or 2 points.*
2. *$\Lambda(\Gamma)$ is a perfect nowhere-dense subset of $\partial\mathbb{H}$.*
3. *$\Lambda(\Gamma) = \partial\mathbb{H}$.*

Proof. Assume that $\Lambda(\Gamma)$ contains more than two points. Our first claim is that then Γ must contain hyperbolic elements. If Γ were purely elliptic, then it's a straightforward exercise to argue that all elements of Γ have the same fixed point. (The product of elliptic transformations with different fixed points is hyperbolic; see, e.g., [108, Theorem 2.4.1]). Then the group would have to be finite cyclic by discreteness, so $\Lambda(\Gamma)$ would be empty.

Now suppose Γ contains a parabolic element. By conjugation assume it's T : $z \mapsto z + 1$ and therefore $\infty \in \Lambda(\Gamma)$. If every element of Γ fixed ∞ then we could easily argue that Γ was parabolic cyclic by discreteness. This would imply $\Lambda(\Gamma) = \{\infty\}$. Hence, under the assumption that $\Lambda(\Gamma)$ has more than two points, Γ must contain in addition to T some transformation that does not fix infinity, say

$$S = \begin{pmatrix} a & b \\ c & d \end{pmatrix},$$

where $c \neq 0$. Observe that

$$\mathrm{tr}\, T^k S = a + kc + d.$$

For k sufficiently large, we have $|\,\mathrm{tr}\, T^k S| > 2$, which implies that $T^k S$ is hyperbolic. Thus Γ contains hyperbolic elements.

Still assuming that $\Lambda(\Gamma)$ has more than two points, we'll next argue that $\Lambda(\Gamma)$ is perfect (every point is a limit point). An arbitrary point in $\Lambda(\Gamma)$ can be moved to 0 by conjugation of the group. So it suffices to assume $0 \in \Lambda(\Gamma)$ and prove that it's a limit point of $\Lambda(\Gamma)$. This is easy if 0 is a hyperbolic fixed point. In this case Γ contains a dilation $T : z \mapsto e^{-\lambda}z$ for $\lambda > 0$. Since $\Lambda(\Gamma)$ contains at least three points, we must have some $q \in \Lambda(\Gamma)$, not equal to 0 or ∞, and then $T^k q \to 0$.

Suppose that $0 \in \Lambda(\Gamma)$ is not a hyperbolic fixed point. Since Γ contains hyperbolic elements by assumption, we can find a hyperbolic $T \in \Gamma$ with fixed points p_1, p_2. Choosing a point w on the axis $\alpha(T)$, let $\{R_j\} \subset \Gamma$ be a sequence such that $R_j w \to 0$. Given $\varepsilon > 0$, we can choose R_j such that $|R_j w| < \varepsilon$. Then because $R_j w$ lies on the half-circle $R_j \alpha(T)$, at least one of the endpoints $R_j p_1$ and $R_j p_2$ must lie in the interval $(-\varepsilon, \varepsilon)$. Since these endpoints are the fixed points of $R_j T R_j^{-1} \in \Gamma$, this shows that 0 is a limit point of hyperbolic fixed points. In particular, 0 is a limit point of $\Lambda(\Gamma)$. This completes the argument that $\Lambda(\Gamma)$ is perfect.

It remains to show that $\Lambda(\Gamma)$ is either $\partial\mathbb{H}$ or nowhere dense. Assume that $\Lambda(\Gamma) \neq \partial\mathbb{H}$. Then we have at least one ordinary point $a \in \partial\mathbb{H} - \Lambda(\Gamma)$. Given $q \in \Lambda(\Gamma)$ and $\varepsilon > 0$, we need to show that there is an ordinary point within ε of q (assuming $q \neq \infty$ without loss of generality). By the arguments above we can find a hyperbolic fixed point p within $\varepsilon/2$ of q. Let $T \in \Gamma$ have p as an attractive hyperbolic fixed point. Then $T^k a$ converges to p as $k \to \infty$. Choosing k so that $|T^k a - p| < \varepsilon/2$ then implies $|T^k a - q| < \varepsilon$. Note that $T^k a$ is ordinary since $\Lambda(\Gamma)$ is Γ-invariant. This shows that there is an ordinary point within every neighborhood of any point of $\Lambda(\Gamma)$. Therefore $\Lambda(\Gamma)$ is nowhere dense in $\partial\mathbb{H}$.   □

With Theorem 2.9 in mind, we introduce some further terminology:

Definition. A Fuchsian group Γ is said to be

1. *elementary* if $\Lambda(\Gamma)$ is finite;
2. *of the first kind* if $\Lambda(\Gamma) = \partial\mathbb{H}$;
3. *of the second kind* if $\Lambda(\Gamma)$ is perfect and nowhere dense.

An alternative defining condition for an elementary group would be the existence of a finite Γ-orbit in $\mathbb{H} \cup \partial\mathbb{H}$. This sounds slightly more general than the definition above, but turns out to be equivalent. Cyclic Fuchsian groups are obviously elementary, with $\Lambda(\Gamma)$ consisting of 0, 1, and 2 points in the elliptic, parabolic, and hyperbolic cases, respectively. The only other elementary possibility is a group conjugate to the group generated by $z \mapsto \lambda z$ and $z \mapsto -1/z$, for which $\Lambda(\Gamma)$ has 2 points also (see, e.g., [108, Theorem 2.4.3]). Since we assume smoothness, the elementary hyperbolic surfaces consist only of $\mathbb{H}$ and its quotients by hyperbolic or parabolic cyclic groups.

If the quotient $\Gamma\backslash\mathbb{H}$ has finite area, then Γ is called *cofinite*. Fuchsian groups of the first kind are precisely the cofinite groups (see, e.g., [108, Section 4.5]). A cofinite Fuchsian group is called *cocompact* if the quotient $\Gamma\backslash\mathbb{H}$ is compact. For most of this book we are concerned with surfaces of infinite area, so our attention will be focused on Fuchsian groups of the second kind.

2.3 Geometrically finite groups

We turn next to the question of the conditions imposed on the group Γ by the assumption of topological finiteness of the quotient $\Gamma\backslash\mathbb{H}$. The answer can be given in terms of a nice geometric condition.

Definition. A Fuchsian group (or corresponding hyperbolic surface) is said to be *geometrically finite* if there exists a fundamental domain that is a finite-sided convex polygon.

There is also an algebraic finiteness condition—we say that Γ is *finitely generated* if there exists a finite list of transformations that generate the group.

Theorem 2.10 (Geometric finiteness). *For a Fuchsian group Γ the following are equivalent:*

1. *$\Gamma\backslash\mathbb{H}$ is topologically finite (i.e., finite Euler characteristic).*
2. *Γ is finitely generated.*
3. *Γ is geometrically finite.*

A related result that we won't prove here is Siegel's theorem, which says that all cofinite Fuchsian groups are geometrically finite (see, e.g., [108, Theorem 4.1.1]).

For the proof of Theorem 2.10 we need to establish some connections between the structure of the group and the geometry of the Dirichlet domain. In Lemma 2.6, we saw that the boundary of $\mathcal{D}_w$ is a union of geodesics. Since $\mathcal{D}_w$ is convex, each geodesic meets $\mathcal{D}_w$ either in a point or in a geodesic segment. The segments in the boundary are called *sides* and must take the form

$$\sigma_w(R) := \{z \in \partial\mathcal{D}_w : d(z, w) = d(z, Rw)\} = \mathcal{D}_w \cap R\mathcal{D}_w, \qquad (2.13)$$

for some $R \in \Gamma$. By Lemma 2.7, the vertices of $\mathcal{D}_w$ are isolated.

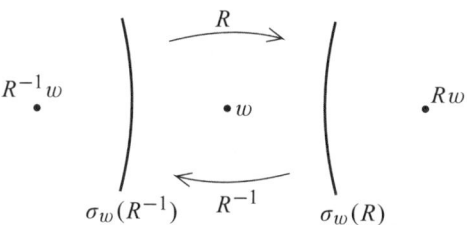

Fig. 2.4. Side-pairing congruences.

Two sides of a Dirichlet domain $\mathcal{D}_w$ are called *congruent* if they are related by an element of Γ. Notice that if a side is given by $\sigma_w(R) \neq \emptyset$, then

$$R^{-1}\sigma_w(R) = (R^{-1}\mathcal{D}_w) \cap \mathcal{D}_w = \sigma_w(R^{-1}).$$

Since this is nonempty as well, $\sigma_w(R^{-1})$ must also be a side of $\mathcal{D}_w$, congruent to the original. This is illustrated in Figure 2.4. It follows that the sides of $\mathcal{D}_w$ come in congruent pairs of the form $\sigma_w(R), \sigma_w(R^{-1})$.

Lemma 2.11. *The side-pairing congruences of a Dirichlet domain generate the group* Γ.

Proof. Suppose $\mathcal{D}_w$ is a Dirichlet domain. Let $\Gamma_s \subset \Gamma$ be the subgroup generated by side-pairing congruences. Clearly, if we define

$$A := \bigcup_{T \in \Gamma_s} T\mathcal{D}_w, \qquad B := \bigcup_{T \in \Gamma - \Gamma_s} T\mathcal{D}_w,$$

then $A \cup B = \mathbb{H}$. Furthermore, A and B are closed, since $\mathcal{D}_w$ is closed and any compact region contains only finitely many copies of $\mathcal{D}_w$ by Lemma 2.7.

Thus if we can show that A and B are disjoint, the connectedness of $\mathbb{H}$ would imply that $B = \emptyset$ (since A is clearly not empty). To prove disjointness, suppose A intersects B. This could happen only at a side or vertex of the Dirichlet tessellation.

To rule out a shared side, suppose $T\mathcal{D}_w$ is adjacent to $R\mathcal{D}_w$ and $T\mathcal{D}_w \subset A$, with $T \in \Gamma_s$. This implies that $R^{-1}T\mathcal{D}_w$ is adjacent to $\mathcal{D}_w$; hence $R^{-1}T$ is a generator of Γ_s; hence $R \in \Gamma_s$. Thus $R\mathcal{D}_w \in A$, which shows that A cannot meet B along a side.

Ruling out a shared vertex is similar. Suppose that $T\mathcal{D}_w$ shares a vertex with $R\mathcal{D}_w$ and $T \in \Gamma_s$. There can be only finitely many sides of the Dirichlet tessellation sharing the same vertex. Therefore $T\mathcal{D}_w$ is connected to $R\mathcal{D}_w$ by a chain of side-sharing faces of the tessellation. We saw above that faces in A share sides only with other faces in A, so we find $R \in \Gamma_s$. Hence A does not intersect B at a vertex, and this finishes the proof that $B = \emptyset$.   □

Proof of Theorem 2.10 (Geometric finiteness). Lemma 2.11 shows in particular that $3 \Rightarrow 2$. The implication $1 \Rightarrow 2$ holds because the fundamental group of a finitely punctured compact surface is finitely generated and $\pi_1(X) \cong \Gamma$. And $3 \Rightarrow 1$ is also clear, because a surface assembled out of a finite-sided polygon by gluing the sides together in pairs must have finite Euler characteristic.

The hard part to prove is $2 \Rightarrow 3$. We will follow the approaches from Beardon [16] and Katok [108]. Assume that Γ is finitely generated, and choose a Dirichlet domain $\mathcal{D}_w$. By Lemma 2.11 we know that the side-pairing transformations of $\mathcal{D}_w$ generate Γ. By assumption, Γ can be generated by finitely many of the side-pairing transformations, say $T_1, \ldots, T_k$.

The strategy is to choose a disk $B(w; r)$ that includes arcs of positive length of the $2k$ sides of $\mathcal{D}_w$ paired by the T_j's. By local finiteness of the sides and vertices (Lemma 2.7), we can choose r so that the boundary circle $\partial B(w; r)$ does not intersect vertices of $\mathcal{D}_w$ and is not tangent to any side. Our goal will be to show that $\mathcal{D}_w - B(w; r)$ is the union of finitely many connected components, each of which meets only finitely many sides of $\mathcal{D}_w$. Thus $\mathcal{D}_w$ has only finitely many sides outside of $B(w; r)$. Since only finitely many sides of $\mathcal{D}_w$ meet the interior of $B(w; r)$, by Lemma 2.7, this will imply that the total number of sides of $\mathcal{D}_w$ is finite.

First we observe that $\Gamma B(w; r)$ is connected. Clearly $B(w; r)$ overlaps $T_j B(w; r)$ for $j = 1, \ldots, k$, since T_j pairs sides of $\mathcal{D}_w$ and $B(w; r)$ was chosen to include arcs of such sides. Then we can argue that $T_j B(w; r)$ overlaps $T_j T_i B(w; r)$, by translation, and so on. Since the T_j's generate Γ, by continuing this process we see that $\Gamma B(w; r)$ is connected.

Let $\eta_1, \ldots, \eta_m$ be the arcs of $\partial B(w; r) \cap \mathcal{D}_w$. If z is an endpoint of η_j then it lies in some side of $\mathcal{D}_w$. Therefore there is a side-pairing $R \in \Gamma$ such that $Rz \in \mathcal{D}_w$ also. By definition of the Dirichlet domain, $z \in \mathcal{D}_w$ implies

$$r = d(z, w) \le d(z, R^{-1}w) = d(Rz, w).$$

On the other hand, $Rz \in \mathcal{D}_w$ implies

$$d(Rz, w) \le d(Rz, Rw) = r.$$

This shows that $d(Rz, w) = r$, so that $Rz \in \partial B(w; r)$ also. Therefore Rz must be an endpoint of some side η_i (possibly the other endpoint of the same η_j).

It suffices to focus on a single arc $\eta = \eta_j$. Given an endpoint of η we can find another endpoint of some η_i congruent to it, and translate that η_i by some element of Γ to add an arc to our original η. Iterating this process in both directions results in a uniquely defined continuous curve β that is a union of arcs each congruent to some η_i. Since there are only finitely many η_i's, β must eventually include two arcs congruent to each other; hence there is some nontrivial $S \in \Gamma$ that preserves β. This setup is illustrated in Figure 2.5.

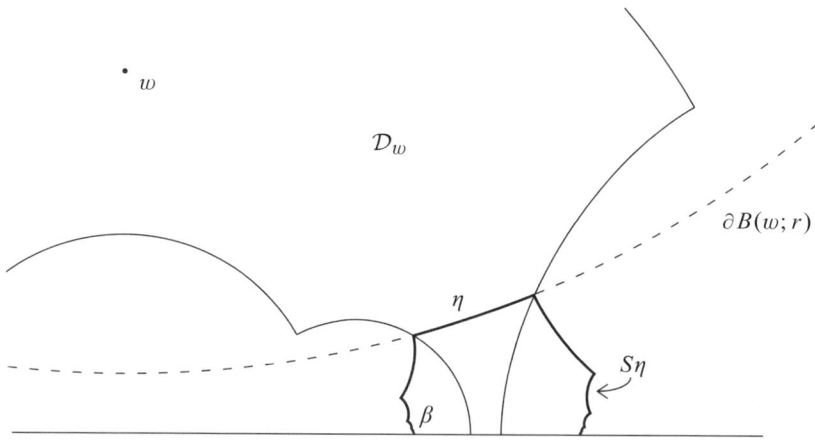

Fig. 2.5. Construction of β.

Let E be the component of $\mathcal{D}_w - B(w; r)$ meeting η. Our goal is to show that E meets $\mathcal{D}_w$ in only finitely many sides. For this purpose it suffices to show that E does not meet $\Lambda(\Gamma)$. Observe that the curve β divides $\mathbb{H}$ into two components, one of which contains E and the other w. Since $\Gamma B(w; r)$ is connected, β separates all of $\Gamma B(w; r)$ from E. The limit points of Γ are all limit points of $\Gamma w \subset \Gamma B(w; r)$. So the only limit points we need to worry about being close to E are the endpoints of β.

Suppose first that S is hyperbolic. Then β must run between its two fixed points. By the definition of a Dirichlet domain, $\mathcal{D}_w$ is contained in the half-planes $\{z : d(z, w) \leq d(z, Sw)\}$ and $\{z : d(z, w) \leq d(z, S^{-1}w)\}$. In the notation (2.13), these half-planes are bounded by arcs $\sigma_w(S)$ and $\sigma_w(S^{-1})$. These two arcs don't intersect (obvious if one conjugates S to a dilation). Since S maps $\sigma_w(S^{-1})$ to $\sigma_w(S)$, neither contains a fixed point of S. Therefore E is separated from the limit points of Γ (in the d_∞ metric). If E met infinitely many sides of $\mathcal{D}_w$ then there would have to be a limit point of Γ on its boundary. Since this doesn't happen, E meets only finitely many sides of $\mathcal{D}_w$.

Now consider the case that S is parabolic. If p denotes the fixed point of S, then the curve β is a closed loop from p to itself. If w lies inside β, then $\Gamma B(w; r)$ does also, and this would imply that p was the only point in $\Lambda(\Gamma)$. Then Γ would be parabolic cyclic and obviously geometrically finite. So assume that w lies on the outside of β, in which case E must be contained inside the loop. If E doesn't meet the boundary $\partial \mathbb{H}$ then it is separated from the limit points of Γ and then we argue as above that E meets only finitely many sides of $\mathcal{D}_w$.

So let us suppose that E lies inside β and meets $\partial \mathbb{H}$ at p. We want to control the shape of $\mathcal{D}_w$ nearby. As in the hyperbolic case above, $\mathcal{D}_w$ lies between the arcs $\sigma_w(S)$ and $\sigma_w(S^{-1})$. These arcs do not intersect in $\mathbb{H}$ but are tangent to each other at p. This implies that a small neighborhood of p meets exactly two sides of $\mathcal{D}_w$. Since otherwise E is bounded away from $\Lambda(\Gamma)$, E meets only finitely many sides of $\mathcal{D}_w$.

Finally, suppose that S is elliptic, in which case β is a closed loop. If E is contained in the interior, then it is obviously bounded away from the limit set. But if w lies in the interior, then $\Lambda(\Gamma) = \emptyset$ and the group is cyclic. $\square$

2.4 Classification of hyperbolic ends

Geometric finiteness imposes strong restrictions on the ends of a hyperbolic surface. We will show that the only possibilities, beyond the hyperbolic plane itself, are the ends of cylinders, i.e., quotients of $\mathbb{H}$ by hyperbolic and parabolic cyclic groups. We start by examining these model cases.

A hyperbolic transformation $T \in \mathrm{PSL}(2, \mathbb{R})$ generates a cyclic hyperbolic group $\langle T \rangle$. The quotient $C_\ell := \langle T \rangle \backslash \mathbb{H}$ is a *hyperbolic cylinder* of diameter $\ell = \ell(T)$. By conjugation, we can identify the generator T with the map $z \mapsto e^\ell z$, and we define Γ_ℓ to be the corresponding cyclic group. A natural fundamental domain for Γ_ℓ would be the region $\mathcal{F}_\ell := \{1 \leq |z| \leq e^\ell\}$. The y-axis is the lift of the only simple closed geodesic on C_ℓ, whose length is ℓ.

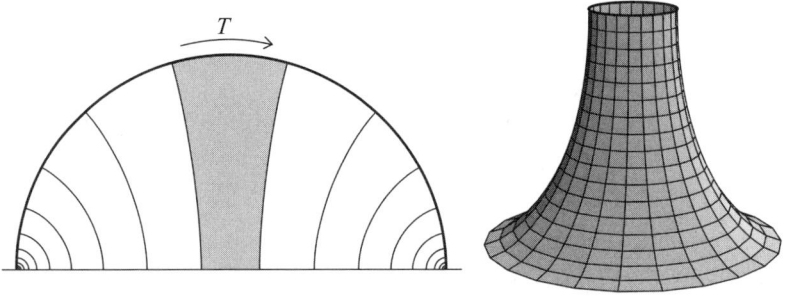

Fig. 2.6. Funnel.

Definition. A *funnel* F_ℓ is half of a hyperbolic cylinder of diameter ℓ, with boundary given by the central geodesic.

Suppose we start with a hyperbolic generator T. If H denotes one of the half-planes of $\mathbb{H}$ bounded by the axis $\alpha(T)$, then $\langle T \rangle \backslash H$ is a funnel of diameter $\ell(T)$. This is illustrated in Figure 2.6, which also shows the Riemannian embedding of a portion the funnel into $\mathbb{R}^3$. By integrating the hyperbolic area form dA given in (2.8) over $\mathcal{F}_\ell \cap \{\operatorname{Re} z \geq 0\}$, we see that

$$\operatorname{area}(F_\ell) = \infty.$$

The quotient of $\mathbb{H}$ by a parabolic cyclic group $\langle T \rangle$ will be called a *parabolic cylinder*. We can always conjugate $\langle T \rangle$ to the group Γ_∞ generated by $z \mapsto z + 1$, so the parabolic cylinder is unique up to isometry. A natural fundamental domain for Γ_∞ is $\mathcal{F}_\infty := \{0 \leq \operatorname{Re} z \leq 1\} \subset \mathbb{H}$. A circle lying in $\mathbb{H}$ and tangent to $\partial \mathbb{H}$ is called a *horocycle*. The curves stabilized by a parabolic transformation, as shown in Figure 2.2, are horocycles tangent at the fixed point.

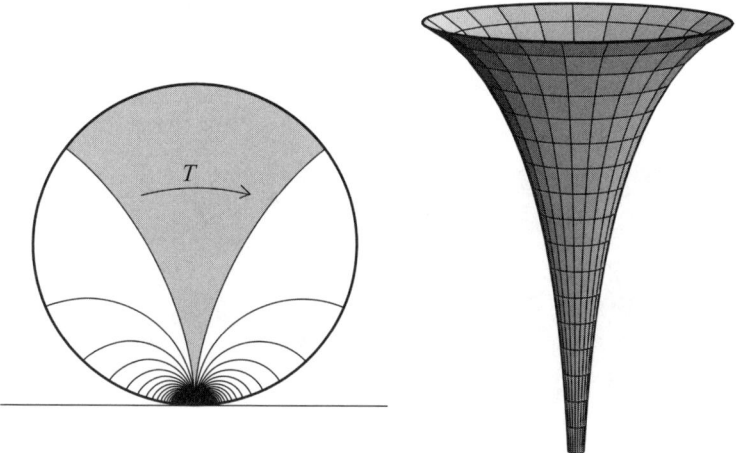

Fig. 2.7. Cusp.

Definition. A *cusp* is the small end of a parabolic cylinder, with boundary the unique closed horocycle of length 1.

There is no canonical choice of boundary for a cusp, but it is convenient to standard-ize the definition by fixing the boundary length. To get a cusp from Γ_∞ as defined above, we first restrict to the subset $\{\operatorname{Im} z > 1\}$ of $\mathcal{F}_\infty$. If we started from a general parabolic generator T with fixed point p, we would first find the unique horocycle σ tangent to $\partial \mathbb{H}$ at p such that $\langle T \rangle \backslash \sigma$ has length one. If O denotes the interior of σ then $\langle T \rangle \backslash O$ is the cusp associated to T. The cusp can be fully embedded into Euclidean $\mathbb{R}^3$, as illustrated in Figure 2.7, where it forms a portion of the classical pseudosphere. Using the fundamental domain for Γ_∞ we compute that

$$\operatorname{area}(\text{cusp}) = \int_1^\infty \int_0^1 \frac{dx\,dy}{y^2} = 1.$$

The one case remaining is the large end of the parabolic cylinder, which is asymptotic to a funnel but not equal to one. We won't give this type of end a separate name because it does not occur in any other hyperbolic surface.

For the rest of this section we'll consider the general case of a hyperbolic surface $X = \Gamma \backslash \mathbb{H}$ where Γ is nonelementary. By Theorem 2.9, $\Lambda(\Gamma)$ is either a perfect nowhere-dense set or equal to $\partial \mathbb{H}$. In the former case, $\partial \mathbb{H} - \Lambda(\Gamma)$ is a countable union of open intervals I_j. Suppose that γ_j is the geodesic whose endpoints are the endpoints of I_j, and H_j the half-plane bounded by γ_j and I_j.

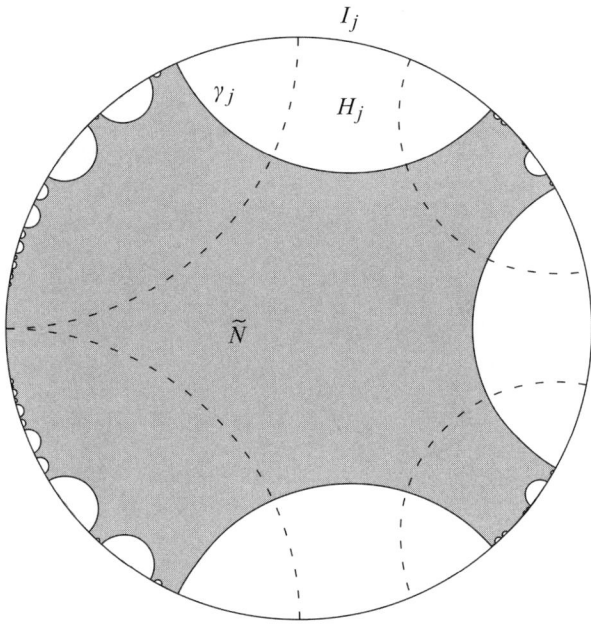

Fig. 2.8. Nielsen region.

Definition. The *Nielsen region* of a Fuchsian group Γ is the set

$$\widetilde{N} := \mathbb{H} - \left(\cup H_j \right). \tag{2.14}$$

Its quotient under Γ,

$$N := \Gamma \backslash \widetilde{N},$$

is called the *convex core* of X.

Figure 2.8 shows a sample construction of the Nielsen region, pictured in the unit disk model for the sake of clarity; the dotted lines mark the boundary of the fundamental domain. The Nielsen region is also commonly (and equivalently) defined as the convex hull of the limit set $\Lambda(\Gamma)$ in $\mathbb{H}$, meaning the union of geodesic arcs $[p, q]$ for all $p, q \in \Lambda(\Gamma)$. The term "convex core" refers to the fact that N is the smallest convex subset of X. If Γ is of the first kind, i.e., $\Lambda(\Gamma) = \partial\mathbb{H}$, then we have $\widetilde{N} := \mathbb{H}$ and $N = X$.

In Theorem 2.13 we'll see that $X - N$ is a finite collection of funnels. But before we get to that, let us develop a way to isolate the cusps also. The basic idea is that each parabolic fixed point in $\Lambda(\Gamma)$ should have a cusp fundamental region attached to it. Given a parabolic fixed point $p \in \partial\mathbb{H}$, let Γ_p be the parabolic cyclic subgroup of Γ fixing p. We take σ_p to be the unique horocycle tangent to $\partial\mathbb{H}$ at p such that $\Gamma_p \backslash \sigma_p$ has length 1. Then let O_p be the open region bounded by σ_p, so that $\Gamma_p \backslash O_p$ is a cusp fitting our convention of boundary length 1.

Lemma 2.12. *With the horocycles σ_p and interiors O_p defined as above, we have the following:*

1. *If two points of O_p are related by $T \in \Gamma$, then T preserves O_p (i.e., $T \in \Gamma_p$).*
2. *The horocycles σ_p are disjoint for different parabolic fixed points.*
3. *The horocycles σ_p do not intersect the half-planes H_j defining the Nielsen region (i.e., each $O_p \subset N$).*

Proof. Let T be the map $z \mapsto z + 1$. By conjugation, we assume that $p = \infty$ and $\Gamma_p = \langle T \rangle$, so that $O_p = \{\operatorname{Im} z > 1\}$. Let $S \in \Gamma$ be an element that does not fix ∞, given by

$$S = \begin{pmatrix} a & b \\ c & d \end{pmatrix},$$

with $ad - bc = 1$ and $c \neq 0$. We claim that in fact $|c| \geq 1$. This will prove the first assertion, because then the inequality

$$\operatorname{Im}(Sz) = \frac{\operatorname{Im} z}{|cz + d|^2} \leq \frac{1}{c^2 \operatorname{Im} z}$$

shows that any point inside O_p is mapped to $\{\operatorname{Im} z < 1\}$ by S.

For the proof that $|c| \geq 1$ we follow Kra [110, Lemma II.2.4]. Suppose we had $|c| < 1$. Define a recursive sequence of $S_n \in \Gamma$ by setting $S_0 := S$ and

$$S_{n+1} := S_n T S_n^{-1}.$$

If the matrix elements of S_n are denoted by a_n, b_n, c_n, d_n, then the recurrence condition becomes

$$\begin{pmatrix} a_{n+1} & b_{n+1} \\ c_{n+1} & d_{n+1} \end{pmatrix} = \begin{pmatrix} 1 - a_n c_n & a_n^2 \\ -c_n^2 & 1 + a_n c_n \end{pmatrix}. \tag{2.15}$$

In particular, under the assumption $|c| < 1$, this shows that

$$c_n = -c^{2^n} \to 0,$$

as $n \to \infty$. Also, from the equation $a_{n+1} = 1 - a_n c_n$, it is easy to prove inductively that a_n is bounded. Hence $c_n \to 0$ implies $a_n \to 1$. Then by (2.15) we conclude immediately that $b_n \to 1$ and $d_n \to 1$ also. We thus have that $S_n T^{-1} \to I$ within Γ, contradicting the discreteness of Γ. We conclude that $|c| \geq 1$, and this proves the first claim.

Still assuming that $p = \infty$, with T and O_p as above, let $q \in \mathbb{R}$ be some other parabolic fixed point of Γ. Then σ_q is a Euclidean circle tangent to $\mathbb{R}$, and the first part of the proof shows that no two points of σ_q could be related by $T : z \mapsto z + 1$. This means in particular that the Euclidean diameter of σ_q must be strictly less than 1, and thus σ_q is too short to intersect O_p.

The proof of the third claim is similar to the second. Suppose H_j is one of the half-planes in question. If ∞ is assumed to be a parabolic fixed point, then H_j cannot include ∞ and so must be a Euclidean half-disk centered on $\mathbb{R}$. The full collection $\cup H_i$ is invariant under Γ by the invariance of $\Lambda(\Gamma)$. The map T clearly does not fix H_j; hence no two points of H_j are related by T. This implies that the Euclidean radius of H_j is less than $1/2$, so H_j cannot intersect O_p. □

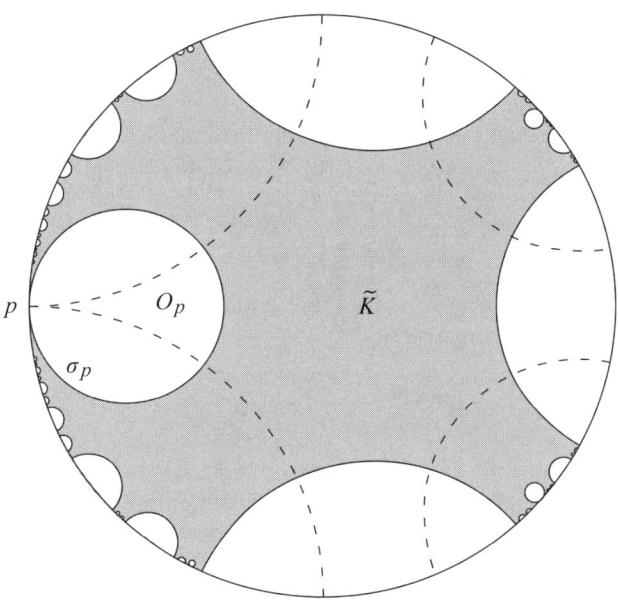

Fig. 2.9. Truncated Nielsen region.

Definition. The *truncated Nielsen region* is

$$\widetilde{K} := \widetilde{N} - \cup O_p,$$

with the union taken over all parabolic fixed points p of Γ. When Γ is geometrically finite, the corresponding quotient region,

$$K := \Gamma \backslash \widetilde{K},$$

is called the *compact core* of X.

Figure 2.9 shows an example of the truncated Nielsen region, for the same Fuchsian group whose Nielsen region was pictured in Figure 2.8. The distinction between the convex core and compact core is illustrated in Figure 2.10.

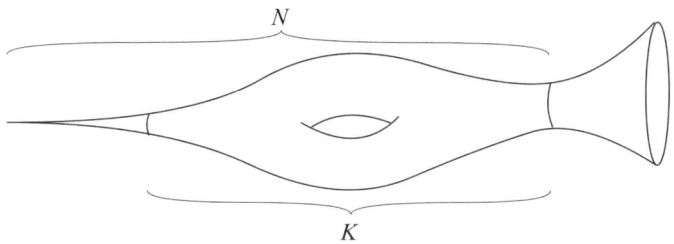

Fig. 2.10. Convex core N and compact core K.

Use of the term "compact core" is justified by the following, which is the main result of this section:

Theorem 2.13 (Classification of hyperbolic ends). *Let $X = \Gamma \backslash \mathbb{H}$ be a nonelementary geometrically finite hyperbolic surface. Then the region K defined above is compact, and $X - K$ is a finite disjoint union of cusps and funnels.*

Proof. Suppose Γ is a geometrically finite, nonelementary Fuchsian group. Let $\mathcal{D}_w$ be a Dirichlet fundamental domain for Γ, which can intersect $\partial \mathbb{H}$ only in a finite number of intervals or isolated points. We saw in the proof of Theorem 2.10 that $\mathcal{D}_w$ could meet $\Lambda(\Gamma)$ only at parabolic fixed points. At such a point p, two sides of $\mathcal{D}_w$ must meet tangentially. So if O_p is the corresponding horocyclic region O_p from Lemma 2.12, then $\mathcal{D}_w - O_p$ is bounded away from p in the d_∞ metric. If $\mathcal{D}_w$ meets $\partial \mathbb{H}$ in an arc η (possibly just a point) consisting of ordinary points, then η must be included in one of the half-planes H_j used to define $\widetilde{N}$ in (2.14). Since the boundary of each H_j meets $\partial \mathbb{H}$ in $\Lambda(\Gamma)$, $\mathcal{D}_w - H_j$ is bounded away from η with respect to d_∞. These arguments show that $\mathcal{D}_w \cap \widetilde{K}$ is bounded away from $\partial \mathbb{H}$ in the d_∞ metric, and therefore compact. Hence K is compact also.

We have shown also that the components of $\mathcal{D}_w - \widetilde{K}$ must be contained either in half-planes H_j or in horocyclic regions O_p. What remains to be seen is that the former give rise to finitely many funnels, and the latter to finitely many cusps.

First the funnel case. Let $\tau_1, \ldots, \tau_k$ denote the geodesic segments of the form $\partial H_j \cap \mathcal{D}_w$. Any point in ∂H_j is congruent to a point in $\mathcal{D}_w$, and these points must lie in some τ_i. In other words, ∂H_j is covered by segments each of which is congruent to one of the τ_i's. Since the τ_i's have finite length, ∂H_j must in fact contain multiple segments congruent to some particular τ_i. Therefore there are hyperbolic elements of

Γ that relate points of ∂H_j, and because the collection $\cup H_j$ is invariant under Γ, such transformations must then preserve ∂H_j. The subgroup $\Gamma_j \subset \Gamma$ that preserves ∂H_j is thus nontrivial, and by discreteness it must be cyclic. Then we have $\Gamma \backslash H_j = \Gamma_j \backslash H_j$, which is by definition a funnel. Because the set of τ_k's was finite to begin with, we conclude that $X - N$ is a finite disjoint union of funnels.

For any parabolic fixed point p, $\Gamma \backslash O_p$ is a cusp bounded by a horocycle of length 1 by Lemma 2.12. Since $\mathcal{D}_w$ meets $\Lambda(\Gamma)$ at only finitely many points, $N - K$ is a finite disjoint union of cusps.   □

Actually, the compactness of K is equivalent to geometric finiteness; see, e.g., [61, Section 15.1]. We will further subdivide the compact core K into a "pants" decomposition in Theorem 13.6.

If the convex core is compact, i.e., $N = K$, then Γ is said to be *convex cocompact*. This is equivalent to "geometrically finite with no cusps."

2.5 Gauss–Bonnet theorem

One of the most satisfying results in differential geometry is the Gauss–Bonnet theorem, which relates the shape of a polygonal region to the curvature of its interior. We present here only the hyperbolic version, for which we can take advantage of the characterization of geodesics given in Proposition 2.2.

The first step in our proof is to compute the area of a triangle in $\mathbb{H}$. Note that since the geodesic arc $[p, q]$ is uniquely defined even for p or q in $\partial \mathbb{H}$, it makes sense to allow "degenerate" triangles with some vertices in $\partial \mathbb{H}$. Two geodesics meeting at a point of $\partial \mathbb{H}$ must be tangent there, so the interior angle at a vertex is zero if and only if it lies in $\partial \mathbb{H}$.

Lemma 2.14 (Triangle area). *Let T be a triangle in $\mathbb{H}$ with interior angles $\alpha, \beta, \gamma \geq 0$. Then*

$$\text{area}(T) = \pi - (\alpha + \beta + \gamma).$$

Proof. This simple proof is taken from Katok [108]. First consider a triangle with at least one point on $\partial \mathbb{H}$. We can apply a Möbius transformation to map this point to ∞ while sending the opposite side to an arc of the unit circle. Let α, β be the interior angles at the other two vertices. From the diagram in Figure 2.11 we can see that the two sides meeting at infinity are the vertical lines $\text{Re}\, z = -\cos \alpha$ and $\text{Re}\, z = \cos \beta$. To compute the area we simply integrate

$$\text{area}(T) = \int_{-\cos \alpha}^{\cos \beta} \int_{\sqrt{1-x^2}}^{\infty} \frac{dy\, dx}{y^2}$$
$$= \int_{-\cos \alpha}^{\cos \beta} \frac{dx}{\sqrt{1 - x^2}}$$
$$= \pi - \alpha - \beta.$$

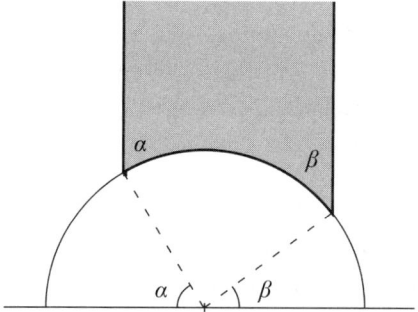

Fig. 2.11. Triangle with vertex at ∞.

If a triangle has all three vertices A, B, C in $\mathbb{H}$, we simply draw an auxiliary triangle by extending the segment $[A, B]$ until it meets $\mathbb{H}$ at a new vertex D, as shown in Figure 2.12. Applying the above computation to the triangles ACD and BCD gives

$$
\begin{aligned}
\text{area}(ABC) &= \text{area}(ACD) - \text{area}(BCD) \\
&= \pi - \alpha - (\gamma - \theta) - (\pi - \theta - (\pi - \beta)) \\
&= \pi - \alpha - \beta - \gamma .
\end{aligned}
\qquad \square
$$

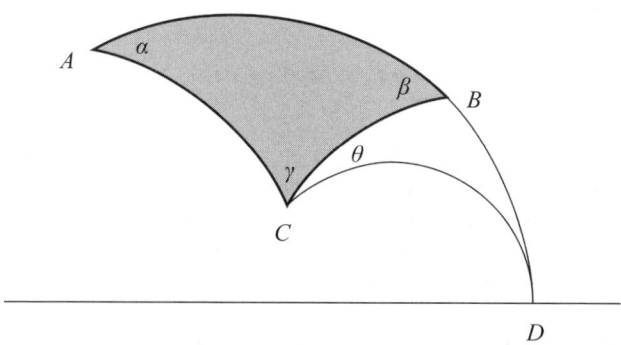

Fig. 2.12. Triangle extension.

The computation of Lemma 2.14 already gives a local version of the Gauss–Bonnet theorem. For the global version we allow a polygonal region with possibly nontrivial topology.

Theorem 2.15 (Gauss–Bonnet). *Suppose Z is a region of finite area in some geometrically finite hyperbolic surface X, with boundary (if any) consisting of n geodesic arcs meeting at interior angles $\alpha_1, \ldots, \alpha_n$. Then*

$$\text{area}(Z) = -2\pi\,\chi(Z) + \sum_{j=1}^{n}(\pi - \alpha_j).$$

In particular, the area of the convex core N of a geometrically finite nonelementary hyperbolic surface X is given by

$$\text{area}(N) = -2\pi\,\chi(X).$$

Proof. First assume Z is compact. If $\mathcal{F}_X$ is a Dirichlet fundamental region for X then Z is represented inside $\mathcal{F}_X$ by a polygonal region $\widetilde{Z}$. By subdividing this polygonal region with geodesic arcs, we can produce a triangulation of Z with all edges geodesic. Let V, E, F be the number of vertices, edges, and faces of the triangulation. By definition, $\chi(Z) = V - E + F$. Note that there are n exterior edges and vertices in the triangulation. Since each interior edge bounds two faces, while an exterior edge bounds a single face, we have

$$3F = 2E - n. \tag{2.16}$$

Applying Lemma 2.14 to the F triangles and summing gives

$$\text{area}(Z) = \pi F - \sum_{i=1}^{3F}\theta_i,$$

where the θ_i are the interior angles of the triangles. For each of the $V - n$ interior vertices the sum of the θ_i's contributes 2π, and of course α_j is the sum of the θ_i's at the exterior vertex j. Hence

$$\text{area}(Z) = \pi F - 2\pi(V - n) - \sum_{j=1}^{n}\alpha_j.$$

By (2.16), $F - 2V + n = -2\chi(Z)$, so this completes the proof for Z compact.

For a noncompact region Z there are two cases to consider. The first is that Z has a vertex at the cusp point, meaning that two sides of Z extend tangentially out the cusp. This case requires no change from the above argument; we simply allow degenerate triangles in the triangulation and assign interior angle zero to any cusp vertices.

The second possibility is that some complete ends of cusps are contained within Z. Suppose that Z encompasses the ends of k cusps. For each of these ends we introduce a geodesic loop (with one new vertex) to cut off the end of the cusp, as shown in Figure 2.13. Let $Z_1, \ldots, Z_k$ be the regions cut off in this way, so that $Y = Z - \cup Z_j$ is a compact region with $n + k$ vertices. Because the regions Y and Z have the same diffeomorphism type, $\chi(Y) = \chi(Z)$. If the interior angles of Y at the added vertices are denoted by $\beta_1, \ldots, \beta_k$, then by the formula above for the compact case we have

$$\text{area}(Y) = -2\pi\,\chi(Z) + \sum_{j=1}^{n}(\pi - \alpha_j) + \sum_{j=1}^{k}(\pi - \beta_j).$$

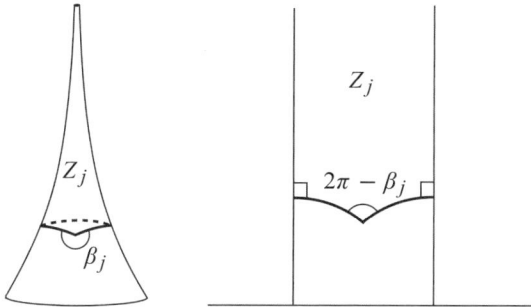

Fig. 2.13. Geodesic loop around a cusp end.

On the other hand, by Lemma 2.14 we can see that

$$\mathrm{area}(Z_j) = \beta_j - \pi,$$

so that the extra terms cancel out in the formula for area(Z). (Note that a fully enclosed cusp end is not counted as a vertex of Z.)   □

2.6 Length spectrum and Selberg's zeta function

The Euler characteristic, genus, and numbers of funnels and cusps, which are related by

$$\chi(X) = 2 - 2g - n_{\mathrm{f}} - n_{\mathrm{c}},$$

are the most basic invariants of a hyperbolic surface. Beyond these, the most natural geometric invariants are the lengths of closed geodesics.

One feature of negative curvature is that by minimizing length within a free homotopy class of closed curves, we expect to find a unique closed geodesic in the class. For a noncompact surface we must be a little careful about this; the horocycle bounding a cusp has no geodesic within its homotopy class. To exclude this exception, we'll say that a curve is *cuspidal* if it is freely homotopic to the horocyclic boundary of a cusp, and noncuspidal otherwise.

Proposition 2.16. *Let η be a homotopically nontrivial curve on a hyperbolic surface X. If η is noncuspidal, then there is a unique closed geodesic γ that is the shortest closed curve freely homotopic to η.*

Proof. Let $\tilde{\eta}$ be a maximal continuous curve in $\mathbb{H}$ obtained by joining successive lifts of η. There is some $T \in \Gamma$ that preserves $\tilde{\eta}$ and corresponds to moving through one period of η. This T must be hyperbolic, or else η would be cuspidal. The axis $\alpha(T)$ descends to a closed geodesic $\gamma = \langle T \rangle \backslash \alpha(T)$, in the free homotopy class of η. Since $\alpha(T)$ is the unique geodesic in $\mathbb{H}$ fixed by T, there is no other geodesic in the free homotopy class of η. It follows from (2.7) that γ is the shortest curve in this class.   □

The proof of Proposition 2.16 reveals an association between closed geodesics and axes of hyperbolic elements of Γ that will turn out to be of great importance to us. We can express this more precisely in the following:

Proposition 2.17. *There is a one-to-one correspondence between the closed, oriented geodesics of a hyperbolic surface $X = \Gamma\backslash\mathbb{H}$ and the conjugacy classes of hyperbolic elements of Γ. The length of the geodesic corresponding to the conjugacy class $[T]$ is the displacement length $\ell(T)$.*

Proof. Suppose $T \in \Gamma$ is hyperbolic. The axis $\alpha(T)$ of T is preserved by T and so projects to a closed geodesic under $\pi : \mathbb{H} \to X$. The length of $\pi(\alpha(T))$ is clearly the distance by which T translates points on $\alpha(T)$, i.e., $\ell(T)$. Note that the axis has a natural orientation because T maps points away from one fixed point toward the other. So the projected geodesic inherits an orientation. The axis of any element of the conjugacy class of T will also project to $\pi(\alpha(T))$, since

$$\alpha(RTR^{-1}) = R(\alpha(T)). \tag{2.17}$$

For the converse, suppose that γ is a closed oriented geodesic in X, with $\gamma(t) = \gamma(t + \ell)$ for some ℓ. We can construct a complete oriented geodesic arc $\widetilde{\gamma}$ in $\mathbb{H}$ by successive lifts of γ. Associated to $\widetilde{\gamma}$ is a unique hyperbolic $T \in \mathrm{PSL}(2, \mathbb{R})$ with axis and displacement length given by $\widetilde{\gamma}$ and ℓ, respectively. To see that T must be an element of Γ, we observe that since $\widetilde{\gamma}(0)$ and $\widetilde{\gamma}(\ell)$ project to the same point of X, we must have $\widetilde{\gamma}(\ell) = R\widetilde{\gamma}(0)$ for some $R \in \Gamma$. In this case $R^{-1}T$ fixes $\widetilde{\gamma}(0)$, implying $T = R$ since Γ acts freely. Hence $T \in \Gamma$. To complete the argument, note that (2.17) shows that any other lift of γ must be the axis of a hyperbolic transformation conjugate to T in Γ. $\qquad\square$

Given a closed geodesic, we can generate a family of iterates that traverse the same path multiple times. We define a *primitive* closed geodesic to be the root element of such a family, a closed geodesic that is not an iterate of a shorter closed geodesic. Similarly, an element of Γ is called primitive if it is not the power of some other element.

It is trivial to see that an oriented closed geodesic γ is uniquely represented as the iterate of a primitive oriented closed geodesic. Starting from $\gamma(0)$, we simply follow the curve and find the first value of $t > 0$ such that $\gamma(t) = \gamma(0)$ and $\gamma'(t) = \gamma'(0)$. The corresponding result for group elements is given in the following lemma:

Lemma 2.18. *Given a Fuchsian group Γ, each element $S \in \Gamma$ can be written uniquely as a power T^k, where $T \in \Gamma$ is primitive and $k \geq 1$. The centralizer Z_S of S in Γ is the cyclic group $\langle T \rangle$.*

Proof. Suppose S is hyperbolic. By conjugation, we can assume that it has the standard form $S : z \mapsto e^{\ell}z$ with $\ell > 0$. The commutation relation

$$\begin{pmatrix} a & b \\ c & d \end{pmatrix} \begin{pmatrix} e^{\ell/2} & 0 \\ 0 & e^{-\ell/2} \end{pmatrix} = \begin{pmatrix} e^{\ell/2} & 0 \\ 0 & e^{-\ell/2} \end{pmatrix} \begin{pmatrix} a & b \\ c & d \end{pmatrix}$$

implies $b = c = 0$, so any element of Z_S is a dilation also. The signed displacement length, given by $\log |R(i)|$ for $R \in Z_S$, therefore defines a homomorphism $Z_S \to \mathbb{R}$. The discreteness of Γ implies that the image must be a lattice $\ell_0 \mathbb{Z}$, for some minimum displacement length $\ell_0 > 0$. The unique choice for T is then $z \mapsto e^{\ell_0}z$, and we let $k = \ell/\ell_0$.

The proof for parabolic S is very similar. $\square$

Definition. The (primitive) *length spectrum* of a hyperbolic surface X is the set

$$\mathcal{L}_X = \{\ell(\gamma) : \gamma \text{ is a primitive oriented closed geodesic on } X\},$$

repeated according to multiplicity.

Note: because the geodesics are oriented, the values of ℓ in $\mathcal{L}_X$ come in pairs, once for each orientation. This might seem redundant, but it proves convenient because of the association with conjugacy classes.

The corresponding *length counting function* is given by

$$\pi_X(t) = \#\{\ell \in \mathcal{L}_X : \ell \leq t\}. \tag{2.18}$$

In Chapter 14 we'll develop a precise asymptotic formula for $\pi_X(t)$, but for the moment we need only a basic bound.

Proposition 2.19. *For X a geometrically finite hyperbolic surface,*

$$\pi_X(t) = O(e^t).$$

Proof. Let K be the compact core of X as introduced in Theorem 2.13. Closed geodesics on X are contained in the convex core N, though not necessarily in K. It is clear, however, that any closed geodesic must at least pass through K, since a cusp cannot contain a closed geodesic completely. Given a realization $X \cong \Gamma \backslash \mathbb{H}$, we also have the truncated Nielsen region $\widetilde{K}$, which covers K.

For some $w \in \widetilde{K}$, let Z be the compact region $\widetilde{K} \cap \mathcal{D}_w$. If γ is a primitive closed geodesic on X, then because γ passes through K it can be covered by a geodesic $\widetilde{\gamma}$ in $\mathbb{H}$ that passes through some point $q \in Z$. This curve $\widetilde{\gamma}$ is the axis of some primitive $T \in \Gamma$. If the length of γ is $\ell = \ell(T)$ and we set a equal to the diameter of Z, then by the triangle inequality,

$$d(w, Tw) \leq d(w, q) + d(q, Tq) + d(Tq, Tw) \leq \ell + 2a.$$

This means that for each primitive closed geodesic of length ℓ there is a transformation T that maps w to a point of distance less than $\ell + 2a$ away. We can therefore bound $\pi_X(t)$ by the number of images of w lying within distance $t + 2a$ of the point w. Alternatively, we could use the number of images of Z lying within distance $t + 3a$ of the point w, which gives

$$\pi_X(t) \leq \frac{\text{area}(B(w; t + 3a))}{\text{area } Z}.$$

The result follows, because $\text{area}(B(w; r)) = O(e^r)$ by (2.11). $\square$

For a Fuchsian group Γ, the function given by summing $e^{-sd(z,Tw)}$ over $T \in \Gamma$ is called the (absolute) *Poincaré series* for Γ. This sum converges for all $\operatorname{Re} s$ above some threshold value, which we single out in the following:

Definition. The *exponent of convergence* of a Fuchsian group Γ is

$$\delta := \inf\left\{s \geq 0 : \sum_{T \in \Gamma} e^{-sd(z,Tw)} < \infty\right\}, \qquad (2.19)$$

for some $z, w \in \mathbb{H}$.

To see that the definition does not depend on the choice of z, w, we simply use the triangle inequality to show that

$$e^{-sd(z,w)}e^{-sd(w,Tw)} \leq e^{-sd(z,Tw)} \leq e^{sd(z,w)}e^{-sd(w,Tw)}. \qquad (2.20)$$

It is easy to check that $\delta = 0$ when Γ is elementary. For Γ geometrically finite, a slight modification of the argument from Proposition 2.19 shows that

$$\#\{T \in \Gamma : d(z, Tw) \leq t\} = O(e^t). \qquad (2.21)$$

This implies in particular that $\delta \leq 1$. It turns out that $\delta = 1$ precisely when $X = \Gamma \backslash \mathbb{H}$ has finite area (Γ is of the first kind).

The interesting case is X nonelementary but of infinite area (Γ is of the second kind). Under this assumption, Beardon [14, 15] established that $0 < \delta < 1$, with $\delta > \frac{1}{2}$ if Γ has parabolic elements (X has cusps). Patterson [156] and Sullivan [202] showed that δ is the Hausdorff dimension of the limit set when Γ is geometrically finite. We'll explore their theory, along with interesting applications to spectral theory, in Chapter 14.

By setting $z = w$ in (2.19) and using the fact that $\ell(T) \leq d(w, Tw)$ for T hyperbolic, we see that

$$\sum_{\ell \in \mathcal{L}_X} e^{-s\ell} < \infty, \qquad \text{for } \operatorname{Re} s > \delta.$$

This gives the range of convergence for the following:

Definition. For a hyperbolic surface X, the *Selberg zeta function* is defined for $\operatorname{Re} s > \delta$ by the product

$$Z_X(s) := \prod_{\ell \in \mathcal{L}_X} \prod_{k=0}^{\infty} \left(1 - e^{-(s+k)\ell}\right). \qquad (2.22)$$

The expression for $Z_X(s)$ is analogous to the Euler product form of the Riemann zeta function, with the role of the prime numbers being played by the primitive length spectrum. Like the Riemann zeta function, the Selberg zeta function admits an analytic continuation to a meromorphic function of $s \in \mathbb{C}$. This can be derived from the Selberg trace formula if X has finite area. For the full geometrically finite case, meromorphic continuation was proven by Guillopé [83]. We will give the proof in Proposition 10.12, essentially following the same route. A simpler proof by dynamical methods is available if X has no cusps (Γ is convex cocompact), and we'll present this in Chapter 15.

Notes

For the basic topology of differentiable manifolds assumed in this chapter (topology of surfaces, Euler characteristic, covering spaces, fundamental group, etc.), see, e.g., Massey [123] or Munkres [142]. The differential geometry needed (metrics, Gaussian curvature, geodesics, etc.) can be found in an introductory book on surfaces, such as do Carmo [51] or Pressley [177]. Anderson [4] covers the basic geometry of the hyperbolic plane.

Our main sources for the theory of hyperbolic surfaces and Fuchsian groups were Beardon [16], Buser [37], Fenchel–Nielsen [61], and Katok [108]. Ratcliffe [179] gives a highly detailed introduction, with extensive historical notes. Milnor summarizes the history of hyperbolic geometry in [137].

Higher-dimensional hyperbolic manifolds are obtained as quotients of $\mathbb{H}^n$ by discrete subgroups of isometries. Isometries of $\mathbb{H}^3$ can be realized by extending the action of Möbius transformations on the Riemann sphere to its interior. Thus in three dimensions the oriented isometry group is $\mathrm{PSL}(2, \mathbb{C})$. A discrete subgroup of $\mathrm{PSL}(2, \mathbb{C})$ is called a *Kleinian group*; see Maskit [122] for the basic theory. Limit sets of Kleinian groups are fascinating objects; see Mumford–Series–Wright [141].

Geometric finiteness is a more complicated issue in higher dimensions; see Bowditch [27] for an account of the various possible definitions.

3

Compact and Finite-Area Surfaces

To set the stage for the development of the spectral theory in the infinite-area case, we will review some details of the spectral theory for compact and finite-area hyperbolic surfaces (Fuchsian groups of the first kind).

As noted in Chapter 1, Selberg's primary interest was automorphic forms, and in this context the important Fuchsian groups are *arithmetic*. The fundamental example is the modular group $\mathrm{PSL}(2, \mathbb{Z})$. A Fuchsian group Γ is arithmetic if it is commensurable with the modular group, meaning that the intersection $\Gamma \cap \mathrm{PSL}(2, \mathbb{Z})$ has finite index in either group (possibly after conjugation of Γ). Arithmetic Fuchsian groups are always of the first kind, and may in general contain elliptic elements of order 2 or 3.

Because of this history the spectral theory of finite-area hyperbolic surfaces has an older and more extensive literature than the infinite-area case. We will not attempt anything like a comprehensive treatment here. Our goal in this chapter is rather to highlight certain results to place our presentation of the the infinite-area theory in context. The results discussed here will not be used directly in later chapters, and many will be stated without proof. (See the notes at the end of the chapter for references.)

3.1 Selberg's trace formula for compact surfaces

In this section we will sketch the arguments leading to the Selberg trace formula [188, 189], following the development by McKean [130] and the exposition by Buser [37]. Suppose $X = \Gamma \backslash \mathbb{H}$ is a compact hyperbolic surface. We'll consider integral operators on X with kernels defined by averaging with respect to Γ. Given $f \in C^\infty([1, \infty))$, consider the sum

$$K(z, w) = \sum_{T \in \Gamma} f(\cosh d(z, Tw)). \qquad (3.1)$$

If we assume that

$$|f(u)| = O(u^{-1-\varepsilon}), \tag{3.2}$$

for some $\varepsilon > 0$, then the estimate (2.21) shows that the sum defining $K(z, w)$ converges uniformly on compact sets. The resulting kernel $K(z, w)$ is invariant under Γ and thus can be viewed as a function on $X \times X$. Since X is compact, using $K(z, w)$ as an integral kernel with respect to the hyperbolic area form dA on X defines a smoothing operator K on $L^2(X)$. In particular, K is trace class with trace given by

$$\operatorname{tr} K := \int_X K(z, z) \, dA(z).$$

The Selberg trace formula arises from the computation of this trace in two different ways.

Let Δ_X denote the (positive) Laplacian defined by the hyperbolic metric on X. The Laplacian is essentially self-adjoint on $C^\infty(X) \subset L^2(X)$, with purely discrete spectrum. (See Section A.3 for a review of these facts, with references.) There exists an orthogonal basis $\{\phi_j\}$ for $L^2(X)$ consisting of eigenvectors for Δ_X, with corresponding eigenvalues,

$$0 = \lambda_0 < \lambda_1 \le \lambda_2 \le \cdots \to \infty.$$

Our first result is that K is also diagonal with respect to the basis $\{\phi_j\}$. In fact, we can actually compute the eigenvalues κ_j explicitly in terms of $\{\lambda_j\}$ and the function f.

Proposition 3.1 (Pretrace formula). *With f, K, ϕ_j, and λ_j as above, we have*

$$K\phi_j = \kappa_j \phi_j,$$

where

$$\kappa_j = h\left(\sqrt{\lambda_j - 1/4}\right)$$

(using the square root with positive imaginary part if $\lambda_j \le 1/4$), where

$$h(t) := \int_{\mathbb{H}} y^{1/2+it} f\left(\frac{1 + x^2 + y^2}{2y}\right) dA(z). \tag{3.3}$$

We will just sketch the proof; see Buser [37, Theorem 9.3.7] for the details. First, lift ϕ_j to a Γ-invariant function on $\mathbb{H}$. Introducing a fundamental domain $\mathcal{F}$ for Γ, we write the eigenvalue equation for κ_j as

$$\kappa_j \phi_j(z) = \int_{\mathcal{F}} K(z, w) \phi_j(w) \, dA(w)$$

$$= \sum_{T \in \Gamma} \int_{\mathcal{F}} f(\cosh d(z, w)) \, \phi_j(w) \, dA(w)$$

$$= \int_{\mathbb{H}} f(\cosh d(z, w)) \, \phi_j(w) \, dA(w). \tag{3.4}$$

Set $z = i$ and let R_θ be the elliptic transformation fixing i with rotation angle θ. We claim that

$$\int_0^{2\pi} \phi_j(R_\theta w)\, d\theta = \phi_j(i) \int_0^{2\pi} (\text{Im}(R_\theta w))^{1/2+it_j}\, d\theta, \qquad (3.5)$$

where $t_j = \sqrt{\lambda_j - 1/4}$. This is shown by noting that both sides are functions of $r = d(i, w)$ that satisfy the radial eigenvalue equation $(-\partial_r^2 + \coth r \partial_r)u = \lambda_j u$, subject to the initial conditions $u(0) = \phi_j(i)$, $\partial_r u(0) = 0$. Because $f(\cosh d(i, w))$ is invariant under R_θ, substitution of (3.5) into (3.4) gives

$$\kappa_j = \int_{\mathbb{H}} f(\cosh d(i, w)) (\text{Im } w)^{1/2+it_j}\, dA(w).$$

The rest of the proof is a direct computation.

The set of eigenvalues $\{\kappa_j\}$ gives us one way to write the trace, namely

$$\text{tr } K = \sum_{j=0}^{\infty} \kappa_j.$$

The other side of the trace formula comes from computing the trace using the definition of $K(z, w)$ as an average the group Γ,

$$\text{tr } K = \sum_{T \in \Gamma} \int_{\mathcal{F}} f(\cosh d(z, Tz))\, dA(z).$$

To bring the length spectrum into the picture, recall that Proposition 2.17 gives a correspondence between $\mathcal{L}_X$ and displacement lengths of conjugacy classes of primitive hyperbolic elements of Γ.

The discreteness of Γ allows us to find a list Π of primitive elements such that any $S \in \Gamma - \{I\}$ is conjugate to T^k for a unique $T \in \Pi$ and $k \in \mathbb{Z} - \{0\}$. In other words,

$$\Gamma - \{I\} = \bigcup_{T \in \Pi} \bigcup_{k \neq 0} [T^k].$$

A general element of $[T^k]$ is represented by $RT^k R^{-1}$ with $R \in \Gamma/\langle T \rangle$. Using this decomposition of Γ in the trace gives

$$\text{tr } K = f(1) \int_{\mathcal{F}} dA + \int_{\mathcal{F}} \left[\sum_{T \in \Pi} \sum_{k=1}^{\infty} \sum_{R \in \Gamma/\langle T \rangle} f\left(\cosh d(z, RT^k R^{-1}z)\right) \right] dA(z).$$

The first term on the right is $f(1)$ times $-2\pi \chi(X)$ by Gauss–Bonnet (Theorem 2.15).

The decay assumption on f justifies switching the order of summation and integration in the remaining terms, so

$$\text{tr } K = -2\pi \chi(X) f(1) + \sum_{T \in \Pi} \sum_{k=1}^{\infty} \sum_{R \in \Gamma/\langle T \rangle} \int_{\mathcal{F}} f(\cosh d(z, RT^k R^{-1}z))\, dA(z).$$

A change of variables $z \to R^{-1}z$ transforms the integral inside the sum,

$$\int_{\mathcal{F}} f(\cosh d(z, RT^k R^{-1}z)) \, dA(z) = \int_{R^{-1}\mathcal{F}} f(\cosh d(z, T^k z)) \, dA(z).$$

This allows us to write

$$\sum_{R \in \Gamma/\langle T \rangle} \int_{\mathcal{F}} f(\cosh d(z, RT^k R^{-1}z)) \, dA(z)$$

$$= \sum_{R \in \Gamma/\langle T \rangle} \int_{R^{-1}\mathcal{F}} f(\cosh d(z, T^k z)) \, dA(z)$$

$$= \int_{\widetilde{\mathcal{F}}} f(\cosh d(z, T^k z)) \, dA(z),$$

where

$$\widetilde{\mathcal{F}} := \bigcup_{R \in \Gamma/\langle T \rangle} R\mathcal{F}.$$

It's not hard to check that $\widetilde{\mathcal{F}}$ is a fundamental domain for the cyclic group $\langle T \rangle$, and clearly $h(d(z, T^k z))$ is invariant under $\langle T \rangle$. Therefore we may express the integral over $\widetilde{\mathcal{F}}$ as an integral over $\langle T \rangle \backslash \mathbb{H}$. To compute it, we conjugate T to $z \mapsto e^{\ell} z$ and replace the fundamental domain $\widetilde{\mathcal{F}}$ by the strip $\{1 \le \operatorname{Im} z \le e^{\ell}\}$. With these transformations we have

$$\sum_{R \in \Gamma/\langle T \rangle} \int_{\mathcal{F}} f(\cosh d(z, RT^k R^{-1}z)) \, dA(z)$$

$$= \int_{\{1 \le \operatorname{Im} z \le e^{\ell}\}} f(\cosh d(z, T^k z) \, dA(z)$$

$$= 2 \int_0^{\infty} \int_1^{e^{\ell}} f\left(1 + \frac{(e^{k\ell} - 1)^2 (x^2 + y^2)}{2 e^{k\ell} y^2}\right) \frac{dx \, dy}{y^2}.$$

Setting u equal to the argument of f in the integral and changing variables leads to the following:

Proposition 3.2 (Length trace formula). *For the operator K defined by (3.1) with f satisfying (3.2), the trace is given by*

$$\operatorname{tr} K = -2\pi \chi(X) f(1)$$

$$+ \sum_{\ell \in \mathcal{L}_X} \sum_{k=1}^{\infty} \frac{\ell}{\sqrt{2} \sinh(k\ell/2)} \int_{\cosh(k\ell)}^{\infty} \frac{f(u)}{\sqrt{u - \cosh(k\ell)}} \, du.$$

Now we can derive the trace formula by combining the pretrace formula and the length trace formula. It is convenient for the applications to switch our perspective from f to the function h defined by (3.3). If we assume that h extends from $\mathbb{R}$ to a holomorphic function on $\{|\operatorname{Im} z| < \frac{1}{2} + \varepsilon\}$, satisfying a uniform bound

$$h(z) = O((1 + |z|^2)^{-1-\varepsilon}),$$

for $\varepsilon > 0$, then the Fourier transform $\widehat{h}(\xi)$ is well-defined and satisfies a bound

$$\widehat{h}(\xi) = O(e^{-(1/2+\varepsilon)|\xi|}).$$

A nontrivial computation involving the Abel transform (see [37, Section 7.3]) can then be used to derive from (3.3) the formulas

$$f(1) = \frac{1}{4\pi} \int_{-\infty}^{\infty} rh(r)\tanh(\pi r)\, dr$$

and

$$\frac{1}{\sqrt{2}} \int_{\cosh\xi}^{\infty} \frac{f(u)}{\sqrt{u - \cosh\xi}}\, du = \widehat{h}(\xi).$$

Substituting these into the length trace formula yields the following:

Theorem 3.3 (Selberg trace formula). *Let X be a compact hyperbolic surface, with $\{\lambda_j\}$ the eigenvalues of Δ_X, and $\mathcal{L}_X$ the primitive length spectrum. For h as above,*

$$\sum_{j=0}^{\infty} h\left(\sqrt{\lambda_j - 1/4}\right) = -\frac{\chi(X)}{2} \int_{-\infty}^{\infty} rh(r)\tanh(\pi r)\, dr$$

$$+ \sum_{\ell \in \mathcal{L}_X} \sum_{k=1}^{\infty} \frac{\ell}{\sinh(k\ell/2)} \widehat{h}(k\ell).$$

The simplest route to application of the trace formula is through the heat kernel. On the eigenvalue side we have the heat trace given by

$$\operatorname{tr} e^{-t\Delta_X} = \sum_{j=0}^{\infty} e^{-t\lambda_j}.$$

Thus we should set

$$h(z) = e^{-t(z^2 + 1/4)}$$

to apply Theorem 3.3. The result, first derived by McKean [130], is the following:

Theorem 3.4 (Heat trace formula). *For $t > 0$,*

$$\sum_{j=0}^{\infty} e^{-t\lambda_j} = -2\pi \chi(X) \frac{e^{-t/4}}{(4\pi t)^{3/2}} \int_{0}^{\infty} \frac{re^{-r^2/4t}}{\sinh(r/2)}\, dr$$

$$+ \frac{1}{2} \frac{e^{-t/4}}{(4\pi t)^{1/2}} \sum_{\ell \in \mathcal{L}_X} \sum_{k=1}^{\infty} \frac{\ell}{\sinh(k\ell/2)} e^{-\ell^2/4t}.$$

3.2 Consequences of the trace formula

By carefully examining both sides of the heat trace formula in Theorem 3.4, we see that from knowledge of $\operatorname{tr} e^{-t\Delta_X}$ (as a function of t), we could deduce all of the pieces $\{\lambda_j\}$, $\mathcal{L}_X$, and $\chi(X)$.

Theorem 3.5 (Huber). *The eigenvalue spectrum $\sigma(\Delta_X)$ and length spectrum $\mathcal{L}_X$ (including multiplicities) determine each other, as well as the Euler characteristic $\chi(X)$.*

Huber's argument [97] did not use the heat trace; this simpler version is due to McKean [130].

In the same paper, McKean applied the equivalence of spectrum and length spectrum to show that the set of surfaces isospectral to a given compact hyperbolic surface X is finite. The starting point is a classic result of Fricke–Klein [64], which says that the single, double, and triple traces of a set of generators of Γ are enough to fix the conjugacy class of Γ in $\mathrm{PSL}(2,\mathbb{R})$ up to a possible reflection. For each $Q \in \Gamma$ there is $R \in \Pi$ such that $Q = R^k$ and therefore we have $|\operatorname{tr} Q| = 2\cosh(k\ell/2)$ for some $\ell \in \mathcal{L}_X$ and $k \geq 1$. Fixing the spectrum (either eigenvalue or length) thus restricts these traces to a countable set of possibilities.

A discrete compact set is finite, so the finiteness of the isospectral classes will follow from a bound on the size of these traces. If we choose generators to be the side-pairing congruences of a Dirichlet domain, then the displacement length of any generator is bounded by the diameter of X. The single, double, and triple traces of generators will then be bounded by $6\cosh(\operatorname{diam}(X))$.

So we must show that fixing the spectrum puts a bound on $\operatorname{diam}(X)$. A result of Mumford [140] (see Lemma 13.11) takes care of this. The minimum length $\ell_0 = \min\{\ell \in \mathcal{L}_X\}$ is obviously a spectral invariant, and Mumford's lemma gives

$$\operatorname{diam}(X) \leq \frac{\operatorname{area}(X)}{\ell_0}.$$

Hence the spectrum puts an upper bound on the diameter, which in turn puts an upper bound on the values of single, double, and triple traces of the generators of the group. This leaves only finitely many choices, proving the following result:

Theorem 3.6 (McKean). *The spectrum $\sigma(\Delta_X)$ determines a compact hyperbolic surface X up to finitely many choices of isometry class.*

Another application of the heat trace formula involves studying the behavior of the heat trace as $t \to 0$. This behavior is in fact universal for compact Riemannian manifolds, depending only on the volume and the dimension. But in the hyperbolic case the heat trace formula allows a very explicit derivation. We can estimate

$$\frac{e^{-t/4}}{(4\pi t)^{1/2}} \sum_{\ell \in \mathcal{L}_X} \sum_{k=1}^{\infty} \frac{\ell}{\sinh(k\ell/2)} e^{-\ell^2/4t} = O(t^{-1/2})$$

and

$$\int_0^\infty \frac{re^{-r^2/4t}}{\sinh(r/2)} \, dr = 2 \int_0^\infty e^{-r^2/4t} \, dr + o(t^{1/2})$$
$$= \sqrt{4\pi t}(1 + o(1)).$$

By the heat trace formula these estimates imply that

$$\sum_{j=0}^\infty e^{-t\lambda_j} \sim \frac{\text{area}(X)}{4\pi t}, \qquad \text{as } t \to 0,$$

which is the general behavior in dimension two. Applying Karamata's Tauberian theorem (see, e.g., [208, Proposition 8.3.2]) to the heat trace asymptotic gives us Weyl's asymptotic law:

$$\lambda_k \sim \frac{4\pi k}{\text{area}(X)} \qquad \text{as } k \to \infty.$$

Analysis of the heat trace as $t \to \infty$ also leads to interesting results, related to the asymptotics of the length counting function $\pi_X(t)$ defined in (2.18). As $t \to \infty$, only the $\lambda_0 = 0$ term survives on the eigenvalue side of the heat trace,

$$\sum_{j=0}^\infty e^{-t\lambda_j} = 1 + o(1).$$

On the length side, the volume term drops out because

$$\int_0^\infty \frac{re^{-r^2/4t}}{\sinh(r/2)} \, dr = O(t),$$

by a simple change of variables. This leaves

$$\frac{1}{2} \frac{e^{-t/4}}{(4\pi t)^{1/2}} \sum_{\ell \in L_X} \sum_{k=1}^\infty \frac{\ell}{\sinh(k\ell/2)} e^{-\ell^2/4t} = 1 + o(1). \qquad (3.6)$$

This formula yields the leading asymptotic of $\pi_X(t)$, by another Tauberian argument. The result is the following theorem of Huber [97]:

Theorem 3.7 (Prime geodesic theorem). *As $t \to \infty$,*

$$\pi_X(t) \sim \frac{e^t}{t}.$$

Of course, the heat trace formula contains more asymptotic information than we used for the right-hand side of (3.6). Huber's result was actually more precise, including error terms corresponding to the eignvalues $\lambda_j < 1/4$.

Not surprisingly, the trace formula also yields important information about the Selberg zeta function $Z_X(s)$. Theorem 3.3 can be applied with the function

$$h(r) = \frac{1}{r^2 + (s - 1/2)^2} - \frac{1}{r^2 + (a - 1/2)^2},$$

for $1 < \text{Re}\, s < a$.

Theorem 3.8 (Zeta function trace formula). *For* $1 < \operatorname{Re} s < a,$

$$\frac{1}{2s-1}\frac{Z'_X}{Z_X}(s) - \frac{1}{2a-1}\frac{Z'_X}{Z_X}(a) = \sum_{k=0}^{\infty}\left[\frac{1}{\lambda_k - s(1-s)} - \frac{1}{\lambda_k - a(1-a)}\right]$$

$$+ \chi(X)\sum_{k=0}^{\infty}\left[\frac{1}{s+k} - \frac{1}{a+k}\right].$$

Since the Weyl asymptotic law implies $\lambda_k \sim ck$, the sum on the right is absolutely convergent for $s \in \mathbb{C}$, except at the obvious poles, where $s(1-s) = \lambda_k$ or $s = -k$. This establishes the analytic continuation of $Z'_X(s)/Z_X(s)$ to a meromorphic function of $s \in \mathbb{C}$. Moreover, the poles are simple with positive integer residues, so we may integrate to obtain the following:

Corollary 3.9. *For a compact hyperbolic surface* X, *Selberg's zeta function* $Z_X(s)$ *is an entire function of order 2. Its divisor consists of spectral zeros at points* s *such that* $s(1-s) \in \sigma(\Delta_X)$ *and topological zeros at* $s = -k$, $k \in \mathbb{N}_0$, *of multiplicity* $-(2k+1)\chi(X)$.

Our final result is a functional equation for the Selberg zeta function. From Theorem 3.8 we obtain

$$\frac{Z'_X}{Z_X}(s) + \frac{Z'_X}{Z_X}(1-s) = (2s-1)\chi(X)\,\pi\cot\pi s, \tag{3.7}$$

using the series expansion

$$\pi\cot\pi s = \frac{1}{s} + \sum_{k=1}^{\infty}\frac{2s}{s^2 - k^2}.$$

To integrate this we introduce the auxiliary entire function

$$G_\infty(s) := (2\pi)^{-s}\Gamma(s)G(s)^2, \tag{3.8}$$

defined in terms of the *Barnes G-function,*

$$G(s+1) := (2\pi)^{s/2}e^{-s/2-(\gamma+1)s^2/2}\prod_{k=1}^{\infty}\left(1+\frac{s}{k}\right)^k e^{-s+s^2/(2k)}.$$

(The inverse $\Gamma_2(s) := 1/G(s)$ is known as the *double gamma function*.) The logarithmic derivative of the Barnes G-function is given by

$$\frac{d}{ds}\log G(s) = \frac{1}{2}\log 2\pi - (s-1/2) + (s-1)\Psi(s),$$

where $\Psi(s)$ is the digamma function $\Gamma'/\Gamma(s)$ (see, e.g., [217]). This implies that

$$\frac{d}{ds}\log G_\infty(s) = (2s-1)(\Psi(s)-1). \tag{3.9}$$

Using the fact that $\Psi(s) - \Psi(1-s) = -\pi \cot \pi s$, we then obtain

$$\frac{d}{ds} \log \frac{G_\infty(s)}{G_\infty(1-s)} = -(2s-1)\pi \cot(\pi s). \tag{3.10}$$

Thus, (3.10) allows us to integrate (3.7) and obtain the following:

Corollary 3.10 (Selberg's functional equation). *For X a compact hyperbolic surface,*

$$\frac{Z_X(s)}{Z_X(1-s)} = \left(\frac{G_\infty(s)}{G_\infty(1-s)}\right)^{-\chi(X)},$$

meromorphically for $s \in \mathbb{C}$.

We will establish the infinite-area analogue of this result in Proposition 10.12.

3.3 Finite-area hyperbolic surfaces

Now consider a hyperbolic surface $X = \Gamma \backslash \mathbb{H}$ that is not compact but has finite area. All cofinite groups are geometrically finite (Siegel's theorem), so X has finitely many cusps. The spectrum of Δ_X is more complicated in this case, no longer purely discrete.

Theorem 3.11 (Lax–Phillips). *For a noncompact finite-volume hyperbolic surface X, the Laplacian has absolutely continuous spectrum $[1/4, \infty)$. The discrete spectrum consists of finitely many eigenvalues in $[0, 1/4)$. There are examples with infinitely many embedded eigenvalues in $[1/4, \infty)$.*

The existence of embedded eigenvalues is a very interesting and delicate point. Selberg refined an earlier conjecture of Roelcke to say (among other things) that a generic finite-area surface should have infinitely embedded eigenvalues. On the other hand, Phillips–Sarnak proved in [174] that certain deformations of Γ destroy embedded eigenvalues. They made the opposite conjecture, that embedded eigenvalues are generically absent and occur only if Γ is arithmetic. The question remains open. (The issue does not arise for infinite-area surfaces; Proposition 7.5 will show that embedded eigenvalues do not occur.)

The continuous spectrum is understood in terms of generalized eigenfunctions. Label the cusps Z_j with $j = 1, \ldots, n_c$. The model for each cusp, as given in Section 2.4, is the quotient of the horocyclic region $\{\operatorname{Im} z \geq 1\} \subset \mathbb{H}$ by the group $\Gamma_\infty = \langle z \mapsto z+1 \rangle$. In Section 2.1 we noted that $\Delta_\mathbb{H} = -y^2(\partial_x^2 + \partial_y^2)$. Thus y^s and y^{1-s} are Γ_∞-invariant solutions of the eigenvalue equation

$$\Delta_\mathbb{H} u = s(1-s)u.$$

For each cusp Z_i, there is a unique solution $E_i(s; z)$ of

$$\Delta_X E_i(s; z) = s(1-s)E_i(s; z),$$

for $\operatorname{Re} s = \frac{1}{2}$, $s \neq \frac{1}{2}$, such that as $y \to \infty$ within cusp Z_j,

$$E_i(s; z) = \delta_{ij} y^s + \phi_{ij}(s) y^{1-s} + O(y^{-\infty}), \tag{3.11}$$

for some $\phi_{ij}(s) \in \mathbb{C}$. These generalized eigenfunctions $E_i(s; z)$ are constructed as *Eisenstein series.* Assuming that Z_i corresponds to the cyclic parabolic subgroup Γ_∞ within Γ, we define

$$E_i(s; z) = \sum_{R \in \Gamma_\infty \backslash \Gamma} (\operatorname{Im} Rz)^s.$$

The sum converges for $\operatorname{Re} s > 1$ and extends meromorphically to $s \in \mathbb{C}$.

The functions $\phi_{ij}(s)$ obtained as coefficients in the asymptotic expansion (3.11) of the $E_i(s)$'s also have meromorphic extensions to $s \in \mathbb{C}$. Collectively they define the *scattering matrix*

$$S_X(s) := \left[\phi_{ij}(s)\right]_{1 \leq i, j \leq m}.$$

(This a rare situation in which the scattering "matrix" is literally a matrix.) The scattering matrix satisfies

$$S_X(s) S_X(1 - s) = I, \qquad \overline{S_X(s)} = S_X(\bar{s}) = S_X(s)^*.$$

The poles of the scattering determinant

$$\tau_X(s) := \det S_X(s)$$

with $\operatorname{Re} s < 1/2$, counted with multiplicity, are called *scattering poles.* Using a Dirichlet series representation for $\tau_X(s)$, Selberg derived the following (see [189]):

Proposition 3.12 (Selberg). *The function $\tau_X(s)$ has order at most 2, and the scattering poles lie in a vertical strip $\{c < \operatorname{Re} s < 1/2\}$.*

The Selberg trace formula extends to the finite-area case, with the contribution from the continuous spectrum expressed in terms of the scattering determinant.

Theorem 3.13 (Selberg trace formula, cofinite version). *Let X be a noncompact finite-area hyperbolic surface. If $g \in C_0^\infty(\mathbb{R})$ is even, then*

$$\sum_j \widehat{g}\left(\sqrt{\lambda_j - 1/4}\right) - \frac{1}{2\pi} \int_{-\infty}^\infty \frac{\tau_X'}{\tau_X}\left(\tfrac{1}{2} + ir\right) \widehat{g}(r) \, dr$$

$$= -\chi(X) \int_{-\infty}^\infty r \tanh(\pi r) \widehat{g}(r) \, dr + \sum_{\ell \in \mathcal{L}_X} \sum_{k=1}^\infty \frac{\ell}{\sinh(k\ell/2)} g(k\ell)$$

$$- \frac{n_c}{\pi} \int_{-\infty}^\infty \Psi(1 + ir) \widehat{g}(r) \, dr + \tfrac{1}{2}(n_c - \operatorname{tr} S_X(\tfrac{1}{2})) \widehat{g}(0) - 2n_c \log 2 g(0),$$

where $\{\lambda_j\}$ are the eigenvalues of Δ_X, n_c is the number of cusps, and $\Psi(z)$ is the digamma function $\Gamma'/\Gamma(z)$.

Selberg also established an analogue of the Weyl law in the finite-area case. In this case the counting function includes a term that could be thought of as the winding number of τ_X:

$$\#\{\lambda_j < t^2\} - \frac{1}{4\pi}\int_{-t}^{t}\frac{\tau_X'}{\tau_X}(\tfrac{1}{2}+ir)\,dr \sim \frac{\text{area}(X)}{4\pi}t^2. \tag{3.12}$$

See Selberg [189] for details.

Just as in the compact case, the trace formula allows us to compute the logarithmic derivative of the Selberg zeta function $Z_X(s)$. The resulting formula includes a sum not over eigenvalues but rather over *resonances*. In this context the precise definition of the resonance set $\mathcal{R}_X$ (quoted from Müller [139]) is:

1. For $\text{Re}\,\zeta \geq 1/2$ and $\zeta \neq 1/2$, $\zeta \in \mathcal{R}_X$ if $\zeta(1-\zeta)$ is an eigenvalue of Δ_X, with multiplicity given by the dimension of the corresponding eigenspace.
2. For $\text{Re}\,\zeta < 1/2$, $\zeta \in \mathcal{R}_X$ if either ζ is a pole of $\tau_X(s)$ or $\zeta(1-\zeta) \in \sigma_d(\Delta_X)$ (the latter possible only when $\zeta \in [0, 1/2)$), with multiplicity equal to the dimension of the eigenspace plus the order of the pole of $\tau_X(s)$ or minus the order of the zero of $\tau_X(s)$ at ζ.
3. The multiplicity of $\zeta = 1/2$ as a resonance is $(\text{tr}\,S_X(\tfrac{1}{2}) + n_c)/2$ plus twice the dimension of the eigenspace of Δ_X at $\lambda = 1/4$.

Theorem 3.14. *For fixed $a > 1$,*

$$\frac{1}{2s-1}\frac{Z_X'}{Z_X}(s) = \frac{1}{2}\sum_{\zeta\in\mathcal{R}_X}\left[\frac{1}{\zeta(1-\zeta)-s(1-s)} - \frac{1}{\zeta(1-\zeta)-a(1-a)}\right]$$
$$-\frac{1}{2s-1}\left[\frac{1}{2}\frac{\tau_X'}{\tau_X}(s) - n_c\Psi(s+1/2)\right]$$
$$+\chi(X)\sum_{k=0}^{\infty}\left(\frac{1}{s+k} - \frac{1}{a+k}\right)$$
$$+\frac{n_c - \text{tr}\,S_X(\tfrac{1}{2})}{(2s-1)^2} - \frac{n_c\log 2}{2s-1} + C.$$

See, e.g., Venkov [210, Chapter 7] for details of the derivation.

One can then deduce from the divisor of $Z_X'/Z_X(s)$ the analytic continuation of $Z_X(s)$ to a meromorphic function of $s \in \mathbb{C}$ and the location of its poles and zeros. The spectral zeros correspond to the resonance set $\mathcal{R}_X$. Topological zeros occur as in the compact case at $s = 0, -1, -2, \ldots$, with orders fixed by $\chi(X)$. In addition, we have have topological poles at $s = -1/2, -3/2, \ldots$, each with order n_c. There is a functional equation, analogous to Corollary 3.10, obtained by integrating the formula for $Z_X'/Z_X(s) + Z_X'/Z_X(1-s)$.

Using the trace formula, Müller [139] extended the results of Huber and McKean to the noncompact finite-area case:

Theorem 3.15 (Müller). *The resonance set $\mathcal{R}_X$ and the length spectrum $\mathcal{L}_X$ determine each other, as well as the genus and number of cusps. Moreover, either the* *resonance set or length spectrum determine the surface X up to finitely many possibilities.*

Notes

Our main source for the spectral theory of compact hyperbolic surfaces was Buser [37]. Venkov [210] gives a full survey of the finite-area theory. Lax–Phillips [117] give a development of the theory from within the framework of Lax–Phillips scattering theory. There are many other expository references for the Selberg trace formula in the compact or finite-area context, including Fischer [62], Hejhal [91, 90], Randol [40, Chapter 11]. Bunke–Olbrich's approach [33] to the theory of Selberg zeta and theta functions generalizes easily to rank-one locally symmetric spaces.

For a guide to the connection between quantum chaos and the spectral theory of hyperbolic surfaces see the lectures by Sarnak [186, 187].

4

Spectral Theory for the Hyperbolic Plane

In our discussion of spectral theory we naturally start with the hyperbolic plane itself, the primary example of a hyperbolic surface of infinite area. In this section we will study the Laplacian on $\mathbb{H}$,

$$\Delta_{\mathbb{H}} = -y^2(\partial_x^2 + \partial_y^2).$$

Not surprisingly, in this case we can write explicit formulas for the basic spectral objects.

For $s \in \mathbb{C}$ the function y^s satisfies an eigenvalue equation,

$$\Delta_{\mathbb{H}} y^s = s(1-s)y^s. \tag{4.1}$$

If $\operatorname{Re} s = \frac{1}{2}$, then by multiplying y^s by appropriate cutoff functions we can construct an orthonormal sequence $\{\phi_n\}$ such that $\|(\Delta_{\mathbb{H}} - \lambda)\phi_n\| \to 0$ for $\lambda = s(1-s) \in [1/4, \infty)$. (The details of this construction can be found later, in the proof of Proposition 7.2.) Weyl's criterion (Theorem A.13) then shows that $[1/4, \infty)$ is contained in the essential spectrum.

We will get a finer picture below by explicit formulas for the resolvent and spectral projections. It turns out (Theorem 4.2) that $[1/4, \infty)$ is the full spectrum of $\Delta_{\mathbb{H}}$ and is absolutely continuous.

4.1 Resolvent

The resolvent of a positive self-adjoint operator A is the bounded operator $(A - z)^{-1}$, defined for $z \in \mathbb{C} - [0, \infty)$ by the spectral theorem. In the case of the operator $\Delta_{\mathbb{H}}$, equation (4.1) hints at the fact that it will be convenient to substitute $z = s(1-s)$ and use s as our spectral parameter.

Definition. For $\operatorname{Re} s > \frac{1}{2}$, $s \notin \left[\frac{1}{2}, 1\right]$, the *resolvent* of $\Delta_{\mathbb{H}}$ is defined by

$$R_{\mathbb{H}}(s) := (\Delta_{\mathbb{H}} - s(1-s))^{-1}.$$

The resolvent kernel $R_{\mathbb{H}}(s; z, z')$ is the integral kernel of $R_{\mathbb{H}}(s)$, with respect to the hyperbolic area element $dA = y^{-2}\, dx\, dy$. With respect to dA, the integral kernel of I is written $y^2\delta(z - z')$, and so the equation

$$(\Delta_{\mathbb{H}} - s(1 - s))R_{\mathbb{H}}(s) = I$$

becomes

$$(\Delta_{\mathbb{H}} - s(1 - s))R_{\mathbb{H}}(s; z, z') = y^2\delta(z - z'), \tag{4.2}$$

with $\Delta_{\mathbb{H}}$ acting on the z coordinate. This shows that the resolvent kernel is essentially the classical *Green's function* for the operator $(\Delta_{\mathbb{H}} - s(1 - s))$.

The symmetry of $\mathbb{H}$ implies that the Green's function depends only on hyperbolic distance, so we can write $R_{\mathbb{H}}(s; z, z') = f_s(d(z, z'))$ for some function f_s. To translate (4.2) into an equation for f_s, we switch to geodesic polar coordinates (r, θ) centered on z', setting $r = d(z, z')$. In (2.10) we noted that the geodesic polar form of the hyperbolic metric is $ds^2 = dr^2 + \sinh^2 r\, d\theta^2$. The corresponding Laplacian is

$$\Delta_{\mathbb{H}} = -\frac{1}{\sinh r}\partial_r(\sinh r\, \partial_r) - \frac{1}{\sinh^2 r}\partial_\theta^2.$$

Thus the homogeneous equation corresponding to (4.2) is

$$\left[-\frac{1}{\sinh r}\partial_r(\sinh r\, \partial_r) - s(1 - s)\right]f_s(r) = 0, \tag{4.3}$$

for $r > 0$.

To solve (4.3), we make one further transformation by setting $g_s(\sigma) = f_s(r)$, where $\sigma := \cosh^2(r/2)$. Then (4.3) becomes

$$\sigma(1 - \sigma)g_s'' + (1 - 2\sigma)g_s' - s(1 - s)g_s = 0, \tag{4.4}$$

for $\sigma > 1$. This is an improvement because equation (4.4) is a special case of the classical hypergeometric equation

$$z(1 - z)h''(z) + (c - (a + b + 1)z)h'(z) - ab\, h(z) = 0. \tag{4.5}$$

The standard solution is the (Gauss) hypergeometric function, defined for $|z| < 1$ by the series

$$F(a, b; c; z) := 1 + \frac{ab}{c}z + \frac{a(a + 1)b(b + 1)}{2!\, c(c + 1)}z^2 + \cdots. \tag{4.6}$$

(This is also commonly denoted by $_2F_1$, where the subscripts refer to the numbers of parameters of each type.) The solution $F(a, b; c; z)$ is regular at $z = 0$, whereas we require a solution regular at $z = \infty$ since $\sigma \in [1, \infty)$. From the Kummer set of solutions (see, e.g., [1, equation (15.5.7)]) we select

$$h(z) = z^{-a}F(a, a + 1 - c; a + 1 - b; z^{-1}).$$

Based on (4.4), we set $a = s$, $b = 1 - s$, and $c = 1$, so that our proposed solution is

$$g_s(\sigma) = c_s \sigma^{-s} F(s, s; 2s; \sigma^{-1}).$$

Euler's integral representation of the hypergeometric function,

$$F(a, b; c; z) = \frac{\Gamma(c)}{\Gamma(b)\Gamma(c-b)} \int_0^1 \frac{t^{b-1}(1-t)^{c-b-1}}{(1-tz)^a} \, dt,$$

allows us to represent the solution for $\operatorname{Re} s > 0$ by

$$g_s(\sigma) = c_s \frac{\Gamma(2s)}{\Gamma(s)^2} \int_0^1 \frac{(t(1-t))^{s-1}}{(\sigma - t)^s} \, dt. \tag{4.7}$$

We claim that the value of c_s can be chosen so that $R_{\mathbb{H}}(s; z, z') = g_s(\sigma(z, z'))$ yields a solution of the inhomogeneous equation (4.2). We need to check the boundary condition as $\sigma \to 1$, which will confirm our choice of this particular Kummer solution and yield the value of c_s. To derive the boundary condition, we integrate (4.2) over the disk $B(0; \varepsilon)$ to obtain

$$1 = -2\pi \int_0^\varepsilon \left[(\sinh r \, f_s'(r))' + s(1-s) \sinh r \, f_s(r)' \right] dr$$

$$= -2\pi \sinh \varepsilon \, f_s'(\varepsilon) - 2\pi s(1-s) \int_0^\varepsilon \sinh r \, f_s(r) \, dr.$$

Under the requirement that the solution be locally L^2, the second term vanishes as $\varepsilon \to 0$. Taking $\varepsilon \to 0$ in the first term then implies

$$f_s'(r) = -\frac{1}{2\pi r} + O(1),$$

as $r \to 0$. Thus the appropriate boundary condition as $r \to 0$ is

$$f_s(r) \sim -\frac{1}{2\pi} \log r. \tag{4.8}$$

We could have carried out the same argument for any metric; the leading singularity of the Green's function is universal, depending only on dimension. Our formula has a minus sign because we used $-\operatorname{div} \operatorname{grad}$ for the Laplacian.

Since $\sigma \sim 1 + r^2/4$ as $r \to 0$, the boundary condition (4.8) translates to

$$g_s(\sigma) \sim -\frac{1}{4\pi} \log(\sigma - 1) \quad \text{as } \sigma \to 1. \tag{4.9}$$

The asymptotics of $g_s(\sigma)$ are easily deduced from (4.7). As $\sigma \to 1$,

$$\int_0^1 \frac{(t(1-t))^{s-1}}{(\sigma - t)^s} \, dt = \int_0^1 \frac{1}{\sigma - t} \, dt + O(1) = -\log(\sigma - 1) + O(1).$$

This shows that we will satisfy the boundary condition (4.9) by setting

$$c_s = \frac{1}{4\pi} \frac{\Gamma(s)^2}{\Gamma(2s)}.$$

Using the Legendre duplication formula,

$$\Gamma(2s) = \pi^{-1/2} 2^{2s-1} \Gamma(s)\Gamma(s + \tfrac{1}{2}), \qquad (4.10)$$

we could rewrite the coefficient as

$$c_s = 2^{-1-2s} \frac{\Gamma(s)}{\sqrt{\pi}\Gamma(s + \tfrac{1}{2})}.$$

These computations express the resolvent kernel of $\mathbb{H}$ as a function of

$$\sigma(z, z') := \cosh^2(d(z, z')/2).$$

By the hyperbolic distance formula (2.6), this function is explicitly given by

$$\sigma(z, z') = \frac{(x - x')^2 + (y + y')^2}{4yy'}. \qquad (4.11)$$

We summarize our results as follows:

Proposition 4.1. *The resolvent kernel for $\Delta_{\mathbb{H}}$ is given by*

$$R_{\mathbb{H}}(s; z, z') = g_s(\sigma(z, z')),$$

where

$$g_s(\sigma) = 2^{-1-2s} \frac{\Gamma(s)}{\sqrt{\pi}\Gamma(s + 1/2)} \sigma^{-s} F(s, s; 2s; \sigma^{-1}). \qquad (4.12)$$

For $\operatorname{Re} s > 0$ this could also be written

$$g_s(\sigma) = \frac{1}{4\pi} \int_0^1 \frac{t^{s-1}(1 - t)^{s-1}}{(\sigma - t)^s} \, dt. \qquad (4.13)$$

It will prove useful later to extend the asymptotic (4.9) to the constant term. From the series expansion of the hypergeometric function $F(s, s; 2s; \sigma^{-1})$ as $\sigma \to 1$ (see, e.g., [57, Section 2.3.1]), we read off that

$$g_s(\sigma) = -\frac{1}{4\pi} \log(\sigma - 1) - \frac{1}{2\pi}(\Psi(s) + \gamma) + o(1), \qquad (4.14)$$

where $\Psi(s)$ is the digamma function $\Gamma'/\Gamma(s)$ and γ is Euler's constant.

4.2 Generalized eigenfunctions

The classical Poisson problem transposed from the unit disk to $\mathbb{H}$ is to find a function u such that

$$\Delta_{\mathbb{H}} u = 0, \qquad u|_{\partial \mathbb{H}} = f,$$

given $f \in C(\partial \mathbb{H})$. The problem is solved by the classical Poisson kernel,

$$u(z) = \frac{1}{\pi} \int_{-\infty}^{\infty} \frac{y}{(x - x')^2 + y^2} f(x') \, dx'.$$

We can use the resolvent kernel to solve the analogous boundary problem for the equation $(\Delta_{\mathbb{H}} - s(1 - s))u = 0$.

Definition. The *generalized eigenfunctions* of $\Delta_{\mathbb{H}}$ are boundary limits of the resolvent kernel, defined for $s \notin -\mathbb{N}_0$ by

$$E_{\mathbb{H}}(s; z, x') := \lim_{y' \to 0} y'^{-s} R_{\mathbb{H}}(s; z, z'),$$

where $z' = (x', y')$.

One should think of these as analogues of the plane waves in Euclidean $\mathbb{R}^n$. By (4.12) and (4.11) we can evaluate the limit explicitly, giving

$$E_{\mathbb{H}}(s; z, x') = \frac{1}{2s - 1} \frac{\Gamma(s)}{\sqrt{\pi} \Gamma(s - \frac{1}{2})} \left[\frac{y}{(x - x')^2 + y^2} \right]^s. \tag{4.15}$$

In particular, the classical Poisson kernel is the generalized eigenfunction $E_{\mathbb{H}}(1; z, x')$. Using (4.15) we can quickly verify that

$$(\Delta_{\mathbb{H}} - s(1 - s))E_{\mathbb{H}}(s; \cdot, x') = 0, \tag{4.16}$$

justifying the term "eigenfunctions." (They are "generalized" in the sense that they do not lie in $L^2(\mathbb{H})$.)

We'll defer the solution of the generalized Poisson problem to the next section. In this section we concentrate on the application to characterizing the spectrum.

Theorem 4.2. *The spectrum of $\Delta_{\mathbb{H}}$ is absolutely continuous and equal to $[1/4, \infty)$.*

We will take an approach to the proof that foreshadows our treatment of the spectrum of general hyperbolic surfaces in Chapter 7. The first step is the following result connecting the resolvent to the generalized eigenfunctions:

Proposition 4.3. *Meromorphically for $s \in \mathbb{C}$, we have*

$$R_{\mathbb{H}}(s; z, w) - R_{\mathbb{H}}(1 - s; z, w) = -(2s - 1) \int_{-\infty}^{\infty} E_{\mathbb{H}}(s; z, x') E_{\mathbb{H}}(1 - s; w, x') \, dx'.$$

Proof. The proof starts with the observation that $R_{\mathbb{H}}(s) = (\Delta_{\mathbb{H}} - s(1-s))^{-1}$ implies, formally, that $R_{\mathbb{H}}(s) - R_{\mathbb{H}}(1 - s)$ should be equal to

$$R_{\mathbb{H}}(s) \Delta_{\mathbb{H}} R_{\mathbb{H}}(1 - s) - \Delta_{\mathbb{H}} R_{\mathbb{H}}(s) R_{\mathbb{H}}(1 - s).$$

This is not correct as stated, because the resolvents $R_{\mathbb{H}}(s)$ and $R_{\mathbb{H}}(1-s)$ cannot be composed. But we can make it into a true statement by regularizing the composition with the limit as $T \to \infty$ of the integral over the rectangular region $\Sigma_T := [-T, T] \times [T^{-1}, T]$:

$$
\begin{aligned}
R_{\mathbb{H}}(s; z, w) - R_{\mathbb{H}}(1-s; z, w) = \lim_{T \to \infty} \int_{\Sigma_T} \Big[R_{\mathbb{H}}(s; z, z') \Delta_{z'} R_{\mathbb{H}}(1-s; z', w) \\
- R_{\mathbb{H}}(1-s; z', w) \Delta_{z'} R_{\mathbb{H}}(s; z, z') \Big] dA(z').
\end{aligned}
$$

Green's formula then turns this into an integral over the boundary,

$$
\begin{aligned}
R_{\mathbb{H}}(s; z, w) - R_{\mathbb{H}}(1-s; z, w) \\
= \lim_{T \to \infty} \int_{\partial \Sigma_T} \Big[\partial_{v'} R_{\mathbb{H}}(s; z, z') \, R_{\mathbb{H}}(1-s; z', w) \\
- R_{\mathbb{H}}(s; z, z') \, \partial_{v'} R_{\mathbb{H}}(1-s; z', w) \Big] ds(z'), \quad (4.17)
\end{aligned}
$$

where $\partial_{v'}$ denotes the unit outward normal vector to Σ_T (acting on the z' variable), and ds is hyperbolic arc length.

Since z is fixed in the interior of $\mathbb{H}$, $\sigma(z, z') \to \infty$ on all edges of $\partial \Sigma_T$, and by Proposition 4.1,

$$
R_{\mathbb{H}}(s; z, z') \sim c_s \sigma(z, z')^{-s}.
$$

Similarly we have $R_{\mathbb{H}}(1-s; z', w) \sim c_{1-s} \sigma(w, z')^{s-1}$. The normal derivatives have the same asymptotics. It then becomes straightforward to bound the integrals along the right, left, and top edges of $\partial \Sigma_T$ by $O(T^{-1})$.

The interesting integral is along the bottom edge, where $y' = T^{-1} \to 0$. By definition of the generalized eigenfunctions,

$$
R_{\mathbb{H}}(s; z, z') = y'^s E_{\mathbb{H}}(s; z, x') + O(y'^{s+1}).
$$

Noting that $\partial_{v'} = -y' \partial_{y'}$ on the bottom edge, we compute

$$
y' \partial_{y'} R_{\mathbb{H}}(s; z, z') = s y'^s E_{\mathbb{H}}(s; z, x') + O(y'^{s+1}).
$$

The length element is $ds(z') = y'^{-1} dx'$, so taking the $T \to \infty$ limit gives, for the first term in (4.17),

$$
\begin{aligned}
\lim_{T \to \infty} \int_{\partial \Sigma_T} \partial_{v'} R_{\mathbb{H}}(s; z, z') \, R_{\mathbb{H}}(1-s; z', w) \, ds(z') \\
= -s \int_{-\infty}^{\infty} E_{\mathbb{H}}(s; z, x') E_{\mathbb{H}}(1-s; w, x') \, dx'.
\end{aligned}
$$

A similar result holds for the second term in the integral on the right-hand side of (4.17), and this yields the claimed formula. □

There is one case for which the expressions in Proposition 4.3 can be evaluated explicitly, namely when $z = w$. One the one hand, by (4.14) we have

$$\left[R_{\mathbb{H}}(s; z, w) - R_{\mathbb{H}}(1 - s; z, w) \right]_{z=w} = \lim_{\sigma \to 1} \left[g_s(\sigma) - g_{1-s}(\sigma) \right]$$

$$= \frac{1}{2\pi} \left[\Psi(1 - s) - \Psi(s) \right].$$

By the relation $\Gamma(s)\Gamma(1 - s) = \pi \csc \pi s$, this simplifies to

$$\left[R_{\mathbb{H}}(s, z, w) - R_{\mathbb{H}}(1 - s, z, w) \right]_{z=w} = \frac{1}{2} \cot \pi s. \tag{4.18}$$

On the other hand, we can simply evaluate

$$- (2s - 1) \int_{-\infty}^{\infty} E_{\mathbb{H}}(s; z, x') E_{\mathbb{H}}(1 - s; z, x') \, dx'.$$

$$= \frac{1}{2s - 1} \frac{\Gamma(s)\Gamma(1 - s)}{\pi \Gamma(s - \frac{1}{2})\Gamma(\frac{1}{2} - s)} \int_{-\infty}^{\infty} \frac{y}{(x - x')^2 + y^2} \, dx'$$

$$= \frac{\Gamma(s)\Gamma(1 - s)}{2\Gamma(s + \frac{1}{2})\Gamma(\frac{1}{2} - s)},$$

yielding the same answer. This function (4.18) will play an important role in our development of the zeta function in Chapter 10.

Proposition 4.3 is related to the proof of Theorem 4.2 because Stone's formula expresses the spectral projectors in terms of $R_{\mathbb{H}}(s) - R_{\mathbb{H}}(1 - s)$.

Proof of Theorem 4.2. Let P_I denote the spectral projector of $\Delta_{\mathbb{H}}$ onto $I \subset \mathbb{R}$. The standard form of the resolvent appearing in Stone's formula (Corollary A.12) corresponds to $R_{\mathbb{H}}(s) = (\Delta_{\mathbb{H}} - s(1 - s))^{-1}$ with $s(1 - s) = \lambda \pm i\varepsilon$ and $\operatorname{Re} s > 1/2$. Figure 4.1 illustrates the limit $\varepsilon \to 0$ needed for Stone's formula, with $z = \lambda + i\varepsilon$ shown on the left and the corresponding $s = \frac{1}{2} \pm \sqrt{1/4 - (\lambda \pm i\varepsilon)}$ on the right.

If $\lambda \leq 1/4$ then the limits of $(\Delta_{\mathbb{H}} - \lambda \mp i\varepsilon)^{-1}$ from either side coincide, as Figure 4.1 demonstrates. Both limits are given by $R_{\mathbb{H}}(s)$ for $s = \frac{1}{2} + \sqrt{1/4 - \lambda}$. We conclude immediately that $P_{[0,1/4)} = 0$; there is no spectrum below $1/4$.

On the other hand, if $\lambda = 1/4 + \xi^2$, then the limits of $(\Delta_{\mathbb{H}} - \lambda \mp i\varepsilon)^{-1}$ are given by $R_{\mathbb{H}}(1/2 \mp i\xi)$. Stone's formula for $b \geq a \geq 1/4$ then becomes

$$\frac{1}{2}(P_{[a,b]} + P_{(a,b)}) = \frac{1}{2\pi i} \int_{\sqrt{a-1/4}}^{\sqrt{b-1/4}} \left[R_{\mathbb{H}}(\tfrac{1}{2} - i\xi) - R_{\mathbb{H}}(\tfrac{1}{2} + i\xi) \right] 2\xi \, d\xi.$$

By Proposition 4.3, the kernel of the spectral measure is given by the restriction to the line $\operatorname{Re} s = \frac{1}{2}$ of

$$(2s - 1) \int_{-\infty}^{\infty} E_{\mathbb{H}}(s; z, x') E_{\mathbb{H}}(1 - s; w, x') \, dx'$$

$$= (2s - 1) c_s c_{1-s} \int_{-\infty}^{\infty} \left[\frac{4y}{(x - x')^2 + y^2} \right]^s \left[\frac{4v}{(u - x')^2 + v^2} \right]^{1-s} dx',$$

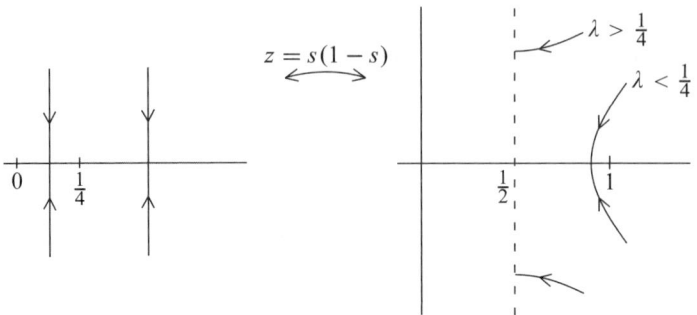

Fig. 4.1. Application of Stone's formula.

where $w = u + iv$. Since this integral defines a continuous function of s, and c_s has no poles in $\operatorname{Re} s > 0$, the spectrum is absolutely continuous. $\square$

4.3 Scattering matrix

Next we consider the problem of using the generalized eigenfunctions to associate solutions of $(\Delta_{\mathbb{H}} - s(1 - s))u = 0$ to functions on $\partial\mathbb{H}$, as in the Poisson problem. Assuming $f \in C_0^\infty(\mathbb{R})$ for convenience, we define

$$E_{\mathbb{H}}(s)f(z) := \int_{-\infty}^{\infty} E_{\mathbb{H}}(s; z, x')f(x')\,dx'.$$

This is analogous to using plane waves to construct the Fourier transform in the Euclidean case. A more general class of functions f could certainly be included here, but this is not necessary for our analysis.

Clearly $u = E_{\mathbb{H}}(s)f$ solves the eigenvalue equation, by (4.16). However, in contrast to the classical Poisson problem, the function f is not the boundary value of u, but rather a leading coefficient in the asymptotic expansion of u at the boundary.

Proposition 4.4. *For $f \in C_0^\infty(\mathbb{R})$ and $s \in \mathbb{C} - \mathbb{Z}/2$, $E_{\mathbb{H}}(s)f \in C^\infty(\mathbb{H})$ has a two-part asymptotic expansion as $y \to 0$. The leading terms are*

$$E_{\mathbb{H}}(s)f \sim \frac{1}{2s - 1}\left[y^{1-s}f + y^s g\right], \tag{4.19}$$

where $g \in C^\infty(\mathbb{R})$ is given for $\operatorname{Re} s < \frac{1}{2}$ by

$$g(x) = \frac{\Gamma(s)}{\sqrt{2}\,\Gamma(s - \frac{1}{2})} \int_{-\infty}^{\infty} \frac{f(x')}{|x - x'|^{2s}}\,dx', \tag{4.20}$$

which extends to define g for $s \notin \frac{1}{2} + \mathbb{N}$ by meromorphic continuation.

Definition. The *scattering matrix* associated to $\Delta_{\mathbb{H}}$ is the map $S_{\mathbb{H}}(s) : f \mapsto g$ defined by Proposition 4.4.

Taking the distributional Fourier transform of the integral kernel appearing in (4.20) shows that $S_{\mathbb{H}}(s)$ is a pseudodifferential operator with symbol

$$2^{1-2s} \frac{\Gamma(\frac{1}{2} - s)}{\Gamma(s - \frac{1}{2})} |\xi|^{2s-1}. \tag{4.21}$$

For $\operatorname{Re} s = \frac{1}{2}$, the two leading terms in the expansion (4.19) are both $O(s^{1/2})$, which is just on the threshold of L^2. The solution $E_{\mathbb{H}}(s) f$ is analogous to a superposition of plane waves in the Euclidean case. To make this analogy more precise, consider the wave equation

$$(\partial_t^2 + \Delta_{\mathbb{H}} - \tfrac{1}{4}) u = 0$$

(with a shift by $1/4$ to account for the bottom of the continuous spectrum). Setting $s = 1/2 + i\alpha$, we can use the generalized eigenfunction to associate to $f \in C_0^\infty(\mathbb{R})$ the solution

$$u(t, z) = 2i\alpha e^{i\alpha t} E_{\mathbb{H}}(s) f(z). \tag{4.22}$$

By Proposition 4.4 the asymptotic behavior of this solution as $y \to 0$ is

$$u(t, z) \sim y^{1/2} e^{i\alpha t + i\alpha \log y} f + y^{1/2} e^{i\alpha t - i\alpha \log y} S_{\mathbb{H}}(s) f.$$

From this we see that f is the coefficient of an incoming wave component, while $S_{\mathbb{H}}(s) f$ is the coefficient of the corresponding outgoing component.

To prove Proposition 4.4, it's apparent from the explicit formula (4.15) that we need to understand the asymptotic behavior of $[w^2 + y^2]^{-s}$ as $y \to 0$. The limit $y \to 0$ exists in the distributional sense, as shown in the following:

Lemma 4.5. *For $\varphi \in C_0^\infty(\mathbb{R})$, let*

$$v(s; x, y) = y^s \int_{-\infty}^{\infty} \frac{\varphi(x')}{\left[(x - x')^2 + y^2\right]^s} \, dx'.$$

Then, for $s \notin \mathbb{Z}/2$, $v(x, y)$ has an asymptotic expansion as $y \to 0$ of the form

$$v(s; x, y) \sim \sum_{k=0}^{\infty} y^{s+k} a_k(s; x) + \sum_{k=0}^{\infty} y^{1-s+k} b_k(s; x),$$

with the coefficients a_k, b_k meromorphic functions $\mathbb{C} \to C^\infty(\mathbb{R})$. The leading coefficients are

$$a_0(s; x) = \int_{-\infty}^{\infty} \frac{\varphi(x')}{|x - x'|^{2s}} \, dx'$$

(convergent for $\operatorname{Re}(s) < 1/2$ and extended analytically to $s \notin \frac{1}{2} + \mathbb{N}$) and

$$b_0(s; x) = \frac{\sqrt{\pi} \, \Gamma(s - \frac{1}{2})}{\Gamma(s)} \varphi(x).$$

Proof. Start by changing variables to $w = x - x'$ and breaking the integral at $|w| = 1$,

$$v(s; x, y) = y^s \int_{-1}^{1} \frac{\varphi(x - w)}{(w^2 + y^2)^s} \, dw + y^s \int_{|w| \geq 1} \frac{\varphi(w - x)}{(w^2 + y^2)^s} \, dw. \qquad (4.23)$$

We will modify the first term on the right by adding and subtracting $\varphi(x)$ in the integrand. For this purpose we first evaluate, for $\operatorname{Re} s > \frac{1}{2}$,

$$\int_{-1}^{1} (w^2 + y^2)^{-s} \, dw = 2 \int_{0}^{\infty} (w^2 + y^2)^{-s} \, dw - 2 \int_{1}^{\infty} (w^2 + y^2)^{-s} \, dw.$$

The first integral is a essentially a beta function, while the second gives a hypergeometric function:

$$\int_{-1}^{1} (w^2 + y^2)^{-s} \, dw = \frac{\sqrt{\pi} \, \Gamma(s - \frac{1}{2})}{\Gamma(s)} y^{1-2s} - \frac{2}{2s - 1} F(s - \tfrac{1}{2}; s; s + 1/2; -y^2).$$

Note that both terms on the right side extend meromorphically to $s \in \mathbb{C}$. Since the left side is analytic in s for $y > 0$, the formula is valid for all $s \in \mathbb{C} - \mathbb{Z}/2$. Using this to insert $\varphi(x)$ in the $|w| \leq 1$ integral in (4.23) gives

$$v(s; x, y) = y^s \int_{-1}^{1} \frac{\varphi(x - w) - \varphi(x)}{(w^2 + y^2)^s} \, dw + y^s \int_{|w| \geq 1} \frac{\varphi(w - x)}{(w^2 + y^2)^s} \, dw$$

$$+ \frac{\sqrt{\pi} \, \Gamma(s - \frac{1}{2})}{\Gamma(s)} y^{1-s} \varphi(x)$$

$$- \frac{2\varphi(x)}{2s - 1} y^s F(s - 1/2; s; s + 1/2; -y^2).$$

For $\operatorname{Re} s < 1$ we find as $y \to 0$ that

$$v(s; x, y) = y^s \left[\int_{-1}^{1} \frac{\varphi(x - w) - \varphi(x)}{|w|^{2s}} \, dw + \int_{|w| \geq 1} \frac{\varphi(w - x)}{|w|^{2s}} \, dw - \frac{2\varphi(x)}{2s - 1} \right]$$

$$+ \frac{\sqrt{\pi} \, \Gamma(s - \frac{1}{2})}{\Gamma(s)} y^{1-s} \varphi(x) + O(y^{\operatorname{Re} s + 2}). \qquad (4.24)$$

This gives the leading terms in the claimed asymptotic expansion.

To demonstrate the expansion to higher order we would apply the same trick but subtract off more terms of the Taylor expansion of $\varphi(x - w)$ in the $|w| \leq 1$ integral, i.e., instead of $\varphi(x - w) - \varphi(x)$ we'd substitute

$$\varphi(x - w) - \sum_{j=0}^{n} \frac{\varphi^{(j)}(x)}{j!} (-w)^j.$$

Since this expression vanishes to order w^n, we can read off the $y \to 0$ asymptotics of the resulting integral for $\operatorname{Re} s < 1 + n/2$.

The claimed expression for $b_0(s; x)$ is obvious from the second term on the right in (4.24). The term in brackets in (4.24) constitutes a standard regularization of the distribution $|w|^{-2s}$. For $\mathrm{Re}\, s < \frac{1}{2}$ the expression collapses to the convergent integral

$$a_0(s; x) = \int_{-\infty}^{\infty} \frac{\varphi(w - x)}{|w|^{2s}} \, dw.$$

One can see the first pole, at $s = \frac{1}{2}$, already in (4.24). Further analytic continuation to the right would be obtained by subtracting off more terms in the Taylor series of φ. This process reveals simple poles at $s \in \frac{1}{2} + \mathbb{N}_0$. These correspond to the poles of $|w|^{-2s}$ as a homogeneous distribution defined by analytic continuation (see, e.g., [207, Section 3.8]).   □

Proof of Proposition 4.4. We start from the formula

$$(2s - 1) E_{\mathbb{H}}(s) f(x) = \frac{\Gamma(s)}{\sqrt{\pi}\,\Gamma(s - \frac{1}{2})} \int_{-\infty}^{\infty} \frac{y^s f(x')}{\left[(x - x')^2 + y^2\right]^s} \, dx'.$$

Then Proposition 4.4 follows immediately from the formulas for a_0 and b_0 given in Lemma 4.5. Note that the first pole in a_0, at $s = \frac{1}{2}$, is canceled by the factor $\Gamma(s - \frac{1}{2})$ in the denominator of the coefficient.   □

Notes

In the appendix we give some additional background and references on spectral theory, distributions and Fourier transforms, and pseudodifferential operators. Hislop [92] gives a general introduction to scattering theory of hyperbolic manifolds, and in particular develops the spectral theory of $\mathbb{H}^n$ as an introductory example.

5

Model Resolvents for Cylinders

In this chapter we'll develop explicit formulas for the resolvent kernels of the other elementary surfaces: the hyperbolic and parabolic cylinders. These explicit formulas will serve as building blocks when we turn to the construction of the resolvent in the general case in Chapter 6. This is because of the decomposition result of Theorem 2.13, which showed that the ends of nonelementary hyperbolic surfaces are funnels and cusps.

5.1 Hyperbolic cylinders

Recall from Section 2.4 the basic model for a hyperbolic cylinder:

$$C_\ell := \Gamma_\ell \backslash \mathbb{H}, \qquad \Gamma_\ell := \langle z \mapsto e^\ell z \rangle.$$

As corresponding fundamental domain we will use $\mathcal{F}_\ell := \{1 \leq |z| \leq e^\ell\}$. This model allows us to conveniently write functions on C_ℓ in terms of their lifts to $\mathbb{H}$.

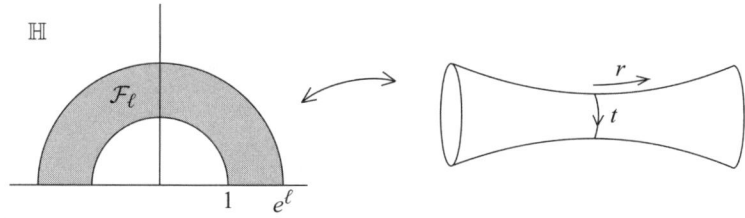

Fig. 5.1. Hyperbolic cylinder coordinates.

We can also introduce a natural set of geodesic coordinates on C_ℓ, based on distance from the central closed geodesic. Let $t \in \mathbb{R}/\ell\mathbb{Z}$ be an arc-length parameter for the central geodesic. For each value of t, we introduce a longitudinal geodesic intersecting the central geodesic orthogonally, and let r denote the signed distance from the central geodesic along these longitudinal geodesics. (Geodesic coordinates constructed in this way are usually called *Fermi coordinates*.)

Figure 5.1 illustrates the two coordinate models for the hyperbolic cylinder. If we identify the central geodesic $r = 0$ with the y-axis and the longitudinal geodesic $t = 0$ with the arc of the unit circle, then the explicit relation between the coordinates is

$$z = e^t \frac{e^r + i}{e^r - i}.$$

We can use this covering map to derive the form of the metric in the (r, t) coordinates,

$$ds^2 = dr^2 + \cosh^2 r \, dt^2. \tag{5.1}$$

(One could also deduce this from the assumptions on r and t and the curvature condition (2.9).)

We will give two separate approaches to construction of the resolvent kernel for C_ℓ. In the first, we construct the lift of the kernel to $\mathbb{H}$ by averaging $R_{\mathbb{H}}(s; z, w)$ over the action of Γ_ℓ. In the second, we will solve the Green's function equation directly in the geodesic coordinates.

In Section 4.1 we found that

$$R_{\mathbb{H}}(s; z, z') = g_s(\sigma(z, z')),$$

where $\sigma(z, z') = \cosh^2(d(z, z')/2)$ and $g_s(\sigma) \sim c_s \sigma^{-s}$ as $\sigma \to \infty$. We thus have

$$R_{\mathbb{H}}(s; z, e^{k\ell} z') = O(e^{-s|k|\ell}),$$

as $k \to \pm\infty$, uniformly for z, z' in compact sets. Hence the sum

$$R_{C_\ell}(s; z, z') := \sum_{k \in \mathbb{Z}} R_{\mathbb{H}}(s; z, e^{k\ell} z') \tag{5.2}$$

converges to an analytic function of s for $\operatorname{Re} s > 0$. This already demonstrates the analytic continuation of $R_{C_\ell}(s; z, w)$ across the continuous spectrum $\operatorname{Re} s = \frac{1}{2}$. For continuation further to the left, we'll follow the argument by Guillopé [83].

Given the definition (4.12) of $g_s(\sigma)$, the hypergeometric series (4.6) yields the expansion

$$g_s(\sigma) = \frac{1}{4\pi} \sum_{n=0}^{\infty} \frac{\Gamma(s + n)^2}{n! \Gamma(2s + n)} \sigma^{-s-n},$$

for $\sigma > 1$ and $s \notin -\mathbb{N}_0$. We can truncate the sum after finitely many terms to obtain

$$g_s(\sigma) = \sum_{n=0}^{N-1} \frac{1}{4\pi} \frac{\Gamma(s + n)^2}{n! \Gamma(2s + n)} \sigma^{-s-n} + F_N(s, \sigma), \tag{5.3}$$

with $F_N(s, \sigma)$ analytic in $\operatorname{Re} s > -N$ and satisfying

$$F_N(s, \sigma) = O(\sigma^{-s-N}) \quad \text{as } \sigma \to \infty. \tag{5.4}$$

Now substitute (5.3) into (5.2). Continuation of the sum over $F_N(s, \sigma)$ will follow from (5.4). The other pieces that we need to analyze are sums of the form

$$\sum_{k \in \mathbb{Z}} \sigma(z, e^{k\ell} z')^{-s-n} = (4yy')^{s+n} \sum_{k \in \mathbb{Z}} \left[a e^{k\ell} + 2b + c e^{-k\ell} \right]^{-s-n},$$

for $n = 0, \ldots, N-1$, where

$$a = x'^2 + y'^2, \quad b = yy' - xx', \quad c = x^2 + y^2. \tag{5.5}$$

Note that the form $q(u, v) := au^2 + 2buv + cv^2$ is positive definite, because the discriminant $ac - b^2 = (xy' + yx')^2 > 0$.

Lemma 5.1. *For $q(u, v) = au^2 + 2buv + cv^2$ a positive definite quadratic form, the series*

$$H(a, b, c, \ell; s) := \sum_{k \in \mathbb{Z}} q(e^{k\ell/2}, e^{-k\ell/2})^{-s}$$

(convergent for $\operatorname{Re} s > 0$) admits a meromorphic continuation in s with poles at $(2\pi i/\ell)\mathbb{Z} - \mathbb{N}_0$.

Proof. Rewrite the sum as

$$q(1, 1)^{-s} + \sum_{k=1}^{\infty} e^{-sk\ell} \left[q(1, e^{-k\ell})^{-s} + q(e^{-k\ell}, 1)^{-s} \right].$$

Because q is positive definite, $q(1, v) \neq 0$ for $|v| \leq \sqrt{a/c}$. Thus $q(1, v)^{-s}$ is analytic in this disk and the Taylor expansion

$$q(1, v)^{-s} = \sum_{j=0}^{\infty} \alpha_j v^j$$

converges uniformly and absolutely for $|v| \leq \sqrt{a/c} - \varepsilon$. Likewise,

$$q(u, 1)^{-s} = \sum_{j=0}^{\infty} \beta_j u^j$$

converges uniformly and absolutely for $|u| \leq \sqrt{c/a} - \varepsilon$.

For k sufficiently large, say $\geq M$, we can interchange the original summation with the Taylor expansion. Breaking the sum at this point gives

$$\sum_{k \in \mathbb{Z}} q(e^{k\ell/2}, e^{-k\ell/2})^{-s} = \sum_{|k| < M} q(e^{k\ell/2}, e^{-k\ell/2})^{-s} + \sum_{j=0}^{\infty} (\alpha_j + \beta_j) \frac{e^{-(s+j)M\ell}}{1 - e^{-(s+j)\ell}}.$$

The meromorphic continuation is now evident, with simple poles occurring at the points where $(s+j)\ell$ is an integer multiple of $2\pi i$. □

This lemma is the crucial step in establishing our next result:

Proposition 5.2. *The resolvent $R_{C_\ell}(s)$ has a meromorphic continuation to $s \in \mathbb{C}$ with poles at $s \in (2\pi i/\ell)\mathbb{Z} - \mathbb{N}_0$.*

Proof. For some fixed $N > 0$ we use (5.3) to write

$$
R_{C_\ell}(s; z, z') = \sum_{n=0}^{N-1} \frac{1}{4\pi} \frac{\Gamma(s+n)^2}{n!\,\Gamma(2s+n)} H(a, b, c, \ell; s+n)(4yy')^{s+n}
$$
$$
+ \sum_{k \in \mathbb{Z}} F_N(s, \sigma(z, e^{k\ell} z')), \tag{5.6}
$$

with a, b, c as in (5.5). By Lemma 5.1, $H(a, b, c, \ell; s+n)$ is meromorphic in s with poles at $s \in (2\pi i/\ell)\mathbb{Z} - \mathbb{N}_0 - n$. By (5.4), the remainder term $F_N(s, \sigma(z, e^{k\ell} z'))$ is $O(e^{-(s+N)k\ell})$ as $k \to \infty$, so the sum over k converges to define an analytic function for $\mathrm{Re}\, s > -N$. $\qquad\square$

Now for the second approach to analyzing $R_{C_\ell}(s)$, which follows Guillopé–Zworski [86], although they computed the scattering matrix rather than the resolvent kernel. The strategy is to solve directly for the Fourier components of the resolvent kernel. For this purpose it's convenient to reparametrize the central geodesic by $\theta = 2\pi t/\ell$. By the expression for the funnel metric (5.1), the Laplacian in the coordinates (r, θ) is

$$
\Delta_{C_\ell} = -\partial_r^2 - \tanh r \, \partial_r - \frac{4\pi^2}{\ell^2 \cosh^2 r} \partial_\theta^2. \tag{5.7}
$$

We will write the Green's function using a Fourier composition in the θ variable,

$$
R_{C_\ell}(s; z, z') = \sum_{k \in \mathbb{Z}} u_k(s; r, r') e^{ik(\theta - \theta')}.
$$

The defining equation $(\Delta_{C_\ell} - s(1-s))R_{C_\ell} = I$ then reduces to an equation for the coefficients u_k,

$$
\left[-\partial_r^2 - \tanh r \, \partial_r - \frac{(2\pi i k)^2}{\ell^2 \cosh^2 r} - s(1-s) \right] u_k(s; r, r') = \frac{\delta(r - r')}{\ell \cosh r}. \tag{5.8}
$$

This effectively reduces the problem to a one-dimensional Schrödinger equation. The particular potential that appears here is of a type known as a *Pöschl–Teller potential*.

For notational convenience, set

$$
\alpha_k := \frac{2\pi k}{\ell}.
$$

The homogeneous equation for u_k is of hypergeometric type. To reduce it to the standard form, we'll change variables to $\eta = -\sinh^2 r$, and introduce

$$f(\eta) = \frac{u_k(s; r, r')}{\sinh r (\cosh r)^{i\alpha_k}},$$

with r' regarded as a fixed parameter. This transforms the homogeneous equation into

$$\eta(1 - \eta) f''(\eta) + \left[\tfrac{3}{2} - (\tfrac{5}{2} + i\alpha_k)\eta\right] f'(\eta)$$
$$- \tfrac{1}{4}\left[s(1 - s) + (i\alpha_k + 1)(i\alpha_k + 2)\right] f(\eta) = 0.$$

This matches the hypergeometric form (4.5), with parameters

$$a = (i\alpha_k + 1 + s)/2, \quad b = (i\alpha_k + 2 - s)/2, \quad c = 3/2.$$

The general solution is

$$f(\eta) = AF(a, b; c; \eta) + B\eta^{1-c} F(a + 1 - c, b + 1 - c; 2 - c; \eta).$$

For the homogeneous equation for u_k this translates back to the solution

$$A \sinh r (\cosh r)^{i\alpha_k} F(\tfrac{i\alpha_k+1+s}{2}, \tfrac{i\alpha_k+2-s}{2}; \tfrac{3}{2}; -\sinh^2 r)$$
$$+ B(\cosh r)^{i\alpha_k} F(\tfrac{i\alpha_k+s}{2}, \tfrac{i\alpha_k+1-s}{2}; \tfrac{1}{2}; -\sinh^2 r). \tag{5.9}$$

To fix the coefficients, consider the asymptotics of the hypergeometric functions as $r \to \infty$, which are evident from the identity (see, e.g., [1, equation (15.3.7)])

$$F(a, b; c; t) = \frac{\Gamma(c)\Gamma(b - a)}{\Gamma(c - a)\Gamma(b)} (-t)^{-a} F(a, a + 1 - c; 1 + a - b; t^{-1})$$
$$+ \frac{\Gamma(c)\Gamma(a - b)}{\Gamma(c - b)\Gamma(a)} (-t)^{-b} F(b, b + 1 - c; 1 + b - a; t^{-1}), \tag{5.10}$$

valid for $t < 0$. From this we can see that the general solution (5.9) has leading asymptotic terms proportional to $(\sinh r)^{-s}$ and/or $(\sinh r)^{s-1}$. Since $R_{C_\ell}(s)$ must be a bounded operator on $L^2(C_\ell)$ for $\mathrm{Re}\, s > 1$, our solution ought to have only $(\sinh r)^{-s}$ as the leading asymptotic as $r \to \infty$. Therefore we take the combination

$$v_k(s; r) := \frac{\sinh r (\cosh r)^{i\alpha_k}}{\Gamma(\tfrac{s-i\alpha_k}{2})\Gamma(\tfrac{s+i\alpha_k}{2})} F(\tfrac{i\alpha_k+1+s}{2}, \tfrac{i\alpha_k+2-s}{2}; \tfrac{3}{2}; -\sinh^2 r)$$
$$- \frac{\cosh^{i\alpha_k} r}{2\Gamma(\tfrac{1+s-i\alpha_k}{2})\Gamma(\tfrac{1+s+i\alpha_k}{2})} F(\tfrac{i\alpha_k+s}{2}, \tfrac{i\alpha_k+1-s}{2}; \tfrac{1}{2}; -\sinh^2 r). \tag{5.11}$$

Since the coefficients in (5.11) are entire, and the hypergeometric function is analytic in its first two parameters, $v_k(s)$ is analytic in s.

Using the identity (5.10), we can rewrite the formula for $v_k(s)$, when $r > 0$ and $s \notin \tfrac{1}{2} - \mathbb{N}$, as

$$v_k(s; r) = -\frac{(\sinh r)^{-s} (\coth r)^{i\alpha_k}}{2\sqrt{\pi}\, \Gamma(s + \tfrac{1}{2})} F(\tfrac{i\alpha_k+1+s}{2}, \tfrac{i\alpha_k+s}{2}; \tfrac{1}{2} + s; -\sinh^{-2} r), \tag{5.12}$$

showing that $v_k(s; r)$ has the desired asymptotic behavior as $r \to +\infty$. If $s = \frac{1}{2} - n$ then the zero in the coefficient cancels the pole caused by $c = \frac{1}{2} + s$ in the hypergeometric function. The effect is to truncate the first n terms from the hypergeometric series (see, e.g., [1, equation (15.2.1)]). This means that $v_k(\frac{1}{2} - n; r)$ decays like $e^{-(n+1/2)r}$ as $r \to \infty$, instead of growing like $e^{(n-1/2)r}$ as expected. The extra decay at these points will play a significant role later, in Section 8.5.

Proposition 5.3. *The resolvent kernel on C_ℓ has the Fourier decomposition*

$$R_{C_\ell}(s; z, z') = \sum_{k \in \mathbb{Z}} u_k(s; r, r') e^{ik(\theta - \theta')},$$

with u_k given in terms of (5.11) by

$$u_k(s; r, r') = \begin{cases} a_k(s) v_k(s; -r) v_k(s; r'), & r \leq r', \\ a_k(s) v_k(s; r) v_k(s; -r'), & r \geq r', \end{cases} \tag{5.13}$$

where

$$a_k(s) = \frac{\pi}{\ell} 4^{1-s} \Gamma\left(s - \frac{2\pi i k}{\ell}\right) \Gamma\left(s + \frac{2\pi i k}{\ell}\right). \tag{5.14}$$

The residue of $R_{C_\ell}(s; z, z')$ at a pole $s \in -\mathbb{N}_0 + (2\pi i/\ell)\mathbb{Z}$ is the kernel of a finite-rank operator.

Proof. To get the proper decay at infinity, $u_k(s; r, r')$ must be a multiple of $v_k(s; r)$ for $r > r'$ and of $v_k(s; -r)$ for $r < r'$. These two solutions must be matched appropriately at r', in order to solve the inhomogeneous equation (5.8).

The deduction of the boundary condition at $r = r'$ is a relatively standard argument from the theory of one-dimensional Schrödinger operators. In general, one would consider a Green's function equation of the form

$$-\left(q(r)u'(r)\right)' + V(r)u(r) = \delta(r - r'). \tag{5.15}$$

Our equation (5.8) fits this model with $q(r) = \ell \cosh r$ and $V(r)$ a smooth potential. The first observation is that the solution $u(r)$ will be continuous across r', because a jump in u would cause a singularity of the form $\delta'(r - r')$ that could not be canceled by the potential term Vu. The same reasoning shows that a jump discontinuity in the derivative of u is needed to produce the delta function on the right. Integrating (5.15) across r' gives

$$\int_{r'-\varepsilon}^{r'+\varepsilon} \left[-\left(q(r)u'(r)\right)' + V(r)u(r)\right] dr = 1.$$

Using integration by parts and taking the limit $\varepsilon \to 0$ (assuming local integrability of $V(r)$), we find the jump condition

$$u'(r'_-) - u'(r'_+) = \frac{1}{q(r')} \tag{5.16}$$

on the derivative, where $r'_\pm$ denotes a left/right limit as $r \to r'$.

Another standard fact about equations of the form (5.15) concerns the Wronskian of two solutions v_1, v_2, of the homogeneous equation, $(qv'_j)' = Vv_j$. By differentiating $q(r)\left[v_1(r)v'_2(r) - v'_1(r)v_2(r)\right]$ we can deduce a formula for the Wronskian:

$$W[v_1, v_2](r) := v_1(r)v'_2(r) - v'_1(r)v_2(r) = \frac{c}{q(r)}, \qquad (5.17)$$

where c is independent of r.

Let us apply these facts to the computation of $u_k(s)$. The function $u_k(s; r, r')$ given in (5.13) clearly has the right decay at infinity and is continuous across $r = r'$. The jump in the derivative at $r = r'$ is given by

$$\partial_r u_k(s; r, r')\big|_{r \to r'_-} - \partial_r u_k(s; r, r')\big|_{r \to r'_+}$$

$$= a_k(s)\left[-v'_k(s; -r')v_k(s; r') - v'_k(s; r')v_k(s; -r')\right].$$

By the Wronskian identity (5.17), with $v_1(r) = v_k(s; r)$, $v_2(r) = v_k(s; -r)$, and $q(r) = \ell \cosh r$, we have

$$-v'_k(s; -r')v_k(s; r') - v'_k(s; r')v_k(s; -r') = \frac{c}{\ell \cosh r'}, \qquad (5.18)$$

for some c independent of r'. This shows that the jump condition (5.16) can be satisfied for all r' by taking $a_k(s) = 1/c$. To compute c, we note that

$$v_k(s; 0) = -\frac{1}{2\Gamma(\frac{1+s-i\alpha_k}{2})\Gamma(\frac{1+s+i\alpha_k}{2})} \qquad (5.19)$$

and

$$v'_k(s; 0) = \frac{1}{\Gamma(\frac{s-i\alpha_k}{2})\Gamma(\frac{s+i\alpha_k}{2})}.$$

With the duplication formula (4.10) for the gamma function, evaluation of (5.18) at $r' = 0$ shows that

$$\frac{c}{\ell} = \frac{4^{s-1}}{\pi \Gamma(s - i\alpha_k)\Gamma(s + i\alpha_k)}.$$

Setting $a_k(s) = 1/c$ yields the formula (5.14).

Since the Green's function must be smooth off the diagonal, the coefficients $u_k(s; r, r')$ decay rapidly as $|k| \to \infty$ for $r \neq r'$. The Fourier series therefore converges uniformly on compact sets bounded away from the diagonal. The $v_\pm(s)$ are analytic, so it follows that the only poles of $R_{C_\ell}(s)$ are the obvious poles in the coefficients $a_k(s)$, which occur at $s \in -\mathbb{N}_0 + (2\pi i/\ell)\mathbb{Z}$. Thus it is easy to check explicitly that the residues have finite rank.   □

The poles of the hyperbolic cylinder resolvent are shown in Figure 5.2. We will show in Section 8.2 that the residue of $R_{C_\ell}(s)$ at any pole has rank 2.

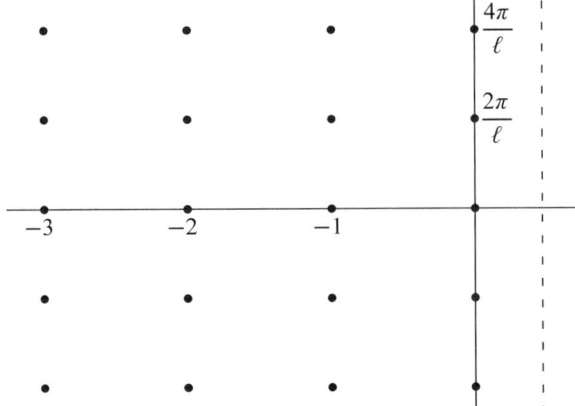

Fig. 5.2. Poles of $R_{C_\ell}(s)$.

5.2 Funnels

While the cylinder resolvent itself would suffice as a model for the resolvent on a funnel end, it is convenient to work out the funnel resolvent separately. Recall that the funnel F_ℓ is half of C_ℓ, with boundary the central closed geodesic. We will impose Dirichlet boundary conditions at the geodesic boundary. (The choice is largely for notational convenience; any boundary condition that defines a self-adjoint extension of Δ_{F_ℓ} would be suitable here.)

In the model $C_\ell = \Gamma_\ell \backslash \mathbb{H}$, F_ℓ corresponds to the region $\operatorname{Re} z \geq 0$, while in geodesic normal coordinates we naturally identify F_ℓ with $\{r \geq 0\}$. Since the hyperbolic cylinder is the double of a funnel across its boundary geodesic, the classical method of images gives a natural way to obtain the funnel Green's function from $R_{C_\ell}(s; z, z')$. Using coordinates $z, w \in \mathbb{H}$, we have

$$R_{F_\ell}(s; z, w) = R_{C_\ell}(s; z, w) - R_{C_\ell}(s; z, -\overline{w}). \tag{5.20}$$

Thus Proposition 5.2 immediately implies the following:

Corollary 5.4. *The funnel resolvent $R_{F_\ell}(s)$ has a meromorphic continuation to $s \in \mathbb{C}$.*

This also implies that the poles of $R_{F_\ell}(s)$ are contained in the set $-\mathbb{N}_0 + (2\pi i/\ell)\mathbb{Z}$ of poles of $R_{C_\ell}(s)$. But we'll see below that half of these poles cancel between the two terms in (5.20).

As with the cylinder, an alternative approach is to solve directly for the Fourier coefficients. If we write the Fourier decomposition,

$$R_{F_\ell}(s; z, z') = \sum_{k \in \mathbb{Z}} u_k(s; r, r') e^{ik(\theta - \theta')},$$

the coefficients $u_k(s; r, r')$ satisfy the same Pöschl–Teller Schrödinger equation (5.8), except that $r, r' \geq 0$ and the coefficients have a boundary condition $u(k; 0, r') = 0$. Thus for $0 \leq r < r'$ we use a homogeneous solution given by

$$v_k^0(s; r) = \sinh r (\cosh r)^{i\alpha_k} F(\tfrac{i\alpha_k + 1 + s}{2}, \tfrac{i\alpha_k + 2 - s}{2}; \tfrac{3}{2}; -\sinh^2 r). \qquad (5.21)$$

For $r > r'$ we continue to use $v_k(s; r)$ as defined in (5.11).

Proposition 5.5. *The Fourier decomposition of the resolvent kernel on F_ℓ is given by*

$$R_{F_\ell}(s; z, z') = \sum_{k \in \mathbb{Z}} u_k(s; r, r') e^{ik(\theta - \theta')},$$

with u_k given by

$$u_k(s; r, r') = \begin{cases} b_k(s) v_k^0(s; r) v_k(s; r'), & 0 \leq r \leq r', \\ b_k(s) v_k(s; r) v_k^0(s; r'), & 0 \leq r' \leq r, \end{cases} \qquad (5.22)$$

where

$$b_k(s) = \frac{2}{\ell} \Gamma\left(\tfrac{1}{2}(1 + s + \tfrac{2\pi ik}{\ell})\right) \Gamma\left(\tfrac{1}{2}(1 + s - \tfrac{2\pi ik}{\ell})\right). \qquad (5.23)$$

For each $n = \mathbb{N}_0$ and $k \in \mathbb{Z}$, $R_{F_\ell}(s)$ has a pole at $s = -1 - 2n + (2\pi i/\ell)k$, with residue an operator of rank two.

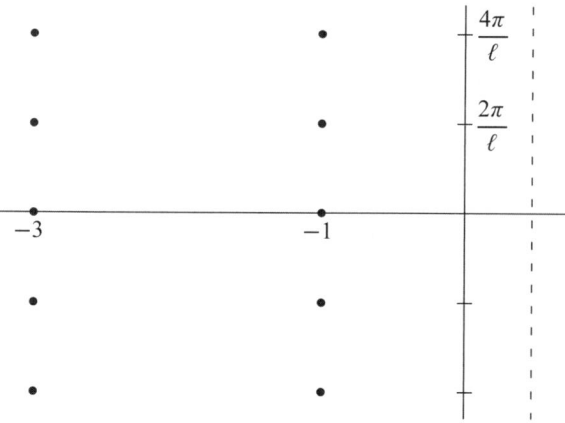

Fig. 5.3. Poles of $R_{F_\ell}(s)$.

    The set of poles of a funnel resolvent, shown in Figure 5.3, thus consists of half of the poles for the corresponding cylinder. The proof of Proposition 5.5 is nearly identical to the proof of Proposition 5.3. After adopting (5.22) as an ansatz, we compute $b_k(s)$ by evaluating the Wronskian of v_k^0, v_k in the limit $r \to 0$, using (5.19) along with the obvious facts $v_k^0(s; 0) = 0$ and $\partial_r v_k^0(s; 0) = 1$.

5.3 Parabolic cylinder

The parabolic cylinder gives us the model for a cusp end. We can work out the formula for its resolvent by methods similar to those used for the hyperbolic cylinder. The standard model is

$$C_\infty := \Gamma_\infty \backslash \mathbb{H}, \qquad \Gamma_\infty = \langle z \mapsto z + 1 \rangle.$$

As for the hyperbolic cylinder, we approach the resolvent in two different ways. The first method, from Guillopé [83], starts with writing the resolvent kernel as an average over the group:

$$R_{C_\infty}(s; z, z') = \sum_{k \in \mathbb{Z}} R_{\mathbb{H}}(s; z, z' - k),$$

convergent for $\operatorname{Re} s > \frac{1}{2}$. By representing $R_{\mathbb{H}}(s; z, w)$ through the truncated sum (5.3), the analytic continuation will be reduced to the following result:

Lemma 5.6. *The sum*

$$J(a, b; s) := \sum_{k \in \mathbb{Z}} \left[(a + k)^2 + b^2 \right]^{-s}$$

has a meromorphic continuation to $s \in \mathbb{C}$. For $s \notin \frac{1}{2} - \mathbb{N}$, this function has the asymptotic behavior

$$J(a, b; s) = \frac{\sqrt{\pi} \Gamma(s - \frac{1}{2})}{\Gamma(s)} b^{1-2s} + o(b^{-\infty}),$$

as $b \to \infty$, uniformly in a.

Proof. A standard trick for handling terms of the form λ^{-s} is introduction of the gamma integral $\int_0^\infty t^{s-1} e^{-\lambda t} dt = \Gamma(s) \lambda^{-s}$. This allows us to write

$$J(a, b; s) = \frac{1}{\Gamma(s)} \sum_{k \in \mathbb{Z}} \int_0^\infty t^{s-1} e^{-((a+k)^2 + b^2)t} dt.$$

Now the sum can be transformed using the Poisson summation formula (Theorem A.6):

$$\sum_{k \in \mathbb{Z}} e^{-(a+k)^2 t} = \sqrt{\frac{\pi}{t}} \sum_{k \in \mathbb{Z}} e^{2\pi i k a - \pi^2 k^2 / t}.$$

This gives

$$J(a, b; s) = \frac{\sqrt{\pi}}{\Gamma(s)} \sum_{k \in \mathbb{Z}} \int_0^\infty t^{s-3/2} e^{2\pi i k a - \pi^2 k^2 / t - b^2 t} dt.$$

Rescaling the integration variable to pull out the b dependence yields

$$J(a, b; s) = \frac{\sqrt{\pi}}{\Gamma(s)} b^{1-2s} \left[\Gamma(s - 1/2) + \sum_{k \neq 0} \int_0^\infty t^{s-3/2} e^{2\pi i k a - \pi^2 b^2 k^2 / t - t} \, dt \right].$$

The remaining integral is analytic in s, and by splitting it at $t = \pi bk$ we can easily estimate

$$\left| \int_0^\infty t^{s-3/2} e^{2\pi i k a - \pi^2 b^2 k^2 / t - t} \, dt \right| \leq C(bk)^{\operatorname{Re} s - 3/2} e^{-\pi bk}. \qquad \square$$

Proposition 5.7. *The resolvent for the parabolic cylinder, $R_{C_\infty}(s)$, has a meromorphic continuation to $s \in \mathbb{C}$.*

Proof. By the decomposition (5.3) we can write

$$R_{C_\infty}(s; z, z') = \sum_{n=0}^{N-1} \frac{1}{4\pi} \frac{\Gamma(s+n)^2}{n! \Gamma(2s+n)} J(a, b; s+n)(4yy')^{s+n}$$
$$+ \sum_{k \in \mathbb{Z}} F_N(s, \sigma(z, z' - k)), \qquad (5.24)$$

with $a = x - x'$ and $b = y + y'$. The decay of F_N as $\sigma \to \infty$ implies that the remainder term is analytic for $\operatorname{Re}(s) > \frac{1}{2} - N$, and Lemma 5.6 gives the meromorphic continuation of the terms $n = 0$ to $N - 1$. $\qquad \square$

Lemma 5.6 indicates that $R_{C_\infty}(s)$ may have poles at $\frac{1}{2} - \mathbb{N}_0$, but in fact we'll see shortly that $s = \frac{1}{2}$ is the only pole.

Our second method for analysis of the kernel of $R_{C_\infty}(s)$ is direct computation of the Fourier decomposition. This is even simpler than in the hyperbolic cylinder case because in the $\Gamma_\infty \backslash \mathbb{H}$ model we already have periodicity in the x-variable. For the Fourier decomposition we simply set $\theta = 2\pi x$ and keep y as the other coordinate. In these coordinates the equation for the Green's function is

$$\left[-y^2 \partial_y^2 - 4\pi^2 y^2 \partial_\theta^2 - s(1-s) \right] u = y^2 \delta(\theta - \theta') \delta(y - y').$$

Setting

$$R_{C_\infty}(s; z, z') = \sum_{k \in \mathbb{Z}} u_k(s; y, y') e^{ik(\theta - \theta')} \qquad (5.25)$$

gives us the coefficient equation

$$\left[-y^2 \partial_y^2 + (2\pi ky)^2 - s(1-s) \right] u_k(s; y, y') = y^2 \delta(y - y').$$

As in the proof of Proposition 5.3, the delta-function singularity is achieved by imposing continuity of the $u_k(s; y, y')$ at $y = y'$ together with a jump condition on the first derivatives:

$$\partial_y u_k(s; y, y')|_{y \to y'_-} - \partial_y u_k(s; y, y')|_{y \to y'_+} = 1.$$

For $k = 0$, the obvious solutions of the homogeneous equation are y^s and y^{1-s}. For $k \neq 0$, the substitution $u_k = \sqrt{y} f(2\pi |k| y)$ transforms the homogeneous equation for u_k into a Bessel form,

$$t^2 f''(t) + t f'(t) - (t^2 + (s - \tfrac{1}{2})^2) f = 0.$$

The two independent solutions are the modified Bessel functions

$$I_{s-\frac{1}{2}}(t) \quad \text{and} \quad K_{s-\frac{1}{2}}(t).$$

Proposition 5.8. *The Fourier decomposition of the parabolic cylinder resolvent kernel is given by*

$$R_{C_\infty}(s; z, z') = \sum_{k \in \mathbb{Z}} u_k(s; y, y') e^{2\pi i k(x - x')},$$

with

$$u_0(s; y, y') = \frac{1}{2s - 1} \begin{cases} y^s y'^{1-s}, & y \leq y', \\ y^{1-s} y'^s, & y \geq y', \end{cases} \tag{5.26}$$

and for $k \neq 0$,

$$u_k(s; y, y') = \begin{cases} \sqrt{yy'} I_{s-\frac{1}{2}}(2\pi |k| y) K_{s-\frac{1}{2}}(2\pi |k| y'), & y \leq y', \\ \sqrt{yy'} K_{s-\frac{1}{2}}(2\pi |k| y) I_{s-\frac{1}{2}}(2\pi |k| y'), & y \geq y', \end{cases} \tag{5.27}$$

for $k \neq 0$. The only pole of $R_{C_\infty}(s)$ is the pole at $s = \frac{1}{2}$ occurring in the u_0 term.

Proof. To obtain the proper decay at infinity in the u_0 term for $\operatorname{Re} s > \frac{1}{2}$, we take the solution y^s for $y < y'$ and y^{1-s} for $y > y'$. Solving for coefficients using the Wronskian as in the proof of Proposition 5.3 is straightforward.

For the $k \neq 0$ terms, note that for $\operatorname{Re} s > 1/2$ the asymptotics of the modified Bessel functions as $t \to 0$ are given by (see, e.g., [1, Sections 9.6–9.7])

$$I_{s-\frac{1}{2}}(t) \sim \frac{(t/2)^{s-\frac{1}{2}}}{\Gamma(s + \frac{1}{2})}, \qquad K_{s-\frac{1}{2}}(t) \sim \frac{1}{2} \Gamma(s - \tfrac{1}{2}) (t/2)^{1/2-s}.$$

As $t \to \infty$ the asymptotics are independent of the parameter s,

$$I_{s-1/2}(t) \sim \frac{e^t}{\sqrt{2\pi t}}, \qquad K_{s-1/2}(t) \sim \sqrt{\frac{\pi}{2t}} e^{-t}.$$

Thus we use the homogeneous solution

$$\begin{cases} \sqrt{y} I_{s-1/2}(2\pi |k| y), & \text{for } y < y', \\ \sqrt{y} K_{s-1/2}(2\pi |k| y), & \text{for } y > y'. \end{cases}$$

The jump condition at y' will be nicely accounted for by the Wronskian formula [1, equation (9.6.5)]

$$K_\nu(t) I_\nu'(t) - K_\nu'(t) I_\nu(t) = \frac{1}{t}.$$

The derivation of (5.27) follows easily.   □

Notes

Ikawa [98] and Gérard [71] showed that for scattering by two convex obstacles, the resonance set is asymptotic to a lattice, analogous to the lattice of resonances for the hyperbolic cylinder as shown in Figure 5.2. As Zworski noted in [226], the underlying classical dynamics of the two systems are the same; both consist of a single closed hyperbolic trajectory.

6

The Resolvent

Now that we have worked out the resolvent kernels for all of the elementary surfaces, we turn to the general case of a geometrically finite hyperbolic surface $X = \Gamma \backslash \mathbb{H}$ of infinite area. In this chapter we will develop a precise picture of the structure of the resolvent kernel and establish its meromorphic continuation. This will lead us to a complete characterization of the spectrum $\sigma(\Delta_X)$ in Chapter 7, and also to the definitions of the scattering matrix and resonances.

Definition. For a hyperbolic surface X the *resolvent* of Δ_X is defined for $\operatorname{Re} s > \frac{1}{2}$, $s \notin \left[\frac{1}{2}, 1\right]$, by

$$R_X(s) := (\Delta_X - s(1-s))^{-1}.$$

To establish the meromorphic continuation to $s \in \mathbb{C}$ we will follow the philosophy of geometric scattering theory developed by Melrose, as outlined in [136]. The general theme of this approach is the introduction of a "radial" compactification of X, after which Δ_X is treated as an elliptic differential operator on $\overline{X}$ that is degenerate at the boundary. Spectral theory for various types of metrics that are asymptotic to standard forms can be handled in this way, e.g., asymptotically Euclidean, cylindrical, or hyperbolic.

6.1 Compactification

In Section 2.1 we defined $\partial \mathbb{H}$ as the boundary of $\mathbb{H}$ in the Riemann-sphere topology. The corresponding compactification, $\overline{\mathbb{H}} = \mathbb{H} \cup \partial \mathbb{H}$, is "radial" in the sense that an endpoint at infinity is added for every geodesic ray emanating from some fixed base point.

We can also derive from the Riemann-sphere topology the radial compactification of a geometrically finite hyperbolic surface $X = \Gamma \backslash \mathbb{H}$, provided we use an appropriate fundamental domain in $\mathbb{H}$. To illustrate this point, suppose we identify the hyperbolic cylinder C_ℓ with the Dirichlet domain $\mathcal{F}_\ell = \{1 \leq |z| \leq e^\ell\}$. Compactification of $\mathcal{F}_\ell$ in the Riemann-sphere topology then adds a boundary circle at either end of the cylinder, giving the appropriate radial compactification. If, on the other hand, we had used the horizontal strip $\{1 \leq \operatorname{Im} z \leq e^{\ell}\}$ as a fundamental domain, the induced compactification would be incorrect, with only a single point added at infinity.

Recall the limit set $\Lambda(\Gamma)$ introduced in Chapter 2. Its complement is the set of ordinary points,

$$\Omega(\Gamma) := \partial\mathbb{H} - \Lambda(\Gamma).$$

For Γ geometrically finite, the analysis in Section 2.4 showed that any Dirichlet fundamental domain $\mathcal{F}$ will meet $\Omega(\Gamma)$ in a finite collection of disjoint arcs. Each arc lies in a half-plane H_j whose quotient $\Gamma\backslash H_j$ corresponds to a funnel end. If we take the closure of $\mathcal{F}\cap(\cup H_j)$ in the Riemann-sphere topology, then the resulting quotient consists of a finite collection of compact cylinders, funnels with one original boundary circle, plus an extra boundary circle "at infinity," as illustrated in Figure 6.1. The collection of extra funnel boundary components is naturally identified with $\Gamma\backslash\Omega(\Gamma)$.

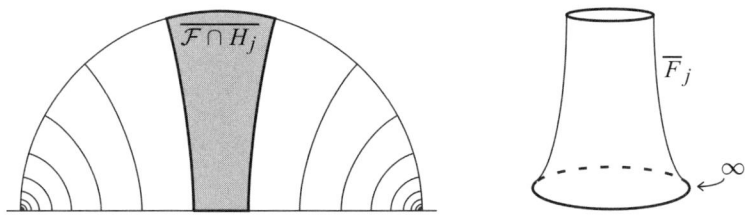

Fig. 6.1. Funnel compactification.

The compactification of a cusp works the same way. Cusps appeared in Section 2.4 as the quotients of horocyclic regions O_p associated to parabolic fixed points $p \in \Lambda(\Gamma)$. If we take the closure of $\mathcal{F}\cap(\cup O_p)$ in the Riemann-sphere topology, the effect is to add a single point at infinity at the end of each cusp, as shown in Figure 6.2. If $\mathcal{P}(\Gamma)$ denotes the set of parabolic fixed points, then $\Gamma\backslash\mathcal{P}(\Gamma)$ is identified with the set of new cusp boundary points.

Fig. 6.2. Cusp compactification.

Definition. The compactification of a geometrically finite hyperbolic surface $X = \Gamma \backslash \mathbb{H}$ is given by the quotient

$$\overline{X} := \Gamma \backslash (\mathbb{H} \cup \Omega(\Gamma) \cup \mathcal{P}(\Gamma)), \tag{6.1}$$

with smooth structure derived from the compactification of a Dirichlet fundamental domain in the Riemann-sphere topology.

The compactification $\overline{X}$ has the structure of a smooth compact surface with boundary and cusp points. The set $C^\infty(\overline{X})$ consists of functions in $C^\infty(X)$ that behave well at infinity, in the following sense. For the funnel ends the usual definition for a smooth manifold with boundary applies: a function is smooth if it is extendible across the boundary to a smooth function on a slightly larger domain. At a cusp point, the condition on $f \in C^\infty(\overline{X})$ is that f and all its derivatives have well-defined limits as the cusp point is approached from the interior of the surface.

Associated with this smooth structure is a class of *boundary-defining functions*, $\rho \in C^\infty(\overline{X})$, characterized by the properties that $\rho > 0$ on X and that ρ vanishes on $\partial \overline{X}$ precisely to first order. For example, where the fundamental domain $\mathcal{F}$ meets $\mathbb{R}$, the coordinate y from $\mathbb{H}$ gives a local boundary-defining function. It is through the boundary-defining function that we will make sense of the notion of asymptotic expansions "at infinity" for objects such as the generalized eigenfunctions. (The usage of the variable y for this purpose was already evident in Proposition 4.4.)

The construction of $\overline{X}$ given above may seem a little ad hoc; we would also like to understand how to arrive at the compactification in a more intrinsic way. This can be formulated in terms of the existence of a boundary-defining function that puts the metric in a standard cusp or funnel form near infinity. The standard form in a funnel is

$$ds^2 = \frac{d\rho^2 + h(\rho, t) \, dt^2}{\rho^2}, \tag{6.2}$$

where t denotes a tangential S^1 variable in the end, and $h(\rho, t)$ is smooth for $\rho \geq 0$. For a cusp the standard form of the metric is

$$ds^2 = \frac{d\rho^2}{\rho^2} + \rho^2 \, h(\rho, t) \, dt^2. \tag{6.3}$$

The funnel metric is a special case of a more general category of metrics.

Definition. Given $\overline{M}, \bar{g}$ a smooth compact Riemannian manifold with boundary, with ρ a defining function for $\partial \overline{M}$, the complete metric $g = \rho^{-2} \bar{g}$ on the interior M is called *conformally compact*.

The hyperbolic plane in the $\mathbb{B}$ model is an obvious example, with $\rho = (1 - |z|^2)/2$. The funnel metric is also of this form. Thus a geometrically finite hyperbolic surface is conformally compact if and only if it has no cusps. In other words,

$$\Gamma \backslash \mathbb{H} \text{ is conformally compact} \iff \Gamma \text{ is convex cocompact.}$$

The coefficients of asymptotic expansions defined with respect to a boundary-defining function ρ are of course dependent on the choice of ρ and not just on the smooth structure of the compactification. It is a relatively simple matter to make invariant definitions of the coefficients as sections of appropriate density bundles on $\partial \overline{X}$. But this becomes cumbersome notationally. Another option is to standardize the choice of ρ, which is easily done in our case because of our simple models for hyperbolic ends. For the sake of notational simplicity, we will take this route. However, one should always keep the dependence on ρ in mind.

Assume that X is a nonelementary geometrically finite hyperbolic surface. Then Theorem 2.13 gives the decomposition of X as a compact core K with a finite number of cusps and funnels attached to it. As illustrated in Figure 6.3, we'll organize this decomposition as

$$X = K \cup F \cup C, \tag{6.4}$$

with the funnels grouped together as the disjoint union

$$F := F_j \cup \cdots \cup F_{n_f},$$

and the cusps,

$$C = C_1 \cup \cdots \cup C_{n_c}.$$

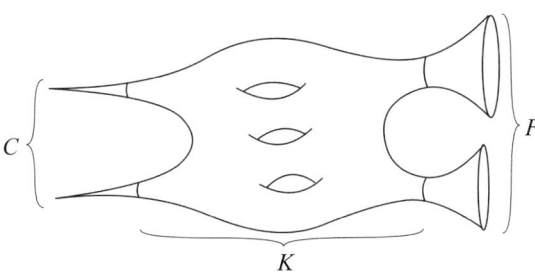

Fig. 6.3. Decomposition of X.

The funnels are bounded by simple closed geodesics, and the cusps are bounded by simple closed horocycles. In either type of end we define the function r as the distance to the compact core K. (This agrees with the geodesic normal coordinate introduced for hyperbolic cylinders in Section 5.1.) Our choice of standard boundary-defining function on $\overline{X}$ is to define

$$\rho(r) := \begin{cases} 2e^{-r} & \text{in } F, \\ e^{-r} & \text{in } C, \end{cases} \tag{6.5}$$

with ρ extended to a smooth nonvanishing function inside K in some arbitrary way.

We will adopt $(\rho, t) \in (0, 2] \times \mathbb{R}/\ell_j \mathbb{Z}$ as the standard coordinates for the funnel F_j, where t is arc length around the central geodesic at $\rho = 2$, in which the metric takes the form

$$ds^2|_{F_j} = \frac{d\rho^2 + (1 + \rho^2/4)^2 \, dt^2}{\rho^2}. \tag{6.6}$$

Note that this is of conformally compact type, with $\bar{g} = d\rho^2 + (1 + \rho^2/4)^2 \, dt^2$. We made the choice $\rho = 2e^{-r}$ in (6.5) so that the metric $\bar{g}$ assigns length ℓ_j, the diameter of the funnel, to the infinite boundary of $\overline{F}_j$. This is simply for notational convenience later on.

For the cusp our standard coordinates $(\rho, t) \in (0, 1] \times \mathbb{R}/\mathbb{Z}$ are based on the model defined by the cyclic group $\Gamma_\infty = \langle z \mapsto z + 1 \rangle$, as in Section 5.3. The cusp boundary is $y = 1$, so that $r = e^y$ and $\rho = 1/y$. We set $t = x$ (mod $\mathbb{Z}$), and the cusp metric becomes

$$ds^2 = \frac{d\rho^2}{\rho^2} + \rho^2 \, dt^2. \tag{6.7}$$

As in the funnel case, our choice of $\rho = e^{-r}$ in the cusp was arbitrary; the motivation was simply to put (6.7) in as simple a form as possible.

Using these standard coordinates we can describe the boundary behavior of functions in $C^\infty(\overline{X})$ a bit more explicitly. For $f \in C^\infty(\overline{X})$, $f|_{F_j}(\rho, t)$ is the restriction to $[0, 1) \times S^1$ of a smooth function defined on $(-\varepsilon, 1) \times S^1$ for some $\varepsilon > 0$. The condition on $f|_{C_j}(r, t)$ is that the function and all its derivatives have limits as $\rho \to 0$ that are independent of t. In particular, the Fourier modes of $f|_{C_j}$ will be smooth functions on $[0, 1)$, with all modes except the zero mode vanishing to infinite order at $\rho = 0$. In any end, a function $f \in C^\infty(\overline{X})$ has a meaningful Taylor expansion at $\rho = 0$, of the form

$$f \sim \sum_{l=0}^{\infty} \rho^l f_l.$$

The existence of such asymptotic expansions could have been taken as the defining property of $C^\infty(\overline{X})$.

6.2 Analytic Fredholm theorem

The method we will use for meromorphic continuation of the resolvent was introduced by Mazzeo–Melrose [128] and is based on the analytic Fredholm theorem. Since this result is crucial to our development, and the version we need is a slight extension of the standard one, we'll give the proof here (adapted from [180, Theorem VI.14]).

For the statement of the theorem let $\mathcal{H}$ be an abstract separable Hilbert space. (In the application this will be a weighted space of the form $\rho^N L^2(X)$.)

Definition. A family of bounded operators $A(s)$ on $\mathcal{H}$, parametrized by $s \in U \subset \mathbb{C}$, is *finitely meromorphic* if for each point $a \in U$ we have a Laurent series representation

$$A(s) = \sum_{k=-m}^{\infty} (s-a)^k A_k$$

converging (in the operator topology) in some neighborhood of a, where the coefficients A_k are finite-rank operators for $k < 0$.

Theorem 6.1 (Analytic Fredholm). *Suppose $E(s)$ is a finitely meromorphic family of compact operators on $\mathcal{H}$, for $s \in U$. If $I - E(s)$ is invertible for at least one $s \in U$, then $(I - E(s))^{-1}$ exists as a finitely meromorphic family on U.*

Proof. It suffices to prove the result in a neighborhood of any point $s_0 \in U$. Hence we can assume that U is small enough to contain only finitely many poles of $E(s)$. With this assumption, we may decompose

$$E(s) = A(s) + F(s),$$

where $F(s)$ is a meromorphic family of finite-rank operators for $s \in U$ and $A(s)$ is a holomorphic family of compact operators. Using the approximation of the compact operator $A(s_0)$ by finite-rank operators, and assuming U is sufficiently small, we can find a fixed finite-rank operator R such that

$$\|A(s) - R\| < 1, \tag{6.8}$$

for all $s \in U$.

Note that (6.8) implies that $I - A(s) + R$ is holomorphically invertible for $s \in U$, by

$$(I - A(s) + R)^{-1} = \sum_{l=0}^{\infty} (A(s) - R)^l.$$

Thus if we set

$$G(s) := (F(s) + R)(I - A(s) + R)^{-1},$$

we can write

$$I - E(s) = (I - G(s))(I - A(s) + R).$$

It's clear that $G(s)$ is of finite rank, since $F(s)$ and R are. We already know that $(I - A(s) + R)^{-1}$ is holomorphic, so the problem is reduced to proving that $I - G(s)$ is meromorphically invertible.

By the Riesz lemma we can write

$$G(s) = \sum_{j,k=1}^{N} \gamma_{jk}(s) \psi_j \langle \phi_k, \cdot \rangle,$$

for some vectors $\psi_j, \phi_k \in \mathcal{H}$, with meromorphic coefficients $\gamma_{jk}(s)$. The ψ_j's are assumed to be independent. To solve

$$(I - G(s))u = w$$

for u given w, we make the ansatz $u = w + \sum_{j=1}^{N} b_j \psi_j$. This reduces the equation to

$$b_j - \sum_{l,k=1}^{N} b_l \gamma_{jk}(s) \langle \phi_k, \psi_l \rangle = \sum_{k=1}^{N} \gamma_{jk}(s) \langle \phi_k, w \rangle.$$

We conclude that inversion of $(I - G(s))$ is possible on the complement of the zero set of the polynomial

$$d(s) := \det \left[\delta_{jl} - \sum_{k=1}^{N} \gamma_{jk}(s) \langle \phi_k, \psi_l \rangle \right].$$

This shows that $I - E(s)$ is meromorphically invertible. The fact that the poles have finite-rank residues follows because they can occur only in the finite-rank operator $G(s)$.   □

6.3 Continuation of the resolvent

Our analysis of the resolvent kernel $R_X(s; z, z')$ will be based on the construction of a *parametrix*, an approximate inverse of $(\Delta_X - s(1 - s))$ whose structure is well understood. It turns out we don't need the full generality of the Mazzeo–Melrose approach; for hyperbolic surfaces a simplified parametrix construction was given by Guillopé–Zworski in [87]. The more explicit construction is significant because it allows growth estimates (in terms of the parameter s) that are not possible in the general case. (We'll develop these in Chapter 9.)

It is the existence of the funnel and cusp models for the ends that allows the more direct parametrix construction. Finding models near the boundary is the main issue, because the structure of the resolvent in the interior can be understood through classical elliptic parametrix theory. The resolvent kernels for funnels and parabolic cylinders developed in Chapter 5 can simply be pulled back to X to give model resolvents on the funnel and cusps ends. These become the principal building blocks of the parametrix. The significance of having exact boundary models is that the patching of model terms occurs only in the interior, making for simpler error terms.

Once we have defined models for the resolvent in different regions of X, we'll paste these pieces together with a simple partition of unity to obtain the parametrix $M(s)$. Then, to see how closely $M(s)$ approximates the actual resolvent, we compose $(\Delta_X - s(1 - s))$ with $M(s)$ and write the result in the form

$$(\Delta_X - s(1 - s))M(s) = I - L(s). \tag{6.9}$$

The trick is then to invert $(I - L(s))$ to obtain

$$R_X(s) = M(s)(I - L(s))^{-1},$$

and this is where analytic Fredholm theory (Theorem 6.1) makes its appearance. To apply it we'll need to establish that $L(s)$ is a finitely meromorphic family of compact operators on the appropriate spaces, which will be obtained by weighting the Hilbert space $L^2(X)$ by powers of ρ.

Theorem 6.2 (Resolvent continuation). *Let X be a geometrically finite hyperbolic surface. For any $N > 0$, the resolvent $R_X(s)$ extends for $\operatorname{Re} s > \frac{1}{2} - N$ to a finitely meromorphic family of operators*

$$R_X(s) : \rho^N L^2(X) \to \rho^{-N} L^2(X).$$

Proof. The elementary cases were dealt with in Chapters 4 and 5, so we assume that X in nonelementary and use the decomposition (6.4) into funnels and cusps. On each funnel F_j we let $R_{F_j}(s)$ denote the resolvent for $\Delta|_{F_j}$ with Dirichlet boundary conditions at the boundary geodesic. Each cusp C_j can be mapped isometrically into the parabolic cylinder C_∞. Then we define $R_{C_j}(s)$ as the pullback to C_j of the model operator $R_{C_\infty}(s)$. To simplify the notation, we'll group the model resolvents together as

$$R_F(s) = \oplus R_{F_j}(s)$$

and

$$R_C(s) = \oplus R_{C_j}(s).$$

It is convenient to regard $R_F(s)$ and $R_C(s)$ as operators on the full surface X that act by zero outside the funnels and cusps, respectively.

In the interior, the resolvent $R_X(s_0)$ evaluated at some fixed $\operatorname{Re} s_0 > 1$ makes a suitable model for $R_X(s)$. To patch $R_F(s)$ and $R_C(s)$ together with $R_X(s_0)$, we introduce a family of cutoff functions. With respect to the coordinate r introduced in Section 6.1 (the geodesic distance from the compact core), we define $\chi_a \in C_0^\infty(X)$ for $a > 0$ such that

$$\chi_a = \begin{cases} 1, & r \le a, \\ 0, & r \ge a + 1. \end{cases} \tag{6.10}$$

Figure 6.4 illustrates the structure of this family of cutoffs.

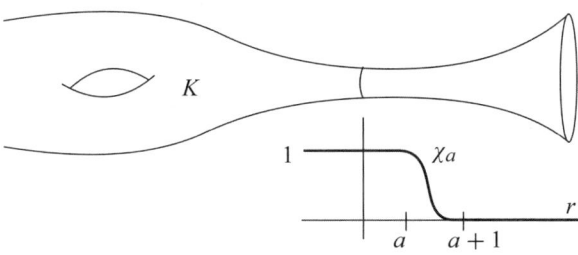

Fig. 6.4. Cutoff function χ_a.

The interior parametrix is defined by

$$M_i := \chi_2 R_X(s_0) \chi_1.$$

For the ends, we set

$$M_f(s) := (1 - \chi_0) R_F(s)(1 - \chi_1),$$
$$M_c(s) := (1 - \chi_0) R_C(s)(1 - \chi_1). \tag{6.11}$$

Together, these pieces make up the full parametrix

$$M(s) := M_i + M_f(s) + M_c(s). \tag{6.12}$$

When we compose $(\Delta_X - s(1-s))$ with $M(s)$, the shift in supports of the cutoffs allows us to exploit the fact that $\chi_a \chi_{a+1} = \chi_a$. The result is

$$(\Delta_X - s(1-s))M(s) = I - L_i(s) - L_f(s) - L_c(s), \tag{6.13}$$

where the error terms are divided into those supported respectively in the interior,

$$L_i(s) = -[\Delta, \chi_2] R_X(s_0)\chi_1 + (s(1-s) - s_0(1-s_0)) M_i, \tag{6.14}$$

funnels,

$$L_f(s) = [\Delta, \chi_0] R_F(s)(1 - \chi_1), \tag{6.15}$$

and cusps,

$$L_c(s) = [\Delta, \chi_0] R_C(s)(1 - \chi_1). \tag{6.16}$$

We must show these error terms to be finitely meromorphic and compact. The interior term $L_i(s)$ is explicitly polynomial in s, so meromorphy is clear enough. As for compactness, the term $[\Delta, \chi_2] R_X(s_0)\chi_1$ is a smoothing operator, because the supports of $[\Delta, \chi_2]$ and χ_1 are compact and do not intersect. For the second term in $L_i(s)$, note that Δ_X is a second-order elliptic differential operator in the interior. The standard elliptic parametrix construction (see, e.g., [208, Section 7.4]) implies that M_i is a compactly supported classical pseudodifferential operator of order -2. Such an operator is compact in particular (see Proposition A.26). Hence $L_i(s)$ is compact on $L^2(X)$ for any s.

The shift in supports of the χ_a's also means that the funnel term, $L_f(s)$, has a smooth kernel. But this does not make it a compact operator, because $(1 - \chi_1)$ is supported near infinity. If $L_f(s; z, z')$ denotes the integral kernel of $L_f(s)$ with respect to the hyperbolic area form dA, we claim that

$$L_f(s; \cdot, \cdot)|_{F_j \times F_j} \in \rho^\infty \rho'^s C^\infty(\overline{F}_j \times \overline{F}_j) \tag{6.17}$$

(and note that $L_f(s; \cdot, \cdot)|_{F_i \times F_j} = 0$ for $i \neq j$). To prove (6.17), it suffices to analyze $R_{C_\ell}(s; z, z')$ by the method of images formula (5.20). Since the shifted supports keep us away from the diagonal, we can read off the asymptotics directly from (5.6).

By (6.6), the area form in standard funnel coordinates is $dA = \rho^{-2} d\rho \, dt + O(1)$, so $\rho^{1/2}$ is the threshold for L^2 asymptotic behavior in a funnel. This implies that L_f is a compact operator for $\operatorname{Re} s > \frac{1}{2}$ and unbounded for $\operatorname{Re} s \leq \frac{1}{2}$. However, extending the region of compactness is easily accomplished by adding a weight to the Hilbert space. For $\operatorname{Re} s > \frac{1}{2} - N$, we have

$$L_f(s) : \rho^N L^2(X) \to C_0^\infty(X);$$

hence $L_f(s)$ is compact as a map $\rho^N L^2(X) \to \rho^{N'} L^2(X)$ for any N'. We observed in Proposition 5.5 that $R_{F_\ell}(s)$ is holomorphic except for poles of finite rank, so $L_f(s)$ is also finitely meromorphic.

A similar analysis applies to $L_c(s)$, whose kernel is supported on $\cup(C_j \times C_j)$, away from the diagonal, with compact support in the first variable. By Proposition 5.8, for z fixed the asymptotics in the second variable are

$$L_c(s; z, \cdot) \in \rho'^{s-1} C^\infty(\overline{X}). \tag{6.18}$$

By (6.7) the cusp area form is $dA = d\rho\, dt$, so borderline L^2 behavior as $\rho \to 0$ is $\rho^{-1/2}$. Hence for $\operatorname{Re} s > \frac{1}{2} - N$,

$$L_c(s) : \rho^N L^2(C) \to C_0^\infty(C).$$

The model resolvent $R_{C\infty}(s)$ had only a single pole of rank one at $s = \frac{1}{2}$, so $L_c(s)$ is finitely meromorphic also.

These arguments show that the full error term, $L(s) = L_i(s) + L_f(s) + L_c(s)$, is a finitely meromorphic family of compact operators on the weighted Hilbert spaces. The remaining ingredient needed for the application of the analytic Fredholm theorem is invertibility of $(I - L(s))^{-1}$ at a single value of s. This can be accomplished by choosing s and s_0 such that $\|L(s)\| < 1$, which is easily done using standard resolvent estimates. In particular, the positivity of Δ_X implies

$$\left| \langle u, (\Delta_X - s(1-s))u \rangle \right| \geq - \operatorname{Re}(s(1-s)) \, \|u\|^2,$$

for u in the domain of Δ_X. Hence for $\operatorname{Re}(s^2 - s) > 1$,

$$\|R_X(s)\| \leq \frac{1}{\operatorname{Re}(s^2 - s)}.$$

By taking both $\operatorname{Re} s$ and $\operatorname{Re}(s_0)$ sufficiently large and using this resolvent estimate in the explicit formulas for L_i, L_f, and L_c, we can guarantee that $\|L(s)\| < 1$. Theorem 6.1 then gives the meromorphic extension of

$$R_X(s) = M(s)(I - L(s))^{-1}. \qquad \square$$

6.4 Structure of the resolvent kernel

The parametrix construction used to prove Theorem 6.2 gives us more than just meromorphic continuation—it is the key to understanding the structure of the resolvent kernel. This structure will be crucial to our development of the spectral and scattering theory in Chapter 7.

Let us recall the convention used earlier that for an operator $A(s)$ we write the integral kernel with respect to the hyperbolic area form as $A(s; z, z')$. To obtain an invariant definition of the integral kernels (i.e., not dependent on the metric), the usual approach would be to introduce half-density bundles and let the operators act on half-densities rather than functions. Such invariance is essential for certain purposes, most importantly the definition of invariant symbols. However, since we wouldn't actually make use of the invariance in our development, we prefer to keep the notation as clean as possible.

In the notation of (6.13), let $M(s)$ be the parametrix of $(\Delta_X - s(1-s))$ and

$$L(s) = L_i(s) + L_f(s) + L_c(s),$$

the remainder term. We set $I + K(s) = (I - L(s))^{-1}$, so that

$$R_X(s) = M(s) + M(s)K(s).$$

Since $M(s)$ is already understood explicitly, we can deduce the structure of $R_X(s)$ by analyzing the composition of $M(s)$ with $K(s)$.

Theorem 6.3 (Resolvent kernel structure). *For X a nonelementary geometrically finite hyperbolic surface, the resolvent admits a decomposition*

$$R_X(s) = M'_i(s) + M_f(s) + M_c(s) + Q(s),$$

where $M'_i(s)$ is a classical pseudodifferential operator of order -2 with kernel compactly supported in the interior of $\overline{X} \times \overline{X}$. The $M_f(s)$ and $M_c(s)$ terms are the components of the parametrix defined in (6.11). The remainder term has the structure

$$Q(s; \cdot, \cdot) \in (\rho_f \rho'_f)^s (\rho_c \rho'_c)^{s-1} C^\infty(\overline{X} \times \overline{X}), \qquad (6.19)$$

where we factor $\rho = \rho_f \rho_c$ so that $\rho_f = \rho$ in F and $\rho_c = \rho$ in C.

Before proceeding with the proof, let us recall the information we already have on the boundary behavior of the kernel of $M(s) = M_i + M_f(s) + M_c(s)$. Of course, the interior term M_i is compactly supported, so we're just considering the funnel and cusp terms.

Lemma 6.4. *For z' fixed within the region $r < a$, we have*

$$(1 - \chi_a(\cdot))M(s; \cdot, z') \in \rho_f^s \rho_c^{s-1} C^\infty(\overline{X}).$$

And for z fixed within $r < a$,

$$M(s; z, \cdot)(1 - \chi_a(\cdot)) \in \rho_f'^s \rho_c'^{s-1} C^\infty(\overline{X}).$$

Proof. Just as in the proof of (6.17), to analyze $M_f(s)$, which is a cutoff version of $R_F(s)$, it suffices to analyze $R_{C_\ell}(s; z, z')$ by the method of images formula (5.20). Our assumptions keep us away from the diagonal, so the boundary behavior follows directly from (5.6).

Similarly, away from the diagonal in a cusp, the boundary behavior of $M_c(s; z, z')$ follows from the explicit form (5.26) of the zeroth Fourier coefficient (with $\rho = 1/y$) and the exponential decay of the higher Fourier modes. □

As a particular corollary of Lemma 6.4, we note that

$$M(s)\varphi \in \rho_f^s \rho_c^{s-1} C^\infty(X), \qquad (6.20)$$

for any $\varphi \in C_0^\infty(X)$.

Proof of Theorem 6.3. The equation $I + K(s) = (I - L(s))^{-1}$ implies that

$$K(s) = L(s) + K(s)L(s) = L(s) + L(s)K(s). \tag{6.21}$$

This identity allows us to deduce the structure of $K(s)$ from that of $L(s)$, and thus to understand the composition $M(s)K(s)$. The interior term $L_i(s)$ is a compactly supported pseudodifferential operator of order -2, and it follows from (6.21) that $K(s; z, z')$ has the same structure in the interior. Since the kernel of $L_i(s)$ is compactly supported in both variables, standard composition theorems for pseudodifferential operators on compact manifolds show that $\chi_2 M(s)K(s)\chi_2$ is a compactly supported pseudodifferential operator of order -4. We write the combined interior term as

$$M_i'(s) := M_i + \chi_2 M(s)K(s)\chi_2.$$

By the definition of $L(s)$,

$$L(s; z, z')(1 - \chi_2(z')) = \left[L_f(s; z, z') + L_c(s; z, z')\right](1 - \chi_2(z')),$$

which is smooth on $X \times X$ and compactly supported in the z variable. As $z' \to \infty$ in either a funnel or cusp, the asymptotics are given by (6.17) and (6.18), respectively. By (6.21), $K(s; z, z')(1 - \chi_2(z'))$ has the same structure. And then the asymptotic behavior of the kernel of

$$Q(s) := M(s)K(s)(1 - \chi_2) + (1 - \chi_2)M(s)K(s)\chi_2$$

is easily deduced from Lemma 6.4. □

In the model cases, one can see clearly that the resolvent kernel is symmetric. This holds true in general and is a simple consequence of the symmetry of the Laplacian itself.

Lemma 6.5. *The resolvent kernel is symmetric in the sense that*

$$R_X(s; z, z') = R_X(s; z', z).$$

Proof. By meromorphic continuation, it suffices to consider $\operatorname{Re} s > \frac{1}{2}$. Suppose $\phi, \psi \in \rho^\infty C^\infty(\overline{X})$. Then by Theorem 6.3 and (6.20), $u = R_X(s)\phi$ and $v = R_X(s)\psi$ are both in $\rho_f^s \rho_c^{s-1} C^\infty(\overline{X})$. This justifies the integration by parts,

$$\int_X v\,(\Delta_X - s(1-s))u\,dA = \int_X u\,(\Delta_X - s(1-s))v\,dA,$$

where as usual dA denotes the hyperbolic area form on X. Substituting in for u and v gives

$$\int_X \phi\,R_X(s)\psi\,dA = \int_X \psi\,R_X(s)\phi\,dA,$$

and the result follows because $\rho^\infty C^\infty(\overline{X})$ is dense in $L^2(X)$. □

The switching of coordinates in the kernel corresponds to the transpose of a map between real Banach spaces. So Lemma 6.5 could be stated as

$$R_X(s)^{\mathrm{t}} = R_X(s).$$

Of course, self-adjointness of Δ_X also implies the relation

$$R_X(s)^* = R_X(\bar{s}),$$

which means that the kernel satisfies

$$\overline{R_X(s; z', z)} = R_X(\bar{s}; z, z'). \qquad (6.22)$$

6.5 The stretched product

Because of our explicit knowledge of the model resolvent terms, Theorem 6.3 gives a sufficient picture of the structure of the resolvent kernel for the subsequent chapters of this book. However, it bypasses the issue of describing the asymptotics of the resolvent kernel at infinity near the diagonal. In this section we will explain this behavior using a framework (the stretched product) provided by Mazzeo–Melrose [128]. This material won't be used elsewhere in this book, but it is an essential component of the theory for hyperbolic metrics in higher dimensions as well as for asymptotically hyperbolic metrics.

For this discussion we will assume that X has no cusps. Proposition 5.8 shows that the behavior of the resolvent kernel in the cusps is essentially one-dimensional. Only the zero mode in the Fourier decomposition is significant in terms of spectral or scattering theory. Because of this very different behavior, cusps are not included in the framework of [128].

To analyze the resolvent kernel at infinity means understanding its behavior near the boundary of the compactification of $X \times X$. We have an obvious candidate for the compactification of $X \times X$, namely $\overline{X} \times \overline{X}$, but unfortunately this choice proves inadequate near the diagonal.

To see the problem, let us first consider a hyperbolic plane with its compactification, $\overline{\mathbb{H}} = \mathbb{H} \cup \partial \mathbb{H}$. The resolvent kernel has the form $R_{\mathbb{H}}(s; z, z') = g_s(\sigma)$, where g_s was given in (4.12), and

$$\sigma(z, z') = \frac{1}{2} + \frac{(x - x')^2 + y^2 + y'^2}{4yy'}.$$

We will continue to use the (x, y) coordinates with $\{y = 0\}$ giving a local picture of the boundary, keeping in mind that these coordinates do not describe the boundary globally. We can immediately see that $\overline{\mathbb{H}} \times \overline{\mathbb{H}}$ is not an appropriate compactification through the fact that $\lim_{y, y' \to 0} \sigma(z, z')$ is not well-defined. The resolvent kernel has no natural extension from $\mathbb{H} \times \mathbb{H}$ to $\overline{\mathbb{H}} \times \overline{\mathbb{H}}$.

Locally we can solve the problem by changing coordinates. In order to give $\sigma(z, z')$ a meaningful extension to the boundary, we set

$$\tau := \sqrt{(x - x')^2 + y^2 + y'^2}, \tag{6.23}$$

and introduce projective coordinates

$$(\omega, \eta, \eta') := \frac{(x - x', y, y')}{\tau} \tag{6.24}$$

(which parametrize a unit quarter-sphere). In the new coordinates,

$$\sigma(z, z') = \frac{1 + 2\eta\eta'}{4\eta\eta'}, \tag{6.25}$$

which has a meaningful limit as η or $\eta' \to 0$. (Indeed, $1/\sigma$ is a smooth function of η, η'.) The change of coordinates $(\tau, x, \omega, \eta, \eta') \mapsto (x, y, x', y')$ is bijective in the interior, but the preimage of the boundary point $(x, 0, x, 0)$ is the quarter-sphere $\{(0, x, \omega, \eta, \eta')\}$. The new coordinates induce (locally) a different compactification of $\overline{\mathbb{H}} \times \overline{\mathbb{H}}$. The direct product $\overline{\mathbb{H}} \times \overline{\mathbb{H}}$ had two boundary faces, $\{y = 0\}$ and $\{y' = 0\}$, intersecting in a corner. In the new coordinates the boundary consists of three faces: $\{\eta = 0\}$, $\{\eta' = 0\}$, and $\{\tau = 0\}$. The corner structure is more complicated, and this extra detail is just what we need to understand the asymptotics of the resolvent kernel.

This new compactification is called a *blowup* of $\mathbb{H} \times \mathbb{H}$. To illustrate this concept in a more familiar situation, consider the first quadrant in $\mathbb{R}^2$. Introducing polar coordinates (r, θ) has the effect of "blowing up" the origin into the quarter-circle $\{r = 0, \theta \in [0, \pi/2]\}$. This represents a change in the smooth structure; a function such as $\sin\theta$ is not smooth in Cartesian coordinates, but is smooth when lifted to the blown-up space.

To describe the blowup procedure in greater generality, suppose M is a manifold with boundary and corners with a submanifold $S \subset \partial M$. We define $\widetilde{S}$ to be the inward-pointing portion of the unit normal bundle of S within M. The blowup $[M; S]$ is defined by excising S from M and gluing $\widetilde{S}$ in its place. There is a natural surjection $b : [M; S] \to M$ (*blowdown*) given by the combination of the identity map on $[M; S] - \widetilde{S}$ and the bundle projection $\widetilde{S} \to S$. In the first-quadrant example given above, the blowdown map is simply the change of coordinates $b(r, \theta) = (r \cos\theta, r \sin\theta)$.

Returning to the case of $\mathbb{H}$, note that $\overline{\mathbb{H}}$ is diffeomorphic to a closed disk. The product $\overline{\mathbb{H}} \times \overline{\mathbb{H}}$ is a manifold with two boundary faces and a corner consisting of the torus $\partial\mathbb{H} \times \partial\mathbb{H}$. By (6.24), we see that the submanifold S to be blown up corresponds to $\tau = 0$, or $x = x', y = y' = 0$. Thus S can be simply described as the diagonal in the corner, and identified with the boundary circle $\partial\mathbb{H}$. The coordinates (x, ω, η, η') give a local parametrization of $\widetilde{S}$, which is diffeomorphic to the product of the quarter-sphere with a circle.

For $\overline{X}$ a smooth compact surface with boundary $\partial\overline{X}$, we let S be the diagonal in the corner $\partial\overline{X} \times \partial\overline{X}$. The blowup of $\overline{X} \times \overline{X}$ along S,

$$\overline{X} \times_0 \overline{X} := [\overline{X} \times \overline{X}; S], \tag{6.26}$$

is called the *stretched product* of $\overline{X}$ with itself. The subscript 0 refers to the "0-calculus" of vector fields vanishing at the boundary. The stretched product proves useful for analyzing differential operators built out of such vector fields, of which the hyperbolic Laplacian is a prime example. See [128, 126, 136] for more details.

The stretched product has three boundary faces, as illustrated in Figure 6.5. If ρ is a boundary-defining function for $\partial\overline{X}$, then using coordinates (z, z') for $\overline{X} \times \overline{X}$, we naturally define $\rho = \rho(z)$ and $\rho' = \rho(z')$. These are boundary-defining functions for the codimension-one boundary faces of $\overline{X} \times \overline{X}$. Let τ be a global defining coordinate for S (patched together from local versions given by (6.23) with a smooth partition of unity). Then the lift of τ to $\overline{X} \times_0 \overline{X}$ becomes a defining coordinate for $\tilde{S}$. Following the model of (6.24), we use $\eta = \rho/\tau$ and $\eta' = \rho'/\tau$ as defining functions for the other faces of $\overline{X} \times_0 \overline{X}$. We can refer to the three faces of $\overline{X} \times_0 \overline{X}$ as the *left face* $\{\eta = 0\}$, the *right face* $\{\eta' = 0\}$, and the *front face* $\tilde{S} = \{\tau = 0\}$. The set $\{\eta = \eta'\} \subset \overline{X} \times_0 \overline{X}$ is called the *lifted diagonal*. Technically it's the closure of the lift of the diagonal of $X \times X$, rather than the lift of the diagonal of $\overline{X} \times \overline{X}$. The lifted diagonal meets the front face in its center and does not intersect the left or right faces.

Let us consider the resolvent kernel in $\mathbb{H}$ in terms of the stretched product. In the blown-up coordinates $(\tau, x, \omega, \eta, \eta')$, the formula (6.25) for σ shows that $R_{\mathbb{H}}$ depends only on η, η'. Since $\sigma = 1$ if and only if $\eta = \eta'$, after blowup the interior singularity of $R_{\mathbb{H}}(s; z, z')$ occurs precisely on the lifted diagonal. Since there is no dependence on r, the defining coordinate for the front face, the singularity behaves uniformly up to the front face and indeed would be extendible across it if the stretched product were doubled at this face. In the interior the diagonal singularity has the standard form for an elliptic parametrix, namely a conormal singularity of order -2.

The class $\mathcal{I}_0^m(\overline{X} \times_0 \overline{X})$, introduced in [128], consists of distributions that have a conormal singularity of order m at the lifted diagonal that is extendible across the front face, are smooth elsewhere in the interior, and vanish to infinite order at the left and right faces.

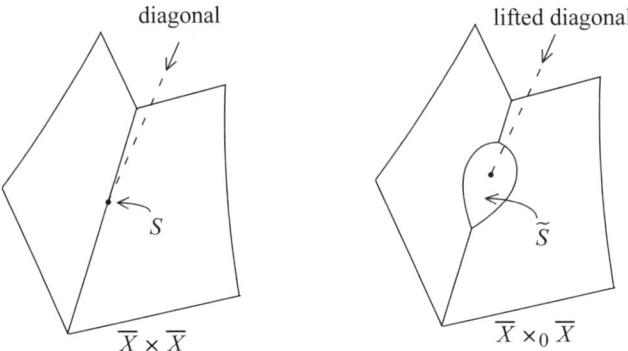

Fig. 6.5. Stretched product.

Proposition 6.6. *For the hyperbolic plane the structure of the resolvent kernel is given by the decomposition*

$$R_{\mathbb{H}}(s; z, z') \in \mathcal{I}_0^{-2}(\overline{\mathbb{H}} \times_0 \overline{\mathbb{H}}) + (\eta\eta')^s C^\infty(\overline{\mathbb{H}} \times_0 \overline{\mathbb{H}}).$$

Proof. Recall that $R_{\mathbb{H}}(s; z, z') = g_s(\sigma(z, z'))$, where

$$g_s(\sigma) = \frac{1}{4\pi} \frac{\Gamma(s)^2}{\Gamma(2s)} \sigma^{-s} F(s, s; 2s; \sigma^{-1}).$$

The decomposition is obtained by considering separately the regions where σ is near 1 versus bounded away from 1. The singularity on the lifted diagonal $\{\sigma = 1\}$ is conormal as discussed above.

Since $F(s, s; 2s; \sigma^{-1})$ is a smooth function of σ^{-1} for $\sigma > 1$, the structure of the second term follows immediately from (6.25).   □

To understand the structure of the resolvent in the general case, using the parametrix construction, we need first to consider the funnel model term R_{F_ℓ}. We let $\overline{F}_\ell$ denote the compactification of F_ℓ induced by the coordinate ρ, as in Section 5.2. Let $\overline{F}_\ell \times_0 \overline{F}_\ell$ denote the product blown up at the infinite end $\rho = 0$ (but not at the original geodesic boundary $\rho = 1$).

Proposition 6.7. *The model funnel resolvent has the structure*

$$R_{F_\ell}(s; \cdot, \cdot) \in \mathcal{I}_0^{-2}(\overline{F}_\ell \times_0 \overline{F}_\ell) + (\eta\eta')^s C^\infty(\overline{F}_\ell \times_0 \overline{F}_\ell) + (\rho\rho')^s C^\infty(\overline{F}_\ell \times \overline{F}_\ell).$$

Proof. The funnel resolvent was obtained from that of the hyperbolic cylinder by the method of images formula:

$$R_{F_\ell}(s; z, z') = R_{C_\ell}(s; z, z') - R_{C_\ell}(s; z, Tz'), \tag{6.27}$$

where T denotes reflection across the central geodesic. So it suffices to understand $R_{C_\ell}(s; z, z')$. This is most conveniently done using the formulas developed for the lift of $R_{C_\ell}(s; z, z')$ to a function on $\mathbb{H} \times \mathbb{H}$, written as an average of $R_{\mathbb{H}}(s; z, z')$ over Γ_ℓ.

Let $\mathcal{F}$ be the fundamental domain $\{1 \le |z| \le e^\ell\}$ for the action of Γ_ℓ on $\mathbb{H}$. If z, z' are restricted to $\mathcal{F}$, then using (5.3) we can decompose for $\operatorname{Re} s > -N$,

$$R_{C_\ell}(s; z, z') = R_{\mathbb{H}}(s; z, z') + \sum_{n=0}^{N-1} \frac{1}{4\pi} \frac{\Gamma(s+n)^2}{n!\Gamma(2s+n)} \sum_{k\neq 0} \sigma(z, e^{k\ell}z')^{-s-n}$$
$$+ \sum_{k\neq 0} F_N(s, \sigma(z, e^{k\ell}z')), \tag{6.28}$$

where $F_N(s, \sigma) \sim c_{s,N}\sigma^{-s-N}$ as $\sigma \to \infty$. This is virtually identical to (5.6) except that we have split off the $k = 0$ term as $R_{\mathbb{H}}$. For $k \neq 0$, $\sigma(z, e^{k\ell}z')^{-1}$ is equal to yy' times a nonvanishing smooth function on $\overline{\mathbb{H}} \times \overline{\mathbb{H}}$. Thus the sums over $k \neq 0$ contribute an element of $(\rho\rho')^s C^\infty(\overline{\mathbb{H}} \times \overline{\mathbb{H}})$.

Since Proposition 6.6 describes the local structure $k = 0$ term, we see that the lift of $R_{C_\ell}(s'; z, z')$ to $\mathbb{H} \times \mathbb{H}$ lies in the combination of spaces

$$\mathcal{I}_0^{-2}(\overline{\mathbb{H}} \times_0 \overline{\mathbb{H}}) + (\eta\eta')^s C^\infty(\overline{\mathbb{H}} \times_0 \overline{\mathbb{H}}) + (\rho\rho')^s C^\infty(\overline{\mathbb{H}} \times \overline{\mathbb{H}}).$$

To apply this result to R_{F_ℓ} we interpret (6.27) as a statement about the lifts of the kernels to $\mathbb{H} \times \mathbb{H}$, with T the reflection across the imaginary axis. With $z, z' \in \mathcal{F}_+ = \mathcal{F}_\ell \cap \{\mathrm{Re}(z) \geq 0\}$, the second term is evaluated away from the diagonal. Hence we deduce the stated result for R_{F_ℓ}. $\qquad\qquad\qquad\qquad\qquad\qquad\qquad\qquad\qquad\qquad\qquad$ □

For a surface without cusps, the singularity on the diagonal extends to infinity only in the $M_f(s)$ term, whose structure is covered by Proposition 6.7. Thus, combining this proposition with Theorem 6.3 gives the following:

Corollary 6.8 (Mazzeo–Melrose). *For X a conformally compact hyperbolic surface (geometrically finite with no cusps),*

$$R_X(s; \cdot, \cdot) \in \mathcal{I}_0^{-2}(\overline{X} \times_0 \overline{X}) + (\eta\eta')^s C^\infty(\overline{X} \times_0 \overline{X}) + (\rho\rho')^s C^\infty(\overline{X} \times \overline{X}).$$

Notes

In the proof of continuation of the resolvent, Theorem 6.2, we used the assumption that the surface was hyperbolic only to characterize the ends. For the interior term in the parametrix, we needed only the property that $\chi_2 R_X(s_0) \chi_1$ was pseudodifferential, which would hold for any metric. Thus the proof implies meromorphic continuation of the resolvent on a general Riemannian surface that is hyperbolic only outside a compact set. This is the context of Guillopé–Zworski [87]. One could also include compactly support potentials, or even potentials that vanish to infinite order at $\rho = 0$, without changing the structure of the argument.

An *asymptotically hyperbolic* manifold X, g is conformally compact with the added restriction that $|d\rho|_{\bar{g}} = 1$ on $\overline{X}$. This implies in particular that all sectional curvatures approach -1 at infinity. Mazzeo–Melrose [128] proved meromorphic continuation of the resolvent for asymptotically hyperbolic metrics, under certain restrictions that were clarified by Guillarmou [78]. The method extends to the more general conformally compact case (see Mazzeo [125, 127] and Borthwick [22]).

For the case of conformally compact hyperbolic manifolds in higher dimensions (convex cocompact groups), alternative methods are available. Perry [161, 162] applied Schrödinger operator techniques to prove meromorphic continuation in this case. (See also Agmon [2] and Hislop [92] for background on this approach.) Guillopé–Zworski [85] give a higher-dimensional version of the construction we've presented here, based on patching of model resolvents.

Cusps of hyperbolic manifolds present additional complications in higher dimensions, but meromorphic continuation of the resolvent and/or scattering matrix has nevertheless been established in such cases by Froese–Hislop–Perry [69, 68], Bunke–Olbrich [36], and Guillarmou [80].

7

Spectral and Scattering Theory

The basic spectral theory of the Laplacian on a geometrically finite hyperbolic manifold was worked out by Lax–Phillips [113, 114], in the abstract framework of Lax–Phillips scattering theory [115]. In particular, they established the following:

Theorem 7.1 (Lax–Phillips). *For a geometrically finite hyperbolic surface X of infinite area, the spectrum of Δ_X consists of at most finitely many eigenvalues in the interval $(0, 1/4)$, and absolutely continuous spectrum $[1/4, \infty)$, with no embedded eigenvalues.*

In this chapter we will use the the structure of the resolvent kernel, as given in Theorem 6.3, to establish these properties of the spectrum. In the process we will introduce the generalized eigenfunctions, which will lead us to the definition of the scattering matrix.

7.1 Essential and discrete spectrum

We begin by locating the essential spectrum of Δ_X, using a standard method. Weyl's criterion (Theorem A.13) says that $\lambda \in \sigma_{\text{ess}}(\Delta_X)$ if and only if there exists an orthonormal sequence $\{\phi_n\} \subset L^2(X)$ with

$$\|(\Delta_X - \lambda)\phi_n\| \to 0.$$

Constructing such a sequence is usually much easier than examining the spectral projections directly.

Proposition 7.2 (Essential spectrum). *For X a geometrically finite hyperbolic surface of infinite area, the interval $[1/4, \infty)$ is contained in $\sigma_{\text{ess}}(\Delta_X)$.*

Proof. Since X must contain at least one funnel, it can be represented by a fundamental domain $\mathcal{F}$ that intersects $\partial \mathbb{H}$ in at least one interval of nonzero Euclidean length. By conjugating the group, we can shift and dilate $\mathcal{F}$ to assume that $\{|\operatorname{Re} z| < 1, \operatorname{Im} z < 1\} \subset \mathcal{F}$. The strategy is to start with y^s, which satisfies $(\Delta_X - s(1 - s))y^s = 0$, and apply cutoff functions supported in this region to construct the ϕ_n.

For each $n \in \mathbb{N}$, we define a cutoff function $\psi_n \in C^\infty(\mathbb{R}^2)$ such that

$$\psi_n(x, t) = \begin{cases} 0 & \text{if } |x| \geq 1 \text{ and } t \notin [0, n], \\ 1 & \text{if } |x| \leq \frac{1}{2} \text{ and } t \in [1, n-1]. \end{cases}$$

We can assume that ψ_n is constructed so that its derivatives to second order are bounded independently of n. For $\operatorname{Re} s = \frac{1}{2}$ we set

$$u_n(z) = y^s \psi_n(x, -\log y).$$

The L^2 norm of u_n is easily estimated from below:

$$\|u_n\|^2 = \int_{\mathcal{F}} y \, \psi_n(x, -\log y)^2 \, \frac{dx \, dy}{y^2}$$

$$\geq \int_{-1/2}^{1/2} \int_{e^{1-n}}^{e^{-1}} y^{-1} \, dx \, dy$$

$$= n - 2.$$

Computing $(\Delta_X - s(1 - s))u_n$ explicitly gives

$$(\Delta_X - s(1 - s))u_n(z) = -y^{2+s} \, \partial_x^2 \psi_n(x, -\log y) - y^s \, \partial_t^2 \psi_n(x, -\log y)$$
$$+ (1 - s)y^s \, \partial_t \psi_n(x, -\log y).$$

With the assumption of a uniform bound on the derivatives of ψ_n, it's easy to check that

$$\|(\Delta_X - s(1 - s))u_n\| = O(1),$$

as $n \to \infty$. Hence by normalizing $\phi_n = u_n / \|u_n\|$, we get a sequence of unit vectors with

$$\|(\Delta_X - s(1 - s))\phi_n\| = O(n^{-1}). \tag{7.1}$$

The simplest way to make the sequence orthonormal is to avoid overlaps of support, as illustrated in Figure 7.1. Noting that dilations are isometries, we can replace ϕ_n by $\phi_n^*(z) = \phi_n(e^{\alpha_n} z)$ for some sequence $\alpha_n > 0$ and still preserve the property (7.1). The support of ϕ_n^* is contained in $\{e^{-\alpha_n - n} \leq y \leq e^{-\alpha_n}\}$. Choosing $\alpha_n = n^2$, for example, will make the sequence ϕ_n^* orthogonal. $\square$

Proposition 7.2 narrows down the possible discrete spectrum of Δ_X to the range $(0, 1/4)$. We can analyze the spectrum in this range by a simple application of Stone's formula, as in the proof of Theorem 4.2.

Proposition 7.3 (Discrete spectrum). *For a geometrically finite hyperbolic surface X with infinite area, the discrete spectrum $\sigma_d(\Delta_X)$ consists of finitely many (possibly zero) eigenvalues in the interval $(0, 1/4)$.*

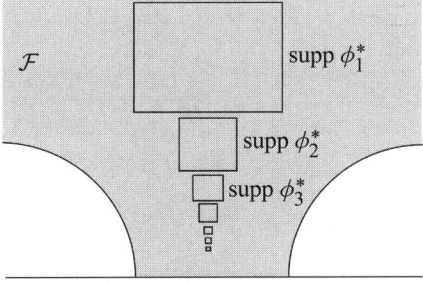

Fig. 7.1. Disjoint supports.

Proof. We start from Stone's formula:

$$\frac{1}{2}(P_{[a,b]} + P_{(a,b)}) = \lim_{\varepsilon \to 0^+} \frac{1}{2\pi i} \int_a^b \left[(\Delta_X - \lambda - i\varepsilon)^{-1} - (\Delta_X - \lambda + i\varepsilon)^{-1} \right] d\lambda,$$

where P_I is the spectral projector for Δ_X onto $I \subset \mathbb{R}$ (see Corollary A.12). As in the proof of Theorem 4.2, we note that if $\lambda \leq 1/4$ then the limits from either side are equal:

$$\lim_{\varepsilon \to 0} (\Delta_X - \lambda \mp i\varepsilon)^{-1} = R_X(s),$$

with $s = \frac{1}{2} + \sqrt{1/4 - \lambda}$, as long as this is not a pole of $R_X(s)$. Hence the spectral projectors are zero away from points where $\lambda = s(1-s)$ with s a pole of $R_X(s)$. The finite meromorphy of $R_X(s)$ then shows that there are only finitely many discrete points in the spectrum below $1/4$. $\square$

7.2 Absence of embedded eigenvalues

The next piece of Theorem 7.1 to consider is the absence of eigenvalues above $1/4$. In proving this we will establish slightly more general conditions under which solutions of $(\Delta_X - s(1-s))u = 0$ fail to exist. These results will have other applications too. For example, at the end of the section we'll show that $R_X(s)$ has no poles on the critical line $\text{Re}\, s = \frac{1}{2}$, except possibly at $s = \frac{1}{2}$.

In their proof of the absence of embedded eigenvalues, Lax–Phillips [113] used a hyperbolic adaptation of the Fourier transform. Here we will prove the result using the unique continuation method as applied to asymptotically hyperbolic manifolds by Mazzeo [127]. In general, "unique continuation" refers to a property possessed by differential equations under certain ellipticity conditions (see, e.g., Hörmander [95]), which says that two solutions that agree to infinite order at a single point must be equal everywhere.

We will establish a unique continuation property for the eigenvalue equation of Δ_X at the infinite ($\rho = 0$) boundary of a funnel. To set up the statement, consider the indicial equation for $(\Delta_{F_\ell} - s(1-s))u = 0$ at $\rho = 0$. Using the standard coordinates (ρ, t) for F_ℓ, with the metric given by (6.6), we have

$$\Delta_{F_\ell} \rho^\gamma = \gamma(1-\gamma)\rho^\gamma + O(\rho^{\gamma+1}).$$

Thus the indicial equation is

$$\gamma(1-\gamma) - s(1-s) = 0, \tag{7.2}$$

with the obvious indicial roots $\gamma = s$ and $1-s$. Of course, our substitution of $z = s(1-s)$ for the spectral parameter is closely related to the form of the indicial equation.

For the unique continuation property we don't need to assume that a solution of $(\Delta_X - s(1-s))u = 0$ vanishes to infinite order at $\rho = 0$. To prove $u = 0$ it suffices to assume that u has an asymptotic expansion based on one of the indicial roots, such that the leading term is zero.

Proposition 7.4 (Unique continuation). *Let X be a nonelementary geometrically finite hyperbolic surface with infinite area. Suppose $u \in C^\infty(X)$ satisfies $(\Delta_X - s(1-s))u = 0$ for $s \notin \pm - \mathbb{N}_0/2$. If in some funnel F_j of X we have*

$$u|_{F_j} \in \rho^{s+1} C^\infty(\overline{F}_j),$$

then $u = 0$.

Before proving this, let us see how it leads to the immediate spectral application:

Proposition 7.5 (No embedded eigenvalues). *For a nonelementary geometrically finite hyperbolic surface X with infinite area, Δ_X has no L^2-eigenvalues in $[1/4, \infty)$.*

Proof. By assumption, X has at least one funnel, say F_j, and this is where we will focus our attention. Suppose we have an eigenfunction $u \in L^2(X)$ with eigenvalue $\lambda \geq 1/4$. Let $s = \frac{1}{2} + i\sqrt{\lambda - 1/4}$, so that $s(1-s) = \lambda$. In order to apply Proposition 7.4, we need to show that u has the right sort of asymptotic expansion at $\rho = 0$ in F_j. The strategy is based on the parametrix equation (6.9). Taking a test function $\psi \in C_0^\infty(X)$, we want to pair u with

$$(\Delta_X - s(1-s))M(s)\psi = \psi - L(s)\psi$$

and integrate. By integrating by parts and using $(\Delta_X - s(1-s))u = 0$, our goal is to derive the equation $u = L(s)^t u$ and read off the asymptotics from the known properties of $L(s)$.

There are two obstacles to be dealt with. First, if there are cusps then $M(s)$ and $L(s)$ have poles at $s = \frac{1}{2}$ from the pole of $R_{C_\infty}(s)$. To avoid this we simply assume that ψ has support within $\{r \geq 2\}$ in the funnel F_j, so that the equation becomes

$$(\Delta_X - s(1-s))M_f(s)\psi = \psi - L_f(s)\psi, \tag{7.3}$$

valid for all $\operatorname{Re} s > 0$.

The second issue is that integration by parts must be justified by showing that $M_f(s)\psi$ lies in the domain of the Laplacian. From (6.20) we see that

$$M_f(s)\psi \in \rho^s C^\infty(\overline{F}_j), \tag{7.4}$$

implying that the desired integration by parts is valid if and only if $\operatorname{Re} s > \frac{1}{2}$. We must therefore integrate by parts first for $s + \varepsilon$, with $\varepsilon > 0$, and then take a limit.

Our argument starts from the formula

$$\int_{F_j} u \left(\Delta_X - (s+\varepsilon)(1-s-\varepsilon)\right) M_f(s+\varepsilon)\psi \, dA$$
$$= \varepsilon(1 - 2s - \varepsilon) \int_{F_j} u \, M_f(s+\varepsilon)\psi \, dA,$$

derived by integrating by parts and using the eigenvalue equation for u. Together with the parametrix equation (7.3) this yields

$$\varepsilon(1 - 2s - \varepsilon) \int_{F_j} u \, M_f(s)\psi \, dA = \int_{F_j} u \, (I - L_f(s))\psi \, dA. \tag{7.5}$$

Now consider the limit $\varepsilon \to 0$. On the left-hand side of (7.5), we must be careful because $M_f(s)\psi$ leaves L^2 in this limit. To handle this, first we estimate for $\varepsilon > 0$ with the Schwarz inequality:

$$\left| \int_{F_j} u \, M_f(s+\varepsilon)\psi \, dA \right| \leq \sup_{F_j} \left| \rho^{-s-\varepsilon} M_f(s+\varepsilon)\psi \right| \, \|\rho^{s+\varepsilon}\|_{F_j} \, \|u\|.$$

By (7.4), the first term on the right is bounded as $\varepsilon \to 0$. A simple calculation using the form of the metric (6.6) shows that

$$\|\rho^{s+\varepsilon}\|_{F_j} = \left[\ell_j \int_0^2 \rho^{1+2\varepsilon} \frac{(1 + \rho^2/4)}{\rho^2} \, d\rho \right]^{1/2} = O(\varepsilon^{-1/2}).$$

Thus the left-hand side of (7.5) is $O(\varepsilon^{1/2})$ and vanishes as $\varepsilon \to 0$.

For the integral on the right-hand side of (7.5) we note that $\psi - L_f(s)\psi \in C_0^\infty(X)$ and $L_f(s)\psi$ varies analytically near $\operatorname{Re} s = \frac{1}{2}$. Thus taking the limit $\varepsilon \to 0$ is justified on this side as well and we deduce that

$$0 = \int_{F_j} u \, (I - L_f(s))\psi \, dA,$$

for $s(1 - s) = \lambda$. Since the test function ψ was arbitrary, this implies that

$$u(z) = \int_X u(w) L_f(s; w, z) \, dA(w), \qquad z \in F_j \cap \{r \geq 2\}.$$

Then, from the structure of the smooth kernel $L_f(s; w, z)$ given in (6.17), we see that

$$u|_{F_j} \in \rho^s C^\infty(\overline{F}_j).$$

Thus u has the desired expansion, and in particular $u|_{F_j} = \rho^s u_0$ for some $u_0 \in C^\infty(\overline{F}_j)$. However, since $\mathrm{Re}\, s = \frac{1}{2}$, u would not be in $L^2(X)$ unless $u_0|_{\rho=0} = 0$. Therefore the expansion must start one order higher,

$$u|_{F_j} \in \rho^{s+1} C^\infty(\overline{F}_j),$$

and then Proposition 7.4 shows that $u = 0$.   □

The proof of Proposition 7.4 rests on a Carleman estimate, which is the bound of a certain weighted L^2 norm of u by a weighted L^2 norm of Δu. This version is adapted from the more general argument given by Mazzeo [127].

Lemma 7.6 (Carleman estimate). *Suppose that $w \in \rho^\infty C^\infty(\overline{F}_\ell)$ has support in the set $\{r \geq r_0\}$, where r is distance from the geodesic boundary. Then for r_0 sufficiently large and k sufficiently large, there exists C independent of k such that*

$$k^3 \int_{F_\ell} e^{2kr} |w|^2 \, dA + k \int_{F_\ell} e^{2kr} |\nabla w|^2 \, dA \leq C \int_{F_\ell} e^{2kr} |\Delta_{F_\ell} w|^2 \, dA.$$

Proof. The Laplacian in the geodesic (Fermi) coordinates (r, t) takes the form

$$\Delta_{F_\ell} = -\partial_r^2 - \tanh r \, \partial_r - (\cosh r)^{-2} \partial_t^2.$$

Let $v = e^{kr} w$ and

$$Q_k = -(\partial_r - k)^2 - (\cosh r)^{-2} \partial_t^2.$$

With these substitutions,

$$e^{kr} \Delta_{F_\ell} w = Q_k v - \tanh r \, (\partial_r - k) v.$$

For the moment we concentrate on the estimate for $\|Q_k v\|$; the contribution from $\tanh r \, (\partial_r - k)v$ will be easily dealt with at the end.

The desired inequality is a lower bound for

$$\begin{aligned}
\|Q_k v\|^2 = &\, \|\partial_r^2 v + (\cosh r)^{-2} \partial_t^2 v\|^2 + 4k^2 \|\partial_r v\|^2 + k^4 \|u\|^2 \\
&- 4k \, \mathrm{Re}\langle \partial_r^2 v, \partial_r v \rangle + 2k^2 \, \mathrm{Re}\langle \partial_r^2 v, v \rangle - 4k^3 \, \mathrm{Re}\langle \partial_r v, v \rangle \\
&- 4k \, \mathrm{Re}\langle \partial_r v, (\cosh r)^{-2} \partial_t^2 v \rangle + 2k^2 \, \mathrm{Re}\langle v, (\cosh r)^{-2} \partial_t^2 v \rangle.
\end{aligned}$$

To produce an estimate one would naturally try to integrate by parts in the various cross terms and then drop positive terms from the right-hand side. Unfortunately, after integration by parts the last cross term yields $-2k^2 \|(\cosh r)^{-1} \partial_t v\|^2$, which is not canceled by a corresponding positive term.

Therefore we must resort to a slightly more convoluted strategy, separating the odd and even powers of k. To handle the even powers we note that

$$\|Q_k v\|^2 + \|Q_{-k} v\|^2 \geq 8k^2 \|\partial_r v\|^2 + k^4 \|v\|^2$$

$$+ 4k^2 \, \mathrm{Re}\langle \partial_r^2 v, v \rangle + 4k^2 \, \mathrm{Re}\langle v, (\cosh r)^{-2} \partial_t^2 v \rangle, \qquad (7.6)$$

where on the right we dropped the (positive) term of order k^0. Some simple integration by parts will evaluate the cross terms

$$4k^2 \operatorname{Re}\langle \partial_r^2 v, v \rangle = -4k^2 \|\partial_r v\|^2 + 2k^2 \|v\|^2$$

and

$$4k^2 \operatorname{Re}\langle v, (\cosh r)^{-2} \partial_t^2 v \rangle = -4k^2 \|(\cosh r)^{-1} \partial_t v\|^2.$$

(Note that the decay assumption on u enters here to justify the lack of boundary terms.) Thus from (7.6) we derive the inequality

$$k^4 \|v\|^2 + 4k^2 \|\partial_r v\|^2 \le \|Q_k v\|^2 + \|Q_{-k} v\|^2 + 4k^2 \|(\cosh r)^{-1} \partial_t v\|^2. \qquad (7.7)$$

For the odd powers of k we consider

$$\|Q_k v\|^2 - \|Q_{-k} v\|^2 = -8k \operatorname{Re}\langle \partial_r^2 v, \partial_r v \rangle - 8k^3 \operatorname{Re}\langle \partial_r v, v \rangle$$
$$- 8k \operatorname{Re}\langle \partial_r v, (\cosh r)^{-2} \partial_t^2 v \rangle. \qquad (7.8)$$

Once again, we integrate by parts, to obtain

$$-8k \operatorname{Re}\langle \partial_r^2 v, \partial_r v \rangle = 4k \int_F |\partial_r v|^2 \tanh r \, dA \ge 4k \|\partial_r v\|^2 (1 - 2e^{-2r_0}).$$

Similarly,

$$-8k^3 \operatorname{Re}\langle \partial_r v, v \rangle \ge 4k^3 \|v\|^2 (1 - 2e^{-2r_0})$$

and

$$-8k \operatorname{Re}\langle \partial_r v, (\cosh r)^{-2} \partial_t^2 v \rangle \ge 4k \|(\cosh r)^{-1} \partial_t v\|^2 (1 - 2e^{-2r_0}).$$

Applying these estimates to (7.8) gives, for r_0 sufficiently large,

$$c(k^3 \|v\|^2 + k \|\nabla v\|^2) + \|Q_{-k} v\|^2 \le \|Q_k v\|^2. \qquad (7.9)$$

To complete the estimate, first divide (7.7) by k to obtain

$$k^3 \|v\|^2 + 4k \|\nabla v\|^2 \le k^{-1} \|Q_k v\|^2 + k^{-1} \|Q_{-k} v\|^2 + 8k \|(\cosh r)^{-1} \partial_t v\|^2.$$

Now we can use (7.9) to control the second and third terms on the right:

$$\|Q_{-k} v\|^2 \le \|Q_k v\|^2$$

and

$$ck \|(\cosh r)^{-1} \partial_t v\|^2 \le \|Q_k v\|^2.$$

We have thus shown that

$$k^3 \|v\|^2 + 4k \|\nabla v\|^2 \le C \|Q_k v\|^2.$$

The final step is to replace Q_k by the actual Laplacian. The difference,

$$Rv := e^{kr} \Delta_{F_\ell} u - Q_k v = -\tanh r \, (\partial_r - k) v,$$

can be bounded by
$$\|Rv\| \le k\|v\| + \|\nabla v\|.$$

Therefore,
$$\|e^{kr}\Delta_{F_\ell}w\|^2 \ge (\|Q_k v\| - k\|v\| - \|\nabla v\|)^2,$$

and the desired lower bound for $\|e^{kr}\Delta_{F_\ell}w\|^2$ follows easily from the lower bound we established for $\|Q_k v\|$. $\qquad\square$

Use of the Carleman estimate to prove unique continuation is a standard technique. The only novelty in our case is the use of the indicial equation to weaken the assumption on the order of vanishing from ρ^∞ to ρ^{s+1}.

Proof of Proposition 7.4. Let us first assume that $(\Delta - \lambda)u = 0$ and u satisfies

$$u|_{F_j} \in \rho^\infty C^\infty(\overline{F}_j). \tag{7.10}$$

Let $\chi_a \in C^\infty(\mathbb{R}_+)$ be the cutoff used in Section 6.3, i.e., $\chi_a(r) = 1$ for $r \le a$ and vanishing for $r \ge a + 1$. By (7.10), $w = (1 - \chi_{r_0})u|_{F_j}$ satisfies the hypotheses of Lemma 7.6. Dropping the $|\nabla w|^2$ term in the inequality, we obtain

$$k^3 \int_{F_j} \rho^{-2k}(1 - \chi_{r_0})^2|u|^2\, dA \le C \int_{F_j} \rho^{-2k}|\Delta(1 - \chi_{r_0})u|^2\, dA,$$

for k sufficiently large. By the eigenvalue equation for u, we then have

$$k^3 \int_{r \ge r_0+1} \rho^{-2k}|u|^2\, dA \le C|\lambda|^2 \int_{r \ge r_0+1} \rho^{-2k}|u|^2\, dA + C'e^{2kr_0},$$

where C' depends on χ and u but not on k. Letting $k \to \infty$ in this inequality, we find that $u = 0$ for $r \ge r_0 + 1$. The standard uniqueness theory for second-order elliptic differential operators (see Aronszajn [7]) implies that a solution of $(\Delta - \lambda)u = 0$ vanishing on an open set must be zero everywhere.

Now suppose u has the more general form given in the statement of Proposition 7.4. We can use the indicial equation at $\rho = 0$ to show that $u|_{F_j}$ must actually vanish to infinite order at $\rho = 0$. Let $u|_{F_j} = \rho^{s+1}v_1$, with $v_1 \in C^\infty(\overline{F}_j)$. Since

$$(\Delta_X - s(1-s))(\rho^{s+1}v_1) = -2s\rho^{s+1}v_1 + O(\rho^{s+2}),$$

we can deduce from $(\Delta_X - s(1-s))u = 0$, assuming $s \ne 0$, that v_1 vanishes at $\rho = 0$. Hence $u|_{F_j} = \rho^{s+2}v_2$, with $v_2 \in C^\infty(\overline{F}_j)$. The process can be continued inductively, as long as $s \notin -\mathbb{N}_0/2$, because

$$(\Delta_X - s(1-s))(\rho^{s+k}v_k) = k(1 - 2s - k)\rho^{s+k}v_k + O(\rho^{s+k+1}).$$

This inductive argument shows that u satisfies (7.10), and the argument above then shows that $u = 0$. $\qquad\square$

For $\operatorname{Re} s = \frac{1}{2}, s \neq \frac{1}{2}$, we can actually prove a stronger version of Proposition 7.4. This strengthening allows us to rule out poles of $R_X(s)$ on the critical line, and it will also prove important for the definition of the scattering matrix later.

Lemma 7.7. *Suppose X is a nonelementary geometrically finite hyperbolic surface with infinite area. Let $\operatorname{Re} s = \frac{1}{2}$ with $s \neq \frac{1}{2}$. Then*

$$(\Delta_X - s(1-s))u = 0$$

has no solutions such that $u|_{F_j} \in \rho^s C^\infty(\overline{F}_j)$ in some funnel F_j.

Proof. The proof is by a boundary-pairing argument adapted from Melrose [135]. Choose $\psi \in C^\infty(\mathbb{R}_+)$ with $\psi(t) = 0$ for $t \leq 1$, $\psi(t) = 1$ for $t \geq 2$, and $\psi' \geq 0$. We'll define $\psi_\varepsilon \in C^\infty(X)$ for $\varepsilon > 0$ by $\psi_\varepsilon = \psi(\rho/\varepsilon)$ on F_j and $\psi_\varepsilon = 1$ on $X - F_j$.

Using the self-adjointness of Δ_X and the fact that $s(1-s)$ is real by assumption, we have

$$\int_X \overline{u} \, [\Delta_X, \psi_\varepsilon] u \, dA = -2i \operatorname{Im}\left(\psi_\varepsilon u, (\Delta_X - s(1-s))u\right) = 0. \qquad (7.11)$$

By (6.6) the Laplacian in the funnel coordinates is

$$\Delta_X = -\rho^2 \partial_\rho^2 - \frac{2\rho^3}{1+\rho^2/4}\partial_\rho + \frac{\rho^2}{(1+\rho^2/4)^2}\partial_t^2.$$

Thus the commutator has three components:

$$[\Delta_X, \psi_\varepsilon] = -2\rho^2\varepsilon^{-1}\psi'(\rho/\varepsilon)\partial_\rho - \rho^2\varepsilon^{-2}\psi''(\rho/\varepsilon)$$
$$- \frac{2\rho^3}{1+\rho^2/4}\varepsilon^{-1}\psi'(\rho/\varepsilon). \qquad (7.12)$$

By (7.11) and (7.12) we have

$$0 = \int_{F_j} \overline{u} \, [\Delta_X, \psi_\varepsilon] u \, dA$$
$$= -\int_0^{\ell_j} \int_\varepsilon^{2\varepsilon} \left(2\varepsilon^{-1}\psi'(\rho/\varepsilon)\,\overline{u}\,\partial_\rho u + \varepsilon^{-2}|u|^2\psi''(\rho/\varepsilon) \right.$$
$$\left. + 2\varepsilon^{-1}|u|^2\rho\psi'(\rho/\varepsilon)\right)(1+\rho^2/4)\,d\rho\,dt.$$

By assumption, $u_{F_j} = \rho^s v$ for $v \in C^\infty(\overline{F}_j)$. Changing variables to $x = \rho/\varepsilon$, we can see that the leading term as $\varepsilon \to 0$ involves only $v(0, t)$. In fact, since $\int_1^2 \psi'(x)\,dx = 1$, we find that

$$0 = (1 - 2s)\int_0^{\ell_j} |v(0, t)|^2\,dt + O(\varepsilon).$$

For $s \neq \frac{1}{2}$ this implies $v(0, t) = 0$. Hence $u|_{F_j} \in \rho^{s+1}C^\infty(\overline{F}_j)$, and then Proposition 7.4 shows that $u = 0$. $\qquad \square$

Corollary 7.8. *For X a geometrically finite infinite-volume hyperbolic surface, $R_X(s)$ has no poles on the line* $\operatorname{Re} s = \frac{1}{2}$ *except possibly at* $s = \frac{1}{2}$.

Proof. If $R_X(s)$ has a pole at ζ, then, since it is finitely meromorphic, it has a Laurent expansion

$$R_X(s) = \sum_{k=-m}^{\infty} (s-a)^k A_k,$$

valid for s near ζ. In particular, for some $\psi \in C_0^\infty(X)$, we have

$$R_X(s)\psi = (s-\zeta)^{-m} u + (s-\zeta)^{-m+1} v(s), \tag{7.13}$$

where $v(s)$ is analytic near ζ. Using the definition of the resolvent, we obtain

$$(\Delta_X - \zeta(1-\zeta)) R_X(s)\psi = \psi + [s(1-s) - \zeta(1-\zeta)] R_X(s)\psi.$$

Substituting (7.13) and taking the leading term as $s \to \zeta$ then shows that

$$(\Delta_X - \zeta(1-\zeta)) u = 0.$$

By Theorem 6.3, we also know that $u \in \rho_f^\zeta \rho_c^{\zeta-1} C^\infty(\overline{X})$. If $\operatorname{Re}\zeta = \frac{1}{2}$ and $\zeta \neq \frac{1}{2}$, then Lemma 7.7 implies $u = 0$. Hence $R_X(s)$ cannot have a singular part at such a value of ζ. □

7.3 Generalized eigenfunctions

We went through the derivation of generalized eigenfunctions for the model case $\mathbb{H}$ in some detail in Section 4.2, in order to illustrate the role that the generalized eigenfunctions play in linking the resolvent to the scattering theory. Now we'll apply the same philosophy in the general case.

We continue to assume here that X is nonelementary and geometrically finite, and refer to the decomposition (6.4) into compact core plus funnels and cusps. As in the model case, the limit of the resolvent kernel $R_X(s; z, z')$ as $\rho' \to 0$ will yield solutions of $(\Delta_X - s(1-s))E = 0$, provided s is not a pole of $R_X(s)$. These solutions will again be parametrized by points at infinity, but now we must keep track of whether $\rho' \to 0$ in a funnel or cusp end.

For the funnel case, we identify z' with the standard coordinates $(\rho', t') \in (0, 1] \times \mathbb{R}/(\ell_j \mathbb{Z})$ in the funnel F_j, and define

$$E_j^f(s; z, t') := \lim_{\rho' \to 0} \rho'^{-s} R_X(s; z, z'), \tag{7.14}$$

for $j = 1, \ldots, n_f$. In the cusp C_j, with standard coordinates $z' = (\rho', t')$, we set

$$E_j^c(s; z) := \lim_{\rho' \to 0} \rho'^{1-s} R_X(s; z, z'), \tag{7.15}$$

for $j = 1, \ldots, n_{\mathrm{c}}$. Both of these limits exist by Theorem 6.3, which shows in addition that the cusp limit (7.15) is independent of t'.

The generalized eigenfunctions coincide with a more classical construction known as *Eisenstein series*. This term refers to their representation as averages over Γ of the model Poisson kernel $E_{\mathbb{H}}(s; z, x')$. Using the form of $E_{\mathbb{H}}(s; z, x')$ given in (4.15), we can write the lift of $E_j^{\mathrm{f}}(s; \cdot, t')$ to $\mathbb{H}$ as a series, convergent for $\operatorname{Re} s > 1$,

$$
E_j^{\mathrm{f}}(s; z, x') = C(s) \sum_{T \in \Gamma} \left(\frac{\operatorname{Im}(Tz)}{|Tz - x'|^2} \right)^s,
$$

where $x' \in \Omega(\Gamma)$. And similarly for $E_j^{\mathrm{c}}(s)$, with x' set equal to a parabolic fixed point of Γ corresponding to the endpoint of C_j.

A very important relation between the resolvent and the generalized eigenfunctions is given by the following:

Proposition 7.9. *For s not a pole of $R_X(s)$ or $R_X(1 - s)$, we have*

$$
R_X(s; z, w) - R_X(1 - s; z, w)
$$
$$
= -(2s - 1) \sum_{j=1}^{n_{\mathrm{f}}} \int_0^{\ell_j} E_j^{\mathrm{f}}(s; z, t') E_j^{\mathrm{f}}(1 - s; w, t') \, dt'
$$
$$
-(2s - 1) \sum_{j=1}^{n_{\mathrm{c}}} E_j^{\mathrm{c}}(s; z) E_j^{\mathrm{c}}(1 - s; w).
$$

Proof. Except for the notational complication provided by the cusp terms, this is very similar to the proof of Proposition 4.3. Let X_ε denote the closed subset of X defined by $\{\rho \geq \varepsilon\}$. The defining equation for the resolvent, $(\Delta_X - s(1 - s)) R_X(s) = I$, implies that

$$
R_X(s; z, w) - R_X(1 - s; z, w) = \lim_{\varepsilon \to 0} \int_{X_\varepsilon} \Big[R_X(s; z, z') \Delta_{z'} R_X(1 - s; z', w)
$$
$$
- R_X(1 - s; z', w) \Delta_{z'} R_X(s; z, z') \Big] \, dA(z').
$$

We can use Green's formula to turn this into a line integral. Noting that the outward unit normal to X_ε is ∂_r in our standard coordinates, and that our positive Laplacian is $-\operatorname{div} \operatorname{grad}$, this yields

$$
R_X(s; z, w) - R_X(1 - s; z, w) = \lim_{\varepsilon \to 0} \int_{\partial X_\varepsilon} \Big[\partial_{r'} R_X(s; z, z') R_X(1 - s; z', w)
$$
$$
- R_X(s; z, z') \partial_{r'} R_X(1 - s; z', w) \Big] \, d\sigma_\varepsilon(z'),
$$

where $d\sigma_\varepsilon$ denotes the arc-length element on ∂X_ε induced by the hyperbolic metric. The boundary ∂X_ε consists of $n_{\mathrm{f}} + n_{\mathrm{c}}$ circles, one for each funnel or cusp.

To compute the contribution from funnel F_j, note that with z fixed in the interior, Theorem 6.3 and (7.14) imply that

$$R_X(s; z, \cdot)|_{F_j} \in \rho'^s C^\infty(\overline{F}_j),$$

with leading term, as $\rho' \to 0$, given by

$$R_X(s; z, z') = \rho'^s E_j^f(s; z, t') + O(\rho'^{s+1}).$$

An analogous result holds for $R_X(1 - s; z', w)$. Since $\partial_{r'} = -\rho' \partial_{\rho'}$, we derive that

$$\partial_{r'} R_X(s; z, z') = -s\rho'^s E_j^f(s; z, t') + O(\rho'^{s+1}),$$

$$\partial_{r'} R_X(1 - s; z', w) = (s - 1)\rho'^{1-s} E_j^f(s; w, t') + O(\rho'^{s+1}).$$

By (6.6),

$$d\sigma_\varepsilon(z') = \varepsilon^{-1}(1 + \varepsilon^2/4)\, dt'.$$

After tallying up the powers of ε and keeping only $O(\varepsilon^0)$ terms in the integral as $\varepsilon \to 0$, we find that

$$\int_{\partial F_{j,\varepsilon}} \left[\partial_{r'} R_X(s; z, z') R_X(1 - s; z', w) \right.$$

$$\left. - R_X(s; z, z') \partial_{r'} R_X(s; z', w) \right] d\sigma_\varepsilon(z')$$

$$= (1 - 2s) \int_0^{\ell_j} E_j^f(s; z, t') E_j^f(1 - s; w, t')\, dt' + O(\varepsilon).$$

In the cusp C_j the analysis is very similar. By Theorem 6.3 and (7.15) we have

$$R_X(s; z, \cdot)|_{C_j} \in \rho'^{s-1} C^\infty(\overline{C}_j)$$

with

$$R_X(s; z, z') = \rho'^{s-1} E_j^c(s; z) + O(\rho'^\infty).$$

By (6.7) the arc-length element is $d\sigma_\varepsilon(z') = \varepsilon\, dt'$ in the standard coordinates. As above, the generalized eigenfunctions give the leading term in the integral, and since $E_j^c(s; z)$ is independent of t', the t' integral simply evaluates to one:

$$\int_{\partial C_{j,\varepsilon}} \left[\partial_{r'} R_X(s; z, z') R_X(1 - s; z', w) \right.$$

$$\left. - R_X(1 - s; z, z') \partial_{r'} R_X(s; z', w) \right] d\sigma_\varepsilon(z')$$

$$= (1 - 2s) E_j^c(s; z) E_j^c(1 - s; w) + O(\varepsilon^\infty). \qquad \square$$

One of the many uses we will have for Proposition 7.9 is to give a representation of the spectral projectors through Stone's formula, which leads to the following:

Proposition 7.10. *For X a geometrically finite hyperbolic surface of infinite area, the essential spectrum of Δ_X is absolutely continuous.*

Proof. Stone's formula (Corollary A.12) allows us to write the spectral projector P_I of Δ_X in the form

$$\frac{1}{2}(P_{[a,b]} + P_{(a,b)}) = \frac{1}{2\pi i} \int_{\sqrt{a-1/4}}^{\sqrt{b-1/4}} \left[R_X(\tfrac{1}{2} - i\xi) - R_X(\tfrac{1}{2} + i\xi) \right] 2\xi \, d\xi.$$

By Proposition 7.9 the kernel of the integrand can be written

$$4i\xi^2 \sum_{j=1}^{n_f} \int_0^{2\pi} E_j^f(\tfrac{1}{2} + i\xi; z, t) E_j^f(\tfrac{1}{2} - i\xi; w, t) \, dt$$

$$+ 4i\xi^2 \sum_{j=1}^{n_c} \int_{C_j} E_j^c(\tfrac{1}{2} + i\xi; z) E_j^c(\tfrac{1}{2} - i\xi; w).$$

Corollary 7.8 implies that these expressions are analytic in ξ for $\xi > 0$; hence the spectral measures are absolutely continuous. $\qquad\square$

With this result we have completed the proof of Theorem 7.1.

7.4 Scattering matrix

In Melrose's approach to geometric scattering theory, as summarized in [136], the scattering matrix is naturally defined as the interpolation operator between leading terms in the asymptotic expansions of solutions of $(\Delta_X - s(1-s))u = 0$ (analogous to a Dirichlet-to-Neumann map).

Assume that X is a nonelementary, geometrically finite hyperbolic surface of infinite area. If u solves $(\Delta_X - s(1-s))u = 0$ then the indicial equation (7.2) shows that in the funnel ends the most general form for an asymptotic expansion of u in powers of ρ would have leading behavior

$$u \sim \rho^{1-s} f + \rho^s f'. \tag{7.16}$$

The coefficients f, f' are identified with functions on the infinite boundary of $\overline{F}_j$. In the cusps, the crucial term in Δ_X for the indicial equation is $-\partial_\rho(\rho^2 \partial_\rho)$, and the indicial roots are $s - 1$ and $-s$. So the leading terms in an asymptotic expansion in the cusps would be

$$u \sim \rho^{-s} a + \rho^{s-1} a', \tag{7.17}$$

with $a, a' \in \mathbb{C}$.

From Section 6.1, $\partial \overline{X}$ is the union of n_f circles bounding the cusps and n_c points bounding the funnels. We identify

$$C^\infty(\partial \overline{X}) \cong \bigoplus_{j=1}^{n_f} C^\infty(\mathbb{R}/\ell_j \mathbb{Z}) \oplus \mathbb{C}^{n_c}.$$

The infinite boundary of $\overline{F}_j$ will be parametrized by t in our standard coordinates (ρ, t) for F_j. This is equivalent to arc length in the metric $\bar{g} = \rho^2 g$, where g is the hyperbolic metric on X. We define a measure dh on $\partial \overline{X}$ using this induced measure dt in the funnel ends and assigning a unit point measure to each cusp point p_j. This allows us to interpret the set of generalized eigenfunctions as the kernel of a map.

Definition. The *Poisson operator* is the map

$$E_X(s) : C^\infty(\partial \overline{X}) \mapsto C^\infty(X),$$

given by

$$E_X(s)\psi(z) := \int_{\partial \overline{X}} E_X(s; z, q)\psi(q)\, dh(q). \tag{7.18}$$

For $\psi = (f_1, \ldots, f_{n_f}, a_1, \ldots, a_{n_c}) \in C^\infty(\partial \overline{X})$, this definition expands to the explicit form

$$E_X(s)\psi(z) := \sum_{j=1}^{n_f} \int_0^{\ell_j} E_j^f(s; z, t') f_j(t')\, dt' + \sum_{j=1}^{n_c} E_j^c(s; z)a_j.$$

With this definition, we can condense the result of Proposition 7.9 into the formula

$$R_X(s) - R_X(1 - s) = -(2s - 1)E_X(s)E_X(1 - s)^t. \tag{7.19}$$

As in Theorem 6.3, we decompose ρ as $\rho_f \rho_c$, with $\rho_f = \rho$ in the funnels and $\rho_c = \rho$ in the cusps.

Proposition 7.11. *For $\psi \in C^\infty(\partial \overline{X})$ and $s \in \mathbb{C}$ not a pole of $R_X(s)$,*

$$(\Delta_X - s(1 - s))E_X(s)\psi = 0.$$

If $s \notin \mathbb{Z}/2$, we have an asymptotic expansion of the form

$$E_X(s)\psi \in \rho_f^{1-s}\rho_c^{-s} C^\infty(\overline{X}) + \rho_f^s \rho_c^{s-1} C^\infty(\overline{X}),$$

with leading behavior given by

$$E_X(s)\psi \sim \frac{1}{2s - 1}\left[\rho_f^{1-s}\rho_c^{-s}\psi + \rho_f^s \rho_c^{s-1}\psi'\right], \tag{7.20}$$

for some $\psi' \in C^\infty(\partial \overline{X})$.

Theorem 6.3 allows us to deduce the asymptotic expansion by analyzing model terms. Dealing with the model cusp terms is essentially trivial, so the main issue here is to work out the expansion in the funnels. We will start with the theory for the model funnel F_ℓ, and then this will be applied to the general case to prove Proposition 7.11.

Let $E_{F_\ell}(s; \rho, t, t')$ denote the generalized eigenfunction obtained from the model funnel resolvent,

$$E_{F_\ell}(s; \rho, t, t') := \lim_{\rho' \to 0} {\rho'}^{-s} R_{F_\ell}(s; \rho, t, t'),$$

where (ρ, t) are the standard coordinates on F_ℓ. Since $\overline{F}_\ell$ has two boundaries, we'll distinguish the infinite boundary at $\{\rho = 0\}$ by the notation $\partial_0 \overline{F}_\ell$. The funnel Poisson operator

$$E_{F_\ell}(s) : C^\infty(\partial_0 \overline{F}_\ell) \to C^\infty(F_\ell)$$

maps $f \in C^\infty(\partial_0 \overline{F}_\ell)$ to

$$E_{F_\ell}(s) f(\rho, t) := \int_0^\ell E_{F_\ell}(s; \rho, t, t') f(t') \, dt'.$$

Lemma 7.12. *For* $f \in C^\infty(\partial_0 \overline{F}_\ell)$, *the function* $E_{F_\ell}(s) f$ *admits an asymptotic expansion as* $\rho \to 0$,

$$E_{F_\ell}(s) f \in \rho^{1-s} C^\infty(\overline{F}_\ell) + \rho^s C^\infty(\overline{F}_\ell), \tag{7.21}$$

with leading terms

$$E_{F_\ell}(s) f(\rho, \cdot) \sim \frac{1}{2s - 1} \left[\rho^{1-s} f + \rho^s f' \right], \tag{7.22}$$

where $f' \in C^\infty(\partial_0 \overline{F}_\ell)$.

Proof. First note that (7.21) is equivalent to the existence of a two-part asymptotic expansion of the form

$$E_{F_\ell}(s) f(\rho, t) \sim \sum_{k=0}^\infty \rho^{1-s+k} a_k(s; t) + \sum_{k=0}^\infty \rho^{s+k} b_k(s; t), \tag{7.23}$$

where the coefficients are smooth functions of t.

The structure of $R_{F_\ell}(s)$ was revealed in the proof of Proposition 6.7. Let $\mathcal{F}_\ell$ be the fundamental domain $\{1 \le |z| \le e^\ell\}$ for the action of Γ_ℓ on $\mathbb{H}$. We can identify $z \in \mathcal{F}_\ell$ with our standard coordinates (ρ, t) for F_ℓ by the map

$$z = e^t \frac{2 + i\rho}{2 - i\rho}.$$

As $\rho \to 0$ we have

$$\rho \sim \frac{y}{x}. \tag{7.24}$$

By (6.27) and (6.28), we can write the lift of the resolvent kernel to $\mathcal{F}_\ell \times \mathcal{F}_\ell$, for $\mathrm{Re}\, s > -N$, as a sum,

$$R_{F_\ell}(s; z, z') = R_{\mathbb{H}}(s; z, z') + \sum_{n=0}^{N-1} \frac{1}{4\pi} \frac{\Gamma(s+n)^2}{n! \Gamma(2s+n)} \sum_{k \neq 0} \sigma(z, e^{k\ell} z')^{-s-n}$$

$$+ \sum_{k \neq 0} F_N(s, \sigma(z, e^{k\ell} z')) + R_{C_\ell}(s; z, -\overline{z}'). \tag{7.25}$$

By (7.24),

$$\lim_{\rho' \to 0} \rho'^{-s} R_{F_\ell}(s; z, z') = x'^s \lim_{y' \to 0} y'^{-s} R_{F_\ell}(s; z, z').$$

With the other variables fixed,

$$\sigma(z, z') \sim \frac{(x - x')^2 + y^2}{4yy'}, \quad \text{as } y' \to 0.$$

Therefore both the $n > 0$ terms in the first sum and the sum over the remainder term F_N contribute zero in the limit. For $z \in \mathcal{F}_\ell, x' \in (1, e^\ell)$ we have

$$E_{F_\ell}(s; z, x') = x'^s E_{\mathbb{H}}(s; z, x') + A(s; z, x'), \tag{7.26}$$

where A is given explicitly by

$$A(s; z, x') := \frac{\Gamma(s)}{2\sqrt{\pi}\,\Gamma(s + 1/2)} (x'y)^s \left(\sum_{k \neq 0} \left[e^{-k\ell}(x - e^{k\ell}x')^2 + e^{-k\ell}y^2 \right]^{-s} \right.$$

$$\left. + \sum_{k \in \mathbb{Z}} \left[e^{-k\ell}(x + e^{k\ell}x')^2 + e^{-k\ell}y^2 \right]^{-s} \right). \tag{7.27}$$

By Lemma 5.1, $A(s; z, x')$ is meromorphic in s, and $y^{-s} A(s; z, x')$ is easily seen to be a smooth function on $\overline{\mathcal{F}}_\ell \times (1, e^\ell)$ because $x \pm e^{k\ell}x'$ is never zero on this set.

Translating back into the standard coordinates, the singular term can be reduced to

$$x'^s E_{\mathbb{H}}(s; z, x') = \frac{\Gamma(s)}{2\sqrt{\pi}\,\Gamma(s + 1/2)} \rho^s \left[4 \sinh^2(\tfrac{t-t'}{2}) + \rho^2 \cosh^2(\tfrac{t-t'}{2}) \right]^{-s}.$$

Note that in splitting off the singular term in this way, we have broken the periodicity in the t, t' variables. This is not a big issue, because the calculation can be localized. We can assume, without loss of generality, that f has compact support in $(0, \ell)$, and analyze the asymptotics of $E_{F_\ell}(s) f(\rho, t)$ for $t \in \mathbb{R}$ instead of $\mathbb{R}/\ell\mathbb{Z}$.

The contribution of the singular term to $E_{F_\ell}(s) f$ is

$$\frac{\Gamma(s)}{2\sqrt{\pi}\,\Gamma(s + 1/2)} \rho^s \int_0^\ell \frac{f(t')}{\left[4 \sinh^2(\tfrac{t-t'}{2}) + \rho^2 \cosh^2(\tfrac{t-t'}{2}) \right]^{-s}} \, dt'. \tag{7.28}$$

This is similar to the integral analyzed in Section 4.2. Indeed, if we set

$$\varphi(t') = \left[\frac{(t - t')^2 + \rho^2}{4 \sinh^2(\tfrac{t-t'}{2}) + \rho^2 \cosh^2(\tfrac{t-t'}{2})} \right]^s f_j(t')$$

(which is smooth even near $t' = t$), then (7.28) becomes

$$\frac{\Gamma(s)}{2\sqrt{\pi}\,\Gamma(s + 1/2)} \rho^s \int_0^\ell \frac{\varphi(t')}{\left[(t - t')^2 + \rho^2 \right]^s} \, dt'.$$

Lemma 4.5 now applies to show that (7.28) has a two-part asymptotic expansion as $\rho \to 0$ of the form

$$\sum_{k=0}^{\infty} \rho^{1-s+k} a_k(s; t) + \sum_{k=0}^{\infty} \rho^{s+k} c_k(s; t),$$

where

$$a_0(s; t) = \frac{1}{2s - 1} f(t),$$

and $c_0(s; t)$ is defined by

$$c_0(s; t) = \frac{1}{(2s - 1)} \frac{\Gamma(s)}{\sqrt{\pi} \Gamma(s - \frac{1}{2})} \int_0^{\ell} \frac{f(t')}{\left[4 \sinh^2 (\frac{t-t'}{2}) \right]^s} \, dt'. \qquad (7.29)$$

The integral in (7.29) is convergent for $\operatorname{Re} s < \frac{1}{2}$ but extends by analytic continuation to $s \notin \frac{1}{2} + \mathbb{N}_0$. The pole at $s = 0$ is canceled by the $\Gamma(s - \frac{1}{2})$ in the denominator.

The nonsingular term $A(s; \rho, t, t')$ is easily dealt with. Since f is assumed compactly supported, the smoothness of $A(s; \cdot, \cdot, \cdot)$ implies a Taylor expansion,

$$\int_0^{\ell} A(s; \rho, t, t') f(t') \, dt' \sim \sum_{k=0}^{\infty} \rho^{s+k} d_k(s, t),$$

with $d_k(s; t)$ smooth in t. Setting $b_k(s; t) = c_k(s; t) + d_k(s; t)$ gives (7.23). $\qquad \square$

Proof of Proposition 7.11. The fact that $E_X(s)\psi$ satisfies the eigenvalue equation is easily deduced from the definition of generalized eigenfunctions as limits of the resolvent. For example, we could set $\phi_\varepsilon := 2\varepsilon \rho^{-s} \chi_{[\varepsilon, 2\varepsilon]}(\rho)\psi$, for $\varepsilon > 0$, so that $R_X(s)\phi_\varepsilon \to E_X(s)\psi$ uniformly on compact sets as $\varepsilon \to 0$. We also have

$$(\Delta_X - s(1 - s)) R_X(s)\phi_\varepsilon = \phi_\varepsilon,$$

which converges to zero uniformly for $\operatorname{Re} s < \frac{1}{2}$. This shows that

$$(\Delta_X - s(1 - s)) E_X(s)\psi = 0$$

for $\operatorname{Re} s < \frac{1}{2}$ and then the result extends to all s by meromorphic continuation.

The asymptotic behavior of $E_X(s) f$ will be deduced from the decomposition

$$R_X(s) = M_i'(s) + M_f(s) + M_c(s) + Q(s),$$

of Theorem 6.3.

The cusp contributions are straightforward. If $z' = (\rho', t')$ denotes a coordinate in cusp C_j, then clearly

$$E_j^c(s; z) = \lim_{\rho' \to 0} \rho'^{1-s} \left[M_c(s; z, z') + Q(s; z, z') \right],$$

since kernels of the other two terms in the decomposition of $R_X(s)$ vanish in the cusps. By Proposition 5.8, for $z, z' \in C_j$,

$$\lim_{\rho' \to 0} \rho'^{1-s} M_c(s; z, z') = (1 - \chi_0(\rho)) \lim_{r' \to \infty} \rho'^{1-s} u_0(s; 1/\rho, 1/\rho')$$

$$= \frac{1}{2s - 1} \rho^{-s} (1 - \chi_0(\rho)).$$

From (6.19), as $\rho' \to 0$ in C_j we have

$$\lim_{\rho' \to 0} \rho'^{1-s} Q(s; \cdot, z') \in \rho_f^s \rho_c^{s-1} C^\infty(\overline{X}).$$

Putting these limits together yields

$$E_j^c(s; \cdot) - \frac{1}{2s - 1} \rho^{-s}(1 - \chi_0(\rho)) \in \rho_f^s \rho_c^{s-1} C^\infty(\overline{X}). \tag{7.30}$$

The starting point for analysis of the funnel terms is similar: with (ρ', t') standard coordinates in the funnel F_j,

$$E_j^f(s; z, t') = \lim_{\rho' \to 0} \rho'^{-s} \left[M_f(s; z, z') + Q(s; z, z') \right].$$

Since $M_f(s; z, z') = (1 - \chi_0) R_{F_j}(s)(1 - \chi_1)$ in funnel F_j, the contribution of this term to $E_j^f(s; z, t')$ is

$$(1 - \chi_0(\rho)) E_{F_j}(s; z, t'),$$

where E_{F_j} denotes the generalized eigenfunction on F_j lifted from the model case F_ℓ. If we let $Q_j^f(s; z, t')$ denote the contribution of $Q(s; z, z')$ to $E_j^f(s; z, t')$, then by (6.19),

$$Q_j^f(s; \cdot, \cdot) \in \rho_f^s \rho_c^{s-1} C^\infty(\overline{X} \times \partial_0 \overline{F}_j). \tag{7.31}$$

The full contribution from f_j in funnel F_j is

$$\int_0^{\ell_j} E_j^f(s; \cdot, t') f_j(t') \, dt' = (1 - \chi_0) E_{F_j}(s) f_j + Q_j^f(s) f_j. \tag{7.32}$$

The asymptotic expansion of the first term was established in Lemma 7.12. And the asymptotic expansion of $Q_j^f(s) f_j$ follows immediately from (7.31). $\qquad \square$

Proposition 7.11 sets up the following:

Definition. The *scattering matrix* is the map $S_X(s) : C^\infty(\partial \overline{X}) \to C^\infty(\partial \overline{X})$ given by

$$S_X(s) : \psi \mapsto \psi',$$

where ψ' is determined by ψ through (7.20).

Before examining the properties of the scattering matrix, we want to show that ψ' is uniquely determined by ψ, at least for generic s. The crucial fact needed for this purpose was proven in Lemma 7.7.

Proposition 7.13. *Assume X is a nonelementary geometrically finite hyperbolic surface of infinite area. Suppose $\operatorname{Re} s \geq \frac{1}{2}$, $s \notin \mathbb{N}/2$, and $s(1-s) \notin \sigma_{\mathrm{d}}(\Delta_X)$. Given $\psi \in C^\infty(\partial \overline{X})$, there is a unique solution of*

$$(\Delta_X - s(1-s))u = 0$$

of the form

$$u \in \rho_{\mathrm{f}}^{1-s} \rho_{\mathrm{c}}^{-s} C^\infty(\overline{X}) + \rho_{\mathrm{f}}^{s} \rho_{\mathrm{c}}^{s-1} C^\infty(\overline{X}),$$

with leading asymptotic behavior

$$u \sim \rho_{\mathrm{f}}^{1-s} \rho_{\mathrm{c}}^{-s} \psi + \rho_{\mathrm{f}}^{s} \rho_{\mathrm{c}}^{s-1} \psi',$$

for some $\psi' \in C^\infty(\partial \overline{X})$.

Proof. Existence of the solution was shown in Proposition 7.11; we can simply take $u = (2s - 1)E_X(s)\psi$. If we assume that there is more than one solution having ψ as coefficient of the $\rho_{\mathrm{f}}^{1-s} \rho_{\mathrm{c}}^{-s}$ term, then the difference of two such solutions satisfies $(\Delta_X - s(1-s))v = 0$ and

$$v \in \rho_{\mathrm{f}}^{2-s} \rho_{\mathrm{c}}^{1-s} C^\infty(\overline{X}) + \rho_{\mathrm{f}}^{s} \rho_{\mathrm{c}}^{s-1} C^\infty(\overline{X}).$$

We proceed using the indicial equation as in the proof of Proposition 7.4. Because

$$(\Delta_X - s(1-s))\rho_{\mathrm{f}}^{2-s} \rho_{\mathrm{c}}^{1-s} \phi \sim (2s - 2)\rho_{\mathrm{f}}^{2-s} \rho_{\mathrm{c}}^{1-s} \phi|_{\rho=0},$$

as $\rho \to 0$, and the left-hand side is zero, the coefficient of the $\rho_{\mathrm{f}}^{2-s} \rho_{\mathrm{c}}^{1-s}$ term in the expansion must vanish. Repeating this argument inductively knocks out the first half of the expansion, leaving us with $v \in \rho_{\mathrm{f}}^{s} \rho_{\mathrm{c}}^{s-1} C^\infty(\overline{X})$. If $\operatorname{Re} s > 0$ then v would be in $L^2(X)$, implying $v = 0$ by the assumption that $s(1-s)$ is not an eigenvalue. If $\operatorname{Re} s = \frac{1}{2}$, then Lemma 7.7 shows that $v = 0$. $\square$

As an immediate consequence of Proposition 7.13, we can derive some fundamental properties of the scattering matrix. (These could also have been derived directly from Proposition 7.9.)

Corollary 7.14. *The Poisson operator and scattering matrix satisfy the relations*

$$E_X(1 - s)S_X(s) = -E_X(s)$$

and

$$S_X(1 - s)S_X(s) = I.$$

The first of these implies that

$$R_X(s) - R_X(1 - s) = (2s - 1)E_X(1 - s) \, S_X(s) \, E_X(1 - s)^{\mathrm{t}}. \tag{7.33}$$

Proof. By analytic continuation it suffices to prove the identities for $\operatorname{Re} s = \frac{1}{2}, s \neq \frac{1}{2}$. Given $\psi \in C^\infty(\partial \overline{X})$, we obtain a solution of the eigenvalue equation

$$(2s - 1)E_X(s)\psi \sim \rho_f^{1-s}\rho_c^{-s}\psi + \rho_f^s\rho_c^{s-1}S_X(s)\psi.$$

Using $S_X(s)\psi$ as input to $E_X(1 - s)$, we have also the solution

$$-(2s - 1)E_X(1 - s)S_X(s)\psi \sim \rho_f^s\rho_c^{s-1}S_X(s)\psi + \rho_f^{1-s}\rho_c^{-s}S_X(1 - s)S_X(s)\psi.$$

Since the $\rho_f^s\rho_c^{s-1}$ terms in these expansions match, Proposition 7.13 implies that the two solutions are identical for any ψ, yielding the stated identities.

The formula (7.33) follows immediately from (7.19) and the first relation.   □

To describe the scattering matrix in more detail, we break it into blocks according to the funnel/cusp boundary types,

$$S_X(s) = \begin{pmatrix} S^{\mathrm{ff}}(s) & S^{\mathrm{fc}}(s) \\ S^{\mathrm{cf}}(s) & S^{\mathrm{cc}}(s) \end{pmatrix}. \tag{7.34}$$

Breaking down further into the individual components of $\partial \overline{X}$, we set

$$\psi = (f_1, \ldots, f_{n_\mathrm{f}}, a_1, \ldots, a_{n_\mathrm{c}}),$$

and then $\psi' = S_X(s)\psi = (f_1', \ldots, f_{n_\mathrm{f}}', a_1', \ldots, a_{n_\mathrm{c}}')$ is given by

$$f_i' = \sum_{j=1}^{n_\mathrm{f}} S_{ij}^{\mathrm{ff}}(s)f_j + \sum_{l=1}^{n_\mathrm{c}} S_{il}^{\mathrm{fc}}(s; \cdot)a_l$$

and

$$a_k' = \sum_{j=1}^{n_\mathrm{f}} \int_0^{\ell_j} S_{kj}^{\mathrm{cf}}(s; t)f_j(t)\, dt + \sum_{l=1}^{n_\mathrm{c}} S_{kl}^{\mathrm{cc}}(s)a_l.$$

Let $Q(s; z, z')$ be the kernel of the remainder term from Theorem 6.3. Its restriction to the boundary,

$$Q^\sharp(s; \cdot, \cdot) = Q(s; \cdot, \cdot)\big|_{\partial \overline{X} \times \partial \overline{X}} \in C^\infty(\partial \overline{X} \times \partial \overline{X}),$$

gives the integral kernel (with respect to dh) of a smoothing operator $Q^\sharp(s)$ on $\partial \overline{X}$. From the definition of $S_X(s)$ and the proof of Proposition 7.11, we can see that $S_X(s)$ is equal to $(2s - 1)Q^\sharp$ except for diagonal terms in the funnels derived from the model case. To make this statement more precise, let $S_{F_j}(s)$ denote the scattering matrix associated to the funnel component F_j with Dirichlet conditions at the geodesic boundary. Thus $S_{F_j}(s)$ is defined on $C^\infty(\partial_0 \overline{F}_j)$ as the map $f \mapsto f'$ in the context of Lemma 7.12, if we identify F_j with F_ℓ. (We will write $S_{F_\ell}(s)$ quite explicitly below in Section 7.5.) These model operators are collected together into an operator on $C^\infty(\partial_0 \overline{F})$,

$$S_F(s) := \oplus_{j=1}^{n_f} S_{F_j}(s). \tag{7.35}$$

Then the proof of Proposition 7.11 yields

$$S_X(s) = \begin{pmatrix} S_F(s) & 0 \\ 0 & 0 \end{pmatrix} + (2s - 1)Q^\sharp(s). \tag{7.36}$$

Proposition 7.15 (Scattering matrix structure). *Assume that X is a nonelementary geometrically finite hyperbolic surface of infinite area. The scattering matrix satisfies*

$$S_X(s)^t = S_X(s), \qquad S_X(s)^* = S_X(\bar{s}). \tag{7.37}$$

Its components are meromorphic in s and have the following properties:

1. *The diagonal funnel–funnel term $S_{jj}^{ff}(s)$ is a pseudodifferential operator of order $2\,\mathrm{Re}\,s - 1$ on $\partial_0 \overline{F}_j$ with principal symbol*

$$2^{1-2s} \frac{\Gamma(\frac{1}{2} - s)}{\Gamma(s - \frac{1}{2})} |\xi|^{2s-1} \tag{7.38}$$

   (where $|\xi|$ is defined with respect to the metric dt^2 induced on $\partial_0 \overline{F}_j \cong \mathbb{R}/\ell_j \mathbb{Z}$). If $s \notin \frac{1}{2} + \mathbb{N}$ and s is not a pole of $R_X(s)$, we have for $t \neq t'$,

$$S_{jj}^{ff}(s; t, t') = (2s - 1) \lim_{\rho,\rho' \to 0} (\rho\rho')^{-s} R_X(s; z, z')\big|_{z,z' \in F_j}.$$

2. *The off-diagonal terms have smooth kernels. If s is not a pole of $R_X(s)$, then*

$$S_{ij}^{ff}(s; t, t') = (2s - 1) \lim_{\rho,\rho' \to 0} (\rho\rho')^{-s} R_X(s; z, z')\big|_{z \in F_i, z' \in F_j}, \quad i \neq j,$$

   and

$$S_{il}^{fc}(s; t) = S_{li}^{cf}(s; t) = (2s - 1) \lim_{\rho,\rho' \to 0} \rho^{-s} \rho'^{1-s} R_X(s; z, z')\big|_{z \in F_i, z' \in C_l}.$$

3. *The cusp–cusp component is a symmetric $n_c \times n_c$ matrix. If s is not a pole of $R_X(s)$, then*

$$S_{kl}^{cc}(s) = (2s - 1) \lim_{\rho,\rho' \to 0} (\rho\rho')^{1-s} Q(s; z, z')\big|_{z \in C_k, z' \in C_l},$$

   where $Q(s)$ is the remainder term from Theorem 6.3.

Proof. The formulas for the off-diagonal and cusp–cusp terms follow immediately from (7.36) and Theorem 6.3.

It is also apparent in (7.36) that $S_{jj}^{ff}(s)$ is equal to $S_{F_j}(s)$ up to a smoothing operator. To see that $S_{F_j}(s)$ is pseudodifferential and read off its symbol, we reexamine the proof of Lemma 7.12. Up to a correction by a smoothing operator, $S_{F_j}(s)f(t)$ is given by the coefficient $c_0(s, t)$ appearing in (7.29). The term $\left[4 \sinh^2(t - t')/2\right]^{-s}$

can be identified as the integral kernel of a pseudodifferential operator by taking a distributional Fourier transform over $\mathbb{R}$. The leading singularity of the integral kernel on the diagonal is $|t - t'|^{-2s}$, which by the Fourier transform corresponds to the operator with symbol

$$2^{1-2s} \frac{\sqrt{\pi}\,\Gamma(\frac{1}{2} - s)}{\Gamma(s)} |\xi|^{2s-1}.$$

Combining the gamma functions gives the principal term (7.38). The formula for the integral kernel of S_{jj}^{ff} is valid (assuming s is not a pole of $R_X(s)$) as long as the integral kernel appearing in the expression (7.29) for $c_0(s, t)$ is a well-defined distribution. This gives the restriction to $s \notin \frac{1}{2} + \mathbb{N}$.

Because of the connection between the scattering matrix and resolvent kernel, the properties (7.37) follow from Lemma 6.5 and (6.22). $\square$

7.5 Scattering matrices for the funnel and cylinders

The scattering matrix for a funnel F_ℓ was described as a pseudodifferential operator in the proof of Proposition 7.15. But we have another way to understand this operator that is completely explicit. The Fourier decomposition diagonalizes the scattering matrix, and we can compute its eigenvalues.

Recall that we found formulas for the Fourier modes of eigenfunctions on F_ℓ in Section 5.2. Let $\{f_k\}$ denote the Fourier basis in the $t \in \mathbb{R}/\ell\mathbb{Z}$ coordinate,

$$f_k(t) := e^{i\alpha_k t},$$

where

$$\alpha_k = \frac{2\pi k}{\ell}.$$

From the mode $v_k^0(s; r)$ given in (5.21) we can derive solutions

$$\psi_k(r, t) := f_k(t) v_k^0(s, r)$$

of

$$(\Delta_{F_\ell} - s(1 - s))\psi_k = 0$$

(satisfying Dirichlet boundary conditions at $r = 0$). Explicitly, these generalized eigenfunctions are

$$\psi_k(r, t) = e^{i\alpha_k t} \sinh r (\cosh r)^{i\alpha_k} F(\tfrac{i\alpha_k+1+s}{2}, \tfrac{i\alpha_k+2-s}{2}; \tfrac{3}{2}; -\sinh^2 r).$$

Asymptotics as $\rho = 2e^{-r} \to 0$ are immediately obtained from the hypergeometric function identity (5.10):

$$\psi_k(\rho, t) \sim f_k(t) \left[\frac{\Gamma(\frac{3}{2})\Gamma(\frac{1}{2} - s)}{\Gamma(\frac{2-s+i\alpha_k}{2})\Gamma(\frac{2-s-i\alpha_k}{2})} \rho^s \right.$$

$$\left. + \frac{\Gamma(\frac{3}{2})\Gamma(s - \frac{1}{2})}{\Gamma(\frac{s+1+i\alpha_k}{2})\Gamma(\frac{s+1-i\alpha_k}{2})} \rho^{1-s} \right]. \quad (7.39)$$

Note that the uniqueness result of Proposition 7.13 relied ultimately on the Carleman estimate of Lemma 7.6, which was a result for a model funnel. Thus uniqueness certainly holds for the funnel itself (and for the hyperbolic cylinder as well, but not for the parabolic cylinder, as we'll see below). Therefore the diagonal form of $S_{F_\ell}(s)$ can be deduced immediately from (7.39). The f_k are eigenfunctions of $S_{F_\ell}(s)$,

$$S_{F_\ell}(s) f_k = \gamma_k(s) f_k,$$

with eigenvalues

$$\gamma_k(s) = \frac{\Gamma(\frac{1}{2} - s)\Gamma(\frac{s+1+i\alpha_k}{2})\Gamma(\frac{s+1-i\alpha_k}{2})}{\Gamma(s - \frac{1}{2})\Gamma(\frac{2-s+i\alpha_k}{2})\Gamma(\frac{2-s-i\alpha_k}{2})}. \tag{7.40}$$

This formula was worked out by Guillopé–Zworski [86], and it will play an important role in certain estimates in Chapter 9.

We can use (7.40) to check our formula for the principal symbol of $S_{F_\ell}(s)$. Stirling's approximation for the gamma function is

$$\Gamma(z) \sim z^{z-1/2} e^{-z} \sqrt{2\pi}, \tag{7.41}$$

as $|z| \to \infty$ in any sector that excludes the negative real axis. Hence, as $k \to \infty$,

$$\gamma_k(s) \sim 2^{1-2s} \frac{\Gamma(\frac{1}{2} - s)}{\Gamma(s - \frac{1}{2})} |\alpha_k|^{2s-1}. \tag{7.42}$$

Since α_k is the eigenvalue of $-i\partial_t$ for the eigenfunction f_k, the asymptotics of $\gamma_k(s)$ show exactly the principal term (7.38).

The hyperbolic cylinder is of course closely related to the funnel. Instead of decomposing the scattering matrix with respect to the two boundary components, it is natural to decompose it into even/odd subspaces with respect to the involution across the central geodesic, $r \mapsto -r$. Note that the generalized eigenfunctions ψ_k given above for the funnel extend to the full cylinder as odd functions. Hence the odd component of the scattering matrix has eigenvalues $\gamma_k^-(s)$ given by the same formula (7.40).

Finding the even generalized eigenfunctions is equivalent to solving a Neumann boundary problem on the funnel. Conveniently, these eigenfunctions are given by the other independent hypergeometric solution appearing in (5.9),

$$\psi_k^+(r, t) = f_k(t)(\cosh r)^{i\alpha_k} F(\tfrac{i\alpha_k+s}{2}, \tfrac{i\alpha_k+1-s}{2}; \tfrac{1}{2}; -\sinh^2 r).$$

From (5.10) we can read off the asymptotics and find the eigenvalues of the even component of the scattering matrix,

$$\gamma_k^+(s) = \frac{\Gamma(\frac{1}{2} - s)\Gamma(\frac{s-i\alpha_k}{2})\Gamma(\frac{s+i\alpha_k}{2})}{\Gamma(s - \frac{1}{2})\Gamma(\frac{1-s-i\alpha_k}{2})\Gamma(\frac{1-s+i\alpha_k}{2})}.$$

The parabolic cylinder is the one case that does not fit the general pattern. The Poisson kernel $E_{C_\infty}(s)$ decomposes into two pieces, $E^f(s)$ and $E^c(s)$ (one for each end). We're stretching the notation a bit here, because the large end of C_∞ is not a funnel. To obtain these as limits of the resolvent as given in Proposition 5.8, note that $\rho' = y'$ as $y' \to 0$ and $1/y'$ as $y' \to \infty$. We find that

$$E^c_{C_\infty}(s; z) = \frac{y^s}{2s - 1}$$

and

$$E^f_{C_\infty}(s; z, x') = \frac{y^{1-s}}{2s - 1} + \sum_{k \neq 0} \frac{(\pi |k|)^{s-\frac{1}{2}}}{\Gamma(s + \frac{1}{2})} \sqrt{y} K_{s-\frac{1}{2}}(2\pi |k| y) e^{2\pi i k(x - x')}.$$

If we define the scattering matrix from the asymptotics of the generalized eigenfunctions as in the general case, we see that

$$S_{C_\infty}(s) = \begin{pmatrix} 0 & \imath \\ \imath^t & 0 \end{pmatrix},$$

where $\imath : \mathbb{C} \to C^\infty(\mathbb{R}/\mathbb{Z})$ denotes inclusion as a constant function. The properties given in Corollary 7.14 do not hold in this case. In particular, $S_{C_\infty}(s) S_{C_\infty}(1 - s)$ is the projection onto the zero mode rather than the identity map.

Notes

The scope of the treatment by Lax–Phillips [113, 114] covers geometrically finite hyperbolic manifolds in arbitrary dimension. In particular, in dimension $n + 1$ and assuming infinite volume, the discrete spectrum of the Laplacian is finite and contained in $(0, n^2/4)$ and the spectrum $[n^2/4, \infty)$ is absolutely continuous with no embedded eigenvalues. Mazzeo [125, 127] proved that the same picture holds in the general asymptotically hyperbolic case, essentially by the methods we have presented in this chapter.

For conformally compact hyperbolic manifolds (geometrically finite without cusps), Perry [162] showed that the scattering matrix is a pseudodifferential operator and computed its symbol. In the asymptotically hyperbolic case, the definition of the scattering matrix in terms of asymptotic expansion of the generalized eigenfunctions was outlined by Melrose in [136]. A full proof for this case was given by Joshi–Sá Barreto [102]. The structure of the scattering matrix was worked out for the general conformally compact case in Borthwick [22].

8

Resonances and Scattering Poles

The concept of a resonance originates in the physics of a damped harmonic oscillator, describing the very large amplitude of oscillation that results when a driving force is applied with frequency matching the resonant frequency of the system. The simplest equation for such a system could be written $(\partial_t^2 + \gamma\,\partial_t + \omega^2)u = f$, where $f(t)$ is the driving term, ω is the undamped eigenfrequency, and γ is the damping coefficient. If the driving term has frequency ξ, then the complex multiplying factor between the driving term and the solution is $(-\xi^2 + i\xi\gamma + \omega^2)^{-1}$, which has two poles at complex values of ξ near $\pm\omega$. The real values of ξ closest to these poles are the resonant frequencies.

In stationary scattering theory, the damping is accounted for by dissipation to infinity, and it is the resolvent operator that plays the role of multiplier between driving term and excitation. Thus a pole of the resolvent corresponds to a resonance in the same sense as above, and this naturally generalizes the notion of eigenvalue. On the other hand, it would also be natural to define resonances as poles of the scattering matrix, for which we view the incoming wave solution as a driving term and the corresponding outgoing solution as the excitation.

To distinguish between these two possible notions, we introduce the following terminology:

Definition. For a geometrically finite hyperbolic surface X, the poles of the meromorphically continued resolvent $R_X(s)$ are called *resonances*, and the set of resonances will be denoted by $\mathcal{R}_X$.

Roughly speaking, poles in the scattering matrix $S_X(s)$ are called *scattering poles*, but there are some technicalities in the definition, to be addressed in Section 8.3.

In this chapter we will discuss how to define the multiplicities of poles in either case, and we will see that the two types are closely related. In later chapters we will focus mainly on resonances, but the connection between resonances and scattering poles will be an essential tool in the analysis. All of the results presented on distribution of resonances could be rephrased in terms of scattering poles without difficulty.

8.1 Multiplicities of resonances

In order to define the multiplicity of a resonance, we must first understand the Laurent expansion of the resolvent in the vicinity of the pole. When the pole comes from an isolated eigenvalue this is standard spectral theory (see, e.g., [107, Section III.5]). We'll start by reviewing this case.

Assume that $\lambda \in (0, 1/4)$ is an isolated eigenvalue of Δ_X. The self-adjointness of Δ_X implies that

$$\|(\Delta_X - z)u\|^2 = \|(\Delta_X - \operatorname{Re} z)u\|^2 + (\operatorname{Im} z)^2 \|u\|^2,$$

which in turn shows that

$$\|(\Delta_X - z)^{-1}\| \leq (\operatorname{Im} z)^{-1}. \tag{8.1}$$

Therefore the pole of $(\Delta_X - z)^{-1}$ at $z = \lambda$ has order 1.

The Laurent expansion takes the form

$$(\Delta_X - z)^{-1} = \frac{A_1}{(z - \lambda)} + H(z),$$

where $H(z)$ is holomorphic on some neighborhood of λ. The residue, A_1, can therefore be picked off by a contour integral,

$$A_1 = \frac{1}{2\pi i} \int_{\gamma_\lambda} (\Delta_X - z)^{-1} \, dz,$$

where γ_λ denotes a small (positively oriented) circle around λ containing no other resonance. The contour integral is defined as a limit of Riemann sums in the operator topology. Changing coordinates to $z = s(1 - s)$ transforms the contour integral to

$$A_1(\zeta) = \frac{1}{2\pi i} \int_{\gamma_\zeta} (1 - 2s) R_X(s) \, ds, \tag{8.2}$$

where $\operatorname{Re} \zeta > \frac{1}{2}$, $\zeta(1 - \zeta) = \lambda$, and γ_ζ is a small (positively oriented) circle around ζ containing no other pole of $R_X(s)$.

Using this contour integral and standard resolvent formulas one can easily deduce that $-A_1(\zeta)$ is the orthogonal projection onto the eigenspace $\ker(\Delta - \lambda) \subset L^2(X)$. (See the proof of Proposition 8.1 for an explanation.) The multiplicity of λ as an eigenvalue of Δ_X could thus be obtained by

$$\dim \ker(\Delta_X - \lambda) = \operatorname{rank}\left(\int_{\gamma_\zeta} R_X(s) \, ds\right).$$

This formula is the basis for our definition at a general pole:

Definition. For a geometrically finite hyperbolic surface X of infinite area, the *multiplicity* of a resonance $\zeta \in \mathbb{C}$ is given by

$$m_\zeta := \text{rank}\left(\int_{\gamma_\zeta} R_X(s) \, ds \right),\tag{8.3}$$

where γ_ζ is a small positively oriented circle containing ζ and no other resonance.

Note that the factor $(1 - 2s)$ dropped from (8.2) would have affected the definition at $s = \frac{1}{2}$. This point obviously requires special care, and we will treat it separately in Section 8.5. The main point is that (8.3) gives an appropriate definition because $1/4$ was ruled out as an eigenvalue by Proposition 7.5.

8.2 Structure of the resolvent at a resonance

In this section we'll study the behavior of $R_X(s)$ near a resonance at $\zeta \neq \frac{1}{2}$. Here X could be any geometrically finite hyperbolic surface, although we are mainly interested in the infinite-area case. Because $R_X(s)$ is finitely meromorphic, the Laurent expansion near a resonance has the form

$$R_X(s) = \sum_{j=1}^{p} \frac{A_j(\zeta)}{[s(1-s) - \zeta(1-\zeta)]^j} + H(s),\tag{8.4}$$

where the $A_j(\zeta)$ are finite-rank operators and $H(s)$ is holomorphic near ζ. The number p is called the *order* of the pole, which is not the same as the multiplicity. For N such that $\text{Re}\,\zeta > \frac{1}{2} - N$, $A_j(\zeta)$ and $H(s)$ are bounded operators $\rho^N L^2(X) \to \rho^{-N} L^2(X)$.

If γ_ζ denotes a positively oriented circle containing ζ and no other resonance, then

$$A_j(\zeta) = \frac{1}{2\pi i} \int_{\gamma_\zeta} (1 - 2s)\big[s(1-s) - \zeta(1-\zeta)\big]^{j-1} R_X(s) \, ds,\tag{8.5}$$

where the contour integral exists as a uniform limit of Riemann sums in

$$\mathcal{L}(\rho^N L^2(X), \rho^{-N} L^2(X)).$$

In particular, the definition (8.3) is equivalent to

$$m_\zeta = \text{rank}\, A_1(\zeta).$$

It also follows from (8.5) that the $A_j(\zeta)$'s inherit the symmetry of $R_X(s)$, as established in Lemma 6.5:

$$A_j(\zeta; z, z') = A_j(\zeta; z', z).$$

Proposition 8.1. *Suppose $R_X(s)$ has a pole of order p at $\zeta \neq \frac{1}{2}$, with Laurent expansion (8.4). Assuming $\mathrm{Re}\,\zeta > \frac{1}{2} - N$, we can find a basis $\{\phi_k\}_{k=1}^{m_\zeta} \subset \rho^{-N} L^2(X)$ for the range of $A_1(\zeta)$ such that*

$$A_1(\zeta) = \sum_{k=1}^{m_\zeta} \phi_k \otimes \phi_k. \tag{8.6}$$

If D denotes the matrix of the restriction of $(\Delta_X - \zeta(1 - \zeta))$ to range $A_1(\zeta)$, with respect to this basis, then D is symmetric, $D^p = 0$, and

$$A_j(\zeta) = \sum_{k,l=1}^{q} \left[D^{j-1} \right]_{kl} \phi_k \otimes \phi_l.$$

For $\zeta(1-\zeta) \in \sigma_{\mathrm{d}}(\Delta_X)$, the pole has order $p = 1$, and $-A_1(\zeta)$ is the orthogonal projection from $L^2(X)$ to the $\zeta(1 - \zeta)$-eigenspace.

Proof. Existence of a basis for which $A_1(\zeta)$ has the form (8.6) follows because it has rank m_ζ and a symmetric kernel.

We can use the fact that $(\Delta_X - s(1 - s))R_X(s) = I$ on $\rho^N L^2(X)$ to write

$$(\Delta_X - \zeta(1 - \zeta))R_X(s) = I + [s(1 - s) - \zeta(1 - \zeta)]\,R_X(s).$$

Substituting this into (8.5) shows that

$$(\Delta_X - \zeta(1 - \zeta))A_j(\zeta) = A_{j+1}(\zeta), \tag{8.7}$$

with the convention that $A_{p+1}(\zeta) = 0$. Similarly, $R_X(s)(\Delta_X - s(1-s)) = I$ implies that

$$A_j(\zeta)(\Delta_X - \zeta(1 - \zeta)) = A_{j+1}(\zeta). \tag{8.8}$$

In particular, $(\Delta_X - \zeta(1 - \zeta))$ commutes with $A_1(\zeta)$ and so preserves its range. This allows us to write

$$(\Delta_X - \zeta(1 - \zeta))\phi_k = \sum_{j=1}^{q} D_{jk}\phi_j.$$

The relations (8.7) and (8.8) further show that $A_j(\zeta)$ commutes with $A_1(\zeta)$. Hence iterating (8.7) gives the coefficients of $A_j(\zeta)$ in terms of D.

For $\zeta(1 - \zeta) \in \sigma_{\mathrm{d}}(\Delta_X)$, we have $p = 1$ by (8.1). The self-adjointness condition $R_X(s)^* = R_X(\bar{s})$ implies that $A_1(\zeta)$ is self-adjoint. To see that $-A_1(\zeta)$ is the projection onto the eigenspace, assume $(\Delta_X - \zeta(1 - \zeta))\phi = 0$ for $\phi \in L^2(X)$, which implies

$$(\Delta_X - s(1 - s))\phi = [\zeta(1 - \zeta) - s(1 - s)]\,\phi.$$

Since $\phi \in L^2$ we can apply $R_X(s)$ to this equation, for $\mathrm{Re}\,s > \frac{1}{2}$, to obtain

$$\phi = [\zeta(1 - \zeta) - s(1 - s)]\,R_X(s)\phi.$$

Taking the limit as $s \to \zeta$ then gives

$$\phi = -A_1(\zeta)\phi. \qquad \square$$

One consequence of the nilpotency of D is that any $\phi \in \text{range}(A_1(\zeta))$ must satisfy

$$(\Delta_X - \zeta(1 - \zeta))^k \phi = 0, \tag{8.9}$$

for some $k \geq 1$. This suggests that the proper analogy of our notion of multiplicity is the algebraic multiplicity of eigenvalues for a finite-dimensional matrix. Note, however, that the argument that $-A_1(\zeta)$ is a projection breaks down if $\text{Re}\,\zeta < \frac{1}{2}$. Indeed, the composition $-A_1(\zeta)^2$ is not even defined in this case.

Using the decomposition of X into cusps and funnels (assuming that X is nonelementary) and the corresponding structure of the resolvent given in Theorem 6.3, we can characterize the ϕ_k's spanning range$(A_1(\zeta))$ more precisely. There is a complication in deducing the structure of the ϕ_k's from (8.5), arising from the fact that the kernel is decomposed in terms of spaces that themselves depend on s. For instance, $Q(s; \cdot, \cdot) \in (\rho_f \rho_f')^s (\rho_c \rho_c')^{s-1} C^\infty(\overline{X} \times \overline{X})$. To remove this dependence, we define

$$B_j(\zeta) = \frac{1}{2\pi i} \int_{\gamma_\zeta} (1 - 2s) \left[s(1 - s) - \zeta(1 - \zeta) \right]^{j-1}$$
$$\times \rho_f^{-s} \rho_c^{1-s} R_X(s) \rho_f^{-s} \rho_c^{1-s} \, ds. \tag{8.10}$$

Theorem 6.3 and the fact that these operators are of finite rank imply that the kernels are smooth:

$$B_j(\zeta, \cdot, \cdot) \in C^\infty(\overline{X} \times \overline{X}).$$

From this, we deduce a series expansion of the form

$$R_X(s) = \sum_{j=1}^{p} \frac{\rho_f^s \rho_c^{s-1} B_j(\zeta) \rho_f^s \rho_c^{s-1}}{[s(1-s) - \zeta(1-\zeta)]^j} + H_2(s). \tag{8.11}$$

To match terms with the $A_j(\zeta)$'s as defined by (8.4), we need also to expand $\rho_f^s \rho_c^{s-1}$ as a power series in the variable $s(1 - s) - \zeta(1 - \zeta)$. Under the assumption $\zeta \neq \frac{1}{2}$, we have a power-series expansion

$$\rho_f^s \rho_c^{s-1} = \sum_{l=0}^{\infty} [s(1 - s) - \zeta(1 - \zeta)]^l f_l, \tag{8.12}$$

valid for s near ζ, where f_l is a function on X of the form

$$f_l \in \sum_{k=0}^{l} \rho_f^\zeta \rho_c^{\zeta-1} (\log \rho)^k C^\infty(\overline{X}).$$

Then matching terms in the singular parts gives

$$A_k = \sum_{j=k}^{p} \sum_{\substack{l,m: \\ l+m=j-k}} f_l B_j(\zeta) f_m. \tag{8.13}$$

In particular, this shows that

$$\text{range } A_1(\zeta) \subset \sum_{k=0}^{p-1} \rho_f^\zeta \rho_c^{\zeta-1} (\log \rho)^k C^\infty(\overline{X}). \qquad (8.14)$$

It is easy to see the necessity of the log terms appearing in (8.14). Suppose $\phi \in \rho_f^s \rho_c^{s-1} C^\infty(\overline{X})$ and $(\Delta_X - s(1-s))^k \phi = 0$ for some $k \geq 1$. By the indicial equation, $(\Delta_X - s(1-s))$ knocks out the leading term in the asymptotic expansion, giving $(\Delta_X - s(1-s))^{k-1} \phi \in \rho_f^{s+1} \rho_c^s C^\infty(\overline{X})$. Then one can argue from $(\Delta_X - s(1-s))^k \phi = 0$ that all successive terms in the expansion of $(\Delta_X - s(1-s))^{k-1} \phi$ vanish also, leaving $(\Delta_X - s(1-s))^{k-1} \phi \in \rho^\infty C^\infty(\overline{X})$. And since this would be L^2 in particular, $(\Delta_X - s(1-s))^{k-1} \phi = 0$. By repeating this argument k times, we can deduce that $(\Delta_X - s(1-s))\phi = 0$. Thus, in range $A_1(\zeta)$, elements whose asymptotic expansions have no log ρ terms can lie only in the kernel of $(\Delta_X - \zeta(1-\zeta))$. In other words, the logarithmic terms must occur whenever the pole has order $p > 1$.

To conclude this section, let us find the multiplicities of resonances in the elementary cases, including the hyperbolic plane. For $\Delta_\mathbb{H}$, we simply examine the resolvent kernel formula from Proposition 4.1,

$$R_\mathbb{H}(s; z, z') = \frac{2^{-1-2s}}{\sqrt{\pi}} \frac{\Gamma(s)}{\Gamma(s + \frac{1}{2})} \sigma^{-s} F(s, s; 2s; \sigma^{-1}), \qquad (8.15)$$

where $\sigma(z, z') := \cosh^2(d(z, z')/2)$. This makes it clear that the set of poles is $-\mathbb{N}_0$ and each pole has order $p = 1$.

We claim that the multiplicities for $\mathbb{H}$ are given by

$$m_{-n} = 2n + 1, \qquad n \in \mathbb{N}_0. \qquad (8.16)$$

To see this, consider first the eigenvalue equation at $s = -n$:

$$(\Delta_\mathbb{H} + n(n+1))u = 0,$$

in geodesic polar coordinates. With the substitution $u(r, \theta) = e^{ik\theta} f(\cosh r)$ this becomes a Legendre equation for f. The only solutions regular at $r = 0$ are given by associated Legendre polynomials,

$$\phi_k(r, \theta) = e^{ik\theta} P_n^{(k)}(\cosh r), \qquad |k| \leq n.$$

(These solutions are the hyperbolic analogues of spherical harmonics.) Since $p = 1$, the range of $A_1(-n)$ is spanned by such solutions and thus has dimension at most $2n + 1$. On the other hand, we read off from (8.15) that

$$A_1(-n) = c_n \sigma^n F(-n, -n; -2n; \sigma^{-1}).$$

To obtain a lower bound on the rank, we consider the limit $y \to 0$. For $f \in C^\infty(\mathbb{H})$ we have

$$\lim_{y \to 0} y^n A_1(-n) f(z) = c_n \int_{\mathbb{H}} \left[(x - x')^2 + y'^2 \right]^n f(z') y'^{-n-2} \, dx' \, dy'.$$

This is a polynomial of degree $2n$ in x, and by moving the support of f around it is easy to see that we could generate at $2n + 1$ independent polynomials from this formula. Hence the rank of $A_1(-n)$ is at least $2n + 1$.

Now consider the hyperbolic cylinder C_ℓ, whose resonance set was evident in Proposition 5.3. We claim that

$$m_\zeta = 2, \qquad \zeta \in -\mathbb{N}_0 + \frac{2\pi i}{\ell} \mathbb{Z}. \tag{8.17}$$

For the points $\zeta = -n + 2\pi i k / \ell$ where $k \neq 0$ this is easily deduced from the formula (5.13) for the Fourier modes $u_k(s; r, r')$ of $R_{C_\ell}(s; z, z')$. Indeed, given $k \neq 0$, the coefficient $a_k(s)$ has simple poles at $s = -n \pm 2\pi i k$ for each $n \in \mathbb{N}_0$. Thus the poles have order one and the residue is

$$A_1(-n + \tfrac{2\pi i k}{\ell}; z, z') = c_k v_k(-n + \tfrac{2\pi i k}{\ell}; r) v_k(-n + \tfrac{2\pi i k}{\ell}; r') e^{ik(\theta - \theta')}$$
$$+ c_{-k} v_{-k}(-n + \tfrac{2\pi i k}{\ell}; r) v_{-k}(-n + \tfrac{2\pi i k}{\ell}; r') e^{-ik(\theta - \theta')},$$

which clearly has rank 2.

The resonances on the real axis are of order two. For simplicity we'll focus on $\zeta = 0$; the other cases $\zeta = -n$ are similar. All of these resonances occur in the zero mode, given in Proposition 5.3 as

$$u_0(s; z, z') = \frac{\pi}{\ell} 4^{1-s} \Gamma(s)^2 v_0(s; \mp r) v_0(s; \pm r'),$$

where v_k was given explicitly in (5.11), and r is the signed distance from the central geodesic. From this expression it's relatively easy to extract the first two terms of the Laurent series at $s = 0$,

$$u_0(s; z, z') = \frac{1}{\pi \ell} s^{-2} - \frac{1}{\pi \ell} (2\gamma + \log(4 \cosh r \cosh r')) s^{-1} + (\text{holo}).$$

From (8.4) we can read off the coefficients:

$$A_2(0; z, z') = \frac{1}{\pi \ell}, \qquad A_1(0; z, z') = -\frac{1}{\pi \ell} (2 + 2\gamma + \log(4 \cosh r \cosh r')).$$

Since $\rho = 2e^{-r}$ as $r \to +\infty$, we see here the log terms in range $A_1(0)$ that were pointed out in (8.14).

From the explicit form (5.7) of Δ_{C_ℓ}, we compute

$$\Delta_{C_\ell}(- \log \cosh r) = 1,$$

which shows that the identities (8.7) and (8.8) are satisfied. Define a basis for range $A_1(0)$ by

$$\phi_1(z) = \frac{1}{\sqrt{2\pi \ell}} (\gamma + \log(2 \cosh r)), \qquad \phi_2(z) = \frac{i}{\sqrt{2\pi \ell}} (2 + \gamma + \log(2 \cosh r)).$$

Then we can see that

$$A_1(0) = \phi_1 \otimes \phi_1 + \phi_2 \otimes \phi_2,$$

which shows in particular that $m_\zeta = 2$. The corresponding nilpotent symmetric matrix D, as defined in Proposition 8.1, is

$$D = \frac{1}{2}\begin{pmatrix} 1 & i \\ i & -1 \end{pmatrix}.$$

The funnel F_ℓ possesses exactly half the resonances of C_ℓ. The same analysis as above shows that the funnel multiplicities are given by

$$m_\zeta = 2, \qquad \zeta \in 1 - 2\mathbb{N} + \frac{2\pi i}{\ell}\mathbb{Z}. \tag{8.18}$$

For the parabolic cylinder C_∞, the formula for $R_{C_\infty}(s; z, z')$ from Proposition 5.8 shows only a single resonance, at $s = \frac{1}{2}$, which occurs in the zero mode of the Fourier decomposition. The order and multiplicity are both equal to one.

8.3 Scattering poles

The scattering matrix $S_X(s)$ forms a meromorphic family of pseudodifferential operators for $s \in \mathbb{C}$. In defining the multiplicity of its poles, we must account for possible cancellation of poles with zeros. The identity $S_X(1 - s)S_X(s) = I$ shows that a pole at ζ comes paired with a zero at $1 - \zeta$. In particular, the scattering poles within $(1/2, 1)$ coming from the discrete spectrum will give rise to zeros in $(0, 1/2)$ that may cancel poles in that range.

The other major issue in the definition of scattering poles is the fact that $S_X(s)$ has poles of infinite rank at $s \in \frac{1}{2} + \mathbb{N}$, and corresponding zeros of infinite rank in $\frac{1}{2} - \mathbb{N}$. One can see them explicitly in the gamma-function factors in the principal symbol (7.38). They are also apparent in the formula (7.40) for the funnel scattering eigenvalues. The origin of these poles and zeros can be traced back to Lemma 4.5 in the analysis of $\Delta_{\mathbb{H}}$, where the coefficient $a_0(s; x)$ was defined by the distributional kernel $|x - x'|^{-2s}$. The poles at half-integers appear in the analytic continuation of this distributional kernel.

To define the multiplicity of a scattering pole, it is convenient to renormalize to obtain a family of bounded operators on $L^2(\partial X)$, and at the same time we want to remove the infinite-rank poles and zeros from the count. Let $\Delta_{\partial_0 \overline{F}}$ denote the Laplacian on $\partial_0 \overline{F}$ induced by the metric $\rho^2 \bar{g}$ on $\overline{X}$, where g denotes the hyperbolic metric on X. In our standard coordinates, $\partial_0 \overline{F}_j$ is identified with $\mathbb{R}/(\ell_j\mathbb{Z})$, and the Laplacian is simply $\Delta_{\partial_0 \overline{F}} = -\partial_t^2$. For $s \in \mathbb{C}$ we introduce a holomorphic family of pseudodifferential operators given by D^s, where

$$D := \frac{1}{2}\left(\Delta_{\partial_0 \overline{F}} + 1\right)^{1/2}.$$

The complex power D^s can be defined explicitly here, using the natural Fourier basis for $L^2(\partial_0 \overline{F}_j)$,

$$D^s\big|_{\partial_0 \overline{F}_j} (e^{2\pi i k t/\ell_j}) = 2^{-s} \left(\frac{4\pi^2 k^2}{\ell_j^2} + 1\right)^{s/2} e^{2\pi i k t/\ell_j}.$$

From the formula (7.40) for the eigenvalues of $S_{F_\ell}(s)$, with asymptotics given by (7.42), we can deduce that

$$S_F(s) = -\frac{\Gamma(\frac{3}{2} - s)}{\Gamma(s + \frac{1}{2})} D^{s-1/2}(I + K_F(s)) D^{s-1/2}, \tag{8.19}$$

where $K_F(s)$ is compact.

Since normalization is required only in the funnel–funnel block, we introduce the holomorphic family

$$\Lambda(s) := \begin{pmatrix} D^{1/2-s} & 0 \\ 0 & I \end{pmatrix}. \tag{8.20}$$

Then the normalized scattering matrix is defined as

$$\widetilde{S}_X(s) := \Lambda(s) \begin{pmatrix} \frac{\Gamma(s+\frac{1}{2})}{\Gamma(\frac{3}{2}-s)} S^{\mathrm{ff}}(s) & \Gamma(s + \frac{1}{2}) S^{\mathrm{fc}}(s) \\ \frac{1}{\Gamma(\frac{3}{2}-s)} S^{\mathrm{cf}}(s) & S^{\mathrm{cc}}(s) \end{pmatrix} \Lambda(s). \tag{8.21}$$

Note that $\widetilde{S}_X(s)$ obeys the same inversion formula as $S_X(s)$, namely

$$\widetilde{S}_X(s)^{-1} = \widetilde{S}_X(1 - s).$$

With the factor of $1/\Gamma(\frac{3}{2}-s)$ canceling the infinite-rank poles of $S^{\mathrm{ff}}(s)$, it follows from Proposition 7.15 and the fact that $R_X(s)$ is finitely meromorphic (Theorem 6.2) that $\widetilde{S}_X(s)$ is a finitely meromorphic family of bounded operators on $L^2(\partial \overline{X})$. Indeed, by (8.19) we have

$$\widetilde{S}_X(s) = -(I + K(s)),$$

where $K(s)$ is a compact operator, so $\widetilde{S}_X(s)$ is a finitely meromorphic family of Fredholm operators on $L^2(\partial \overline{X})$.

The theory for counting multiplicities of such families was worked out by Gohberg–Sigal [73], who developed an operator version of the "argument principle" for meromorphic functions.

Definition. A *scattering pole* is a pole of $\widetilde{S}_X(s)$. The *multiplicity* of such a pole at $\zeta \in \mathbb{C}$ is given by

$$\nu_\zeta := -\operatorname{tr}\left(\frac{1}{2\pi i} \int_{\gamma_\zeta} \widetilde{S}_X(1 - s) \widetilde{S}'_X(s)\, ds\right). \tag{8.22}$$

Note that the convention here is that poles have positive multiplicity, and zeros negative.

8.4 Operator logarithmic residues

To develop our understanding of the definition of ν_ζ, let us review some details from Gohberg–Sigal [73]. Suppose $A(s)$ is a meromorphic family of operators on the Banach space $\mathcal{B}$, for s in some neighborhood of ζ. The kernel of $A(s)$ at ζ is not well-defined if $A(s)$ has a pole at ζ, and we wish to provide a substitute for this concept.

A *root function* of A at ζ is a holomorphic $\mathcal{B}$-valued function φ with $\varphi(\zeta) \neq 0$, satisfying

$$\lim_{s \to \zeta} A(s)\varphi(s) = 0.$$

The vector $\varphi_\zeta := \varphi(\zeta)$, which plays the role of a null vector of $A(\zeta)$, is called a *root vector*. The *rank* of the root vector φ_ζ is the maximal order of vanishing of $A(s)\varphi(s)$ at ζ, over all choices of root function. The (linear) space of all root vectors of A at ζ is denoted by $\ker_\zeta(A)$. (This is of course equal to $\ker A(\zeta)$ if $A(s)$ is holomorphic at $s = \zeta$.)

To illustrate some of the possibilities, consider

$$A_1(s) = \begin{pmatrix} s & 0 \\ 0 & s^{-2} \end{pmatrix}.$$

Then $A_1(s)$ has a root vector $\varphi_0 = (1, 0)$, of rank one, and $A_1(s)^{-1}$ has a root vector $\varphi(s) = (0, 1)$, of rank two. For the matrix

$$A_2(s) = \begin{pmatrix} 1 & s^{-1} \\ 0 & 1 \end{pmatrix},$$

both $A_2(s)$ and $A_2(s)^{-1}$ have the rank-one root vector $\varphi_0 = (1, 0)$.

If $\ker_\zeta(A)$ is finite-dimensional and the ranks of all root vectors are finite, then we can try to define the multiplicity of ζ as a zero of $A(s)$. The idea is to add up the ranks of the root vectors, but one must be careful to avoid overcounting. A maximal set of root vectors is a basis $\{\psi_j\}_{j=1}^m$ for $\ker_\zeta(A)$ chosen according to the following prescription. The first vector ψ_1 is chosen so that $r_1 := \operatorname{rank} \psi_1$ is maximal among all root vectors. Then, at each stage $j = 2, \ldots, n$, we choose ψ_j so that $r_j := \operatorname{rank} \psi_j$ is maximal among all root vectors in some direct complement of the span of $\{\psi_1, \ldots, \psi_{j-1}\}$ in $\ker_\zeta(A)$.

Definition. Given a maximal set of root vectors for $A(s)$ at ζ, with ranks $r_1 \geq \cdots \geq r_m$, where $m = \dim \ker_\zeta(A)$, the *null-multiplicity* of $A(s)$ at ζ is

$$N_\zeta(A) := \sum_{j=1}^m r_j,$$

A primary result of Gohberg–Sigal [73] is the following operator version of the argument principle:

Theorem 8.2 (Gohberg–Sigal logarithmic residue theorem). *Suppose $A(s)$ and $A(s)^{-1}$ are finitely meromorphic families of Fredholm operators on the Banach space $\mathcal{B}$, defined for s in some neighborhood of ζ. Assume that the nonsingular part of $A(s)$ at ζ has index zero. Then*

$$\mathrm{tr}\left(\frac{1}{2\pi i}\int_{\gamma_\zeta} A(s)^{-1} A'(s)\, ds\right) = N_\zeta(A) - N_\zeta(A^{-1}),$$

where γ_ζ is a small circle around ζ within which $A(s)$ and $A(s)^{-1}$ are holomorphic except possibly at ζ.

Before giving the proof, let's consider the interpretation of this formula in the context of the scattering matrix. The Banach space $\mathcal{B}$ is taken to be $L^2(\partial X)$ (with respect to the measure dh defined on ∂X in Section 7.4). We have already noted that $\widetilde{S}_X(s)$ defines a meromorphically invertible family of Fredholm operators on $L^2(\partial X)$.

In view of the relation $\widetilde{S}_X(s)^{-1} = \widetilde{S}_X(1-s)$, Theorem 8.2 implies that the scattering multiplicity is

$$\nu_\zeta = N_{1-\zeta}(\widetilde{S}_X) - N_\zeta(\widetilde{S}_X). \tag{8.23}$$

For simplicity, assume $\zeta \notin \frac{1}{2} + \mathbb{Z}$, so that we may deal with $S_X(s)$ itself. Then $N_\zeta(\widetilde{S}_X) > 0$ if there is a root function $\psi_s \in L^2(\partial \overline{X})$ such that $S_X(s)\psi_s$ vanishes at $s = \zeta$. We will see below that we can assume that the root functions are smooth on ∂X for all s near ζ. By applying the Poisson operator $E_X(s)$ to ψ_s, we obtain a family u_s of solutions to $(\Delta_X - s(1-s))u_s = 0$ such that

$$u_s \sim \rho_{\mathrm{f}}^{1-s}\rho_{\mathrm{c}}^{-s}\psi_s + \rho_{\mathrm{f}}^s\rho_{\mathrm{c}}^{s-1}S_X(s)\psi_s.$$

The fact that ψ_s is a root function implies the vanishing of the right half of the asymptotic expansion at the point $s = \zeta$. Note that if $\mathrm{Re}\,\zeta < \frac{1}{2}$ then this would imply $u_\zeta \in L^2(X)$, and hence could occur only if $\zeta(1-\zeta) \in \sigma_d(\Delta_X)$.

By the same argument, $N_{1-\zeta}(\widetilde{S}_X) > 0$, for $\zeta \notin \frac{1}{2} + \mathbb{Z}$, would imply existence of families ψ_s, u_s such that

$$u_s \sim \rho_{\mathrm{f}}^s\rho_{\mathrm{c}}^{s-1}\psi_s + \rho_{\mathrm{f}}^{1-s}\rho_{\mathrm{c}}^{-s}S_X(1-s)\psi_s,$$

where the right half of the expansion vanishes at $s = \zeta$. In particular, $N_{1-\zeta}(S_X) = 0$ for $\mathrm{Re}\,\zeta > \frac{1}{2}$ unless $\zeta(1-\zeta) \in \sigma_d(\Delta_X)$.

Thus our interpretation of ν_ζ at a nonexceptional value of ζ (meaning $\mathrm{Re}\,\zeta < \frac{1}{2}$, $\zeta \notin \frac{1}{2} - \mathbb{N}$, and $\zeta(1-\zeta) \notin \sigma_d(\Delta_X)$) is that this multiplicity counts the number of solutions of $(\Delta_X - \zeta(1-\zeta))u = 0$ such that the asymptotic expansion of u is of the form

$$u \in \rho_{\mathrm{f}}^\zeta\rho_{\mathrm{c}}^{\zeta-1}C^\infty(\partial\overline{X}), \tag{8.24}$$

i.e., with the second half of the expansion missing. Note that this implies that the uniqueness result of Proposition 7.13 cannot be extended to $\mathrm{Re}\,s < \frac{1}{2}$ when there are scattering poles; the scattering poles occur precisely at the points where uniqueness fails.

If ζ is a scattering pole with $\operatorname{Re} \zeta < \frac{1}{2}$ and $\zeta(1-\zeta) \in \sigma_d(\Delta_X)$, then both expressions on the right side of (8.23) are nonzero. The scattering multiplicity ν_ζ would be the count of generalized eigenfunctions of the form (8.24) minus the number of actual L^2 eigenfunctions.

Let us denote the trace of the logarithmic residue of A at ζ by

$$M(A;\zeta) := \operatorname{tr}\left(\frac{1}{2\pi i}\int_{\gamma_\zeta} A(s)^{-1}A'(s)\,ds\right). \tag{8.25}$$

The following auxiliary result (taken from [73, Theorem 5.2]) will be useful in the proof of Theorem 8.2:

Lemma 8.3. *If $A(s)$ and $B(s)$ are meromorphically invertible families of Fredholm operators, with possible finite-rank singularities at $s = \zeta$, then*

$$M(AB;\zeta) := M(A;\zeta) + M(B;\zeta).$$

Proof. Let $\Xi(A(s))$ denote the singular part of the Laurent expansion of A at ζ, and denote the holomorphic remainder by $T(A(s)) := A(s) - \Xi(A(s))$. We first note that

$$\Xi(A(s)B(s)) = \Xi(A(s))\Xi(B(s)) + \Xi\Big(\Xi(A(s))T(B(s))\Big)$$
$$+ \Xi\Big(T(A(s))\Xi(B(s))\Big).$$

Because the singular parts are of finite rank, we can use the cyclicity of the ordinary trace to deduce

$$\operatorname{tr}\Xi(A(s))\Xi(B(s)) = \operatorname{tr}\Xi(B(s))\Xi(A(s))$$

and

$$\operatorname{tr}\Xi(A(s))T(B(s)) = \operatorname{tr}T(B(s))\Xi(A(s)),$$

and similarly for $\operatorname{tr}T(A(s))\Xi(B(s))$. This implies that

$$\operatorname{tr}\Xi(A(s)B(s)) = \operatorname{tr}\Xi(B(s)A(s)). \tag{8.26}$$

To evaluate $M(AB;\zeta)$ we must consider the integrand

$$(A(s)B(s))^{-1}(A(s)B(s))' = B(s)^{-1}A(s)^{-1}A'(s)B(s) + B(s)^{-1}B(s)'.$$

When the trace of the residue is taken, the second term on the right gives $M(B;\zeta)$ directly. And by (8.26), the contribution from the first term on the right is

$$\operatorname{tr}\left(\frac{1}{2\pi i}\int_{\gamma_\zeta} B(s)^{-1}A(s)^{-1}A'(s)B(s)\,ds\right) = M(A;\zeta). \qquad \square$$

Proof of Theorem 8.2. Gohberg–Sigal [73] proved this by establishing a certain factorization,

$$A(s) = E(s)D(s)F(s), \tag{8.27}$$

with $E(s)$ and $F(s)$ holomorphically invertible in a neighborhood of ζ. The middle factor takes the form

$$D(s) = P_0 + \sum_{j=1}^{n} (s - \zeta)^{k_j} P_j, \tag{8.28}$$

where the P_j are mutually orthogonal projections, having rank 1 for $j > 0$, with

$$I = \sum_{j=0}^{n} P_j.$$

The k_j are nonzero integers with $k_1 \leq \cdots \leq k_n$.

Before deriving (8.27), we note that the logarithmic residue formula follows easily from it. The form of (8.28) makes it easy to construct maximal sets of root vectors and deduce that

$$N_\zeta(A) = \sum_{k_j > 0} k_j, \qquad N_\zeta(A^{-1}) = - \sum_{k_j < 0} k_j.$$

By Lemma 8.3, and the holomorphic invertibility of $E(s)$ and $F(s)$ near ζ, we have

$$M(A; \zeta) = M(D; \zeta).$$

Then we simply use

$$D(s)^{-1} D'(s) = (s - \zeta)^{-1} \sum_{j=1}^{n} k_j P_j$$

to compute that

$$M(D; \zeta) = \sum_{j=1}^{n} k_j = N_\zeta(A) - N_\zeta(A^{-1}).$$

It remains to establish the factorization (8.27). Out first step is reduction to a finite-dimensional problem. It suffices to prove the case $\zeta = 0$. Start by writing the Laurent expansion of $A(s)$,

$$A(s) = \sum_{j=-m}^{\infty} s^j A_j.$$

By assumption the index of A_0 is zero, so we can find a finite-dimensional operator K_0 such that $K_0 + A_0$ is invertible. Hence the operator

$$E_1(s) := K_0 + \sum_{j=0}^{\infty} s^j A_j$$

will be invertible in some neighborhood of 0. Then we can write

$$A(s) = E_1(s)(I + K(s)),$$

where

$$K(s) = E_1(s)^{-1}\left(\sum_{j=-m}^{-1} s^j A_j - K_0\right).$$

Let W_0 be the intersection of the kernels of the A_j for $j < 0$ and K_0. Then $K(s)|_{W_0} = 0$ for all s near ζ. If $K(s)$ has the Laurent expansion,

$$K(s) = \sum_{j=-m}^{-1} s^j K_j + T_1(s),$$

then we can find a decomposition $\mathcal{B} = V \oplus W$ such that $W \subset W_0$ and V is a finite-dimensional subspace that is invariant with respect to K_j for each $j = -m, \ldots, -1$. Let P be the projection onto V parallel to W, and set $P_0 = 1 - P$. From $K(s)|_W = 0$ and $P_0 K_j = 0$, it follows that

$$K(s) = K(s)P = PK(s)P + P_0 T_1(s)P.$$

Hence if we set $F_1(s) = I + P_0 T_1(s)P$, which is holomorphically invertible, then

$$I + K(s) = (I + PK(s)P)F_1(s).$$

We have thus written $A(s)$ in the form

$$A(s) = E_1(s)(I + PK(s)P)F_1(s). \tag{8.29}$$

This allows us to focus our attention on the finite-rank operator $D_1(s) : V \to V$ given by

$$D_1(s) := P(I + K(s))P.$$

Choose a basis for V so that $D_1(s)$ can be written as a matrix. We claim that $D_1(s)$ can be put in the normal form

$$D_1(s) = G(s)\begin{pmatrix} s^{k_1} & 0 & \cdots & 0 \\ 0 & s^{k_2} & \cdots & 0 \\ \vdots & \vdots & \ddots & \vdots \\ 0 & 0 & \cdots & s^{k_n} \end{pmatrix} H(s), \tag{8.30}$$

with $k_1 \leq \cdots \leq k_n$ and $G(s)$ and $H(s)$ holomorphically invertible maps $V \to V$. This follows from the row-reduction algorithm used to construct the Smith normal form for polynomial matrices (see, e.g., [112, Chapter 4]). In fact, one has such a normal form for matrices over any principal ideal domain (see, e.g., [101, Theorem 3.8]). The ring of meromorphic functions on a neighborhood $U \ni 0$ with zeros or poles only at the origin is a principal ideal domain, because all ideals are generated by the monomials s^k for $k \in \mathbb{Z}$.

The normal form algorithm works as follows. Set $M(s) = s^m D_1(s)$ to cancel any poles, and note that $M(s)$ is invertible for $s \neq 0$ by assumption. Let $m_{ij}(s)$ be the entries of the matrix $M(s)$. By restricting the size of our neighborhood if necessary, we can assume that for each i, j, either $m_{ij}(s)$ vanishes identically or $m_{ij}(s) \neq 0$ for $s \neq 0$. Let $r_1 \geq 0$ be the minimum order of the zeros of the $m_{ij}(s)$'s that don't vanish identically. By rearranging and then performing row and column reduction, we can put the matrix in the form

$$M(s) = G_1(s) \begin{pmatrix} s^{r_1} & 0 & \cdots & 0 \\ 0 & & & \\ \vdots & & M_1(s) & \\ 0 & & & \end{pmatrix} H_1(s).$$

Choosing r_1 as the minimum order ensures that the row reduction can be accomplished with $G_1(s)$ and $H_1(s)$ holomorphically invertible. Iterating this procedure for successively smaller blocks, we eventually obtain

$$M(s) = G(s) \begin{pmatrix} s^{r_1} & 0 & \cdots & 0 \\ 0 & s^{r_2} & \cdots & 0 \\ \vdots & \vdots & \ddots & \vdots \\ 0 & 0 & \cdots & s^{r_n} \end{pmatrix} H(s), \qquad (8.31)$$

where $G(s)$ and $H(s)$ are holomorphically invertible. Then we simply multiply by s^{-m} to get (8.30).

If we extend $G(s)$ and $H(s)$ to $\mathcal{B}$ as operators that commute with P and act by zero on W, then we can rewrite (8.30) as

$$P(I + K(s))P = G(s) \left(\sum_{j=1}^{n} s^{k_j} P_j \right) H(s),$$

where $\sum P_j = P$. By (8.29) we have

$$A(s) = E_1(s)(P_0 + G(s)) \left(P_0 + \sum_{j=1}^{n} s^{k_j} P_j \right) (P_0 + H(s)) F_1(s),$$

which completes the proof of (8.27). $\square$

8.5 Half-integer points

Before establishing the precise relationship between resonances and scattering poles, we need to analyze the special cases when $\zeta \in \frac{1}{2} + \mathbb{Z}$.

Our first observation is that $S_X(\frac{1}{2})$ is self-adjoint and unitary by (7.37) and Corollary 7.14. Since $\widetilde{S}_X(\frac{1}{2}) = S_X(\frac{1}{2})$, we have

$$\nu_{1/2} = 0.$$

We also know that there is no eigenvalue at $\lambda = 1/4$ under the assumption that X has infinite area, by Proposition 7.5. A resonance may still occur at $s = \frac{1}{2}$, however.

Lemma 8.4. *Suppose X is a geometrically finite hyperbolic surface of infinite area. If $R_X(s)$ has a pole at $s = \frac{1}{2}$, then in some neighborhood of $\frac{1}{2}$ the resolvent can be written as*

$$R_X(s) = \frac{1}{2s-1} \sum_{k=1}^{m_{1/2}} \phi_k(s) \otimes \phi_k(s) + H(s),$$

where $H(s)$ is holomorphic near $\frac{1}{2}$, and the $\phi_k(s)$'s are independent functions, depending holomorphically on s, satisfying

$$(\Delta_X - 1/4)\phi_k(\tfrac{1}{2}) = 0, \qquad \phi_k(s) \in \rho_f^s \rho_c^{s-1} C^\infty(\overline{X}).$$

Proof. The self-adjointness of Δ_X implies, for $\varphi \in C_0^\infty(X)$,

$$\left| \int_X \overline{\varphi}\,(\Delta_X - s(1-s))\varphi\, dA \right| \geq \left| \operatorname{Im} \int_X \overline{\varphi}\,(\Delta_X - s(1-s))\varphi\, dA \right|$$

$$= \left| \operatorname{Im}(s^2 - s) \right| \|\varphi\|^2.$$

Observe that $\operatorname{Im}(s^2 - s)$ could also be written $\operatorname{Im}(s - \frac{1}{2})^2$. Applying the Schwarz inequality to the left-hand side, we obtain

$$\|(\Delta_X - s(1-s))\varphi\| \geq \left| \operatorname{Im}(s - \tfrac{1}{2})^2 \right| \|\varphi\|.$$

For $\operatorname{Re} s > \frac{1}{2}$, the resolvent $R_X(s)$ is a bounded operator, and the above inequality implies that

$$\|R_X(s)\| \leq \frac{1}{|\operatorname{Im}(s - \tfrac{1}{2})^2|}. \tag{8.32}$$

Therefore, a pole in $R_X(s)$ at $s = \frac{1}{2}$ has order at most 2.

The singular part of $R_X(s)$ at $s = \frac{1}{2}$ can thus be written as

$$\frac{A_2}{(2s-1)^2} + \frac{A_1}{2s-1},$$

for some finite-rank operators $A_j : \rho^\varepsilon L^2(X) \to \rho^{-\varepsilon} L^2(X)$, with $\varepsilon > 0$. By $(\Delta_X - s(1-s))R_X(s) = I$, we can derive

$$(\Delta_X - 1/4)R_X(s) = I - \frac{(2s-1)^2}{4} R_X(s).$$

Substituting the Laurent series of $R_X(s)$ into this equation shows that

$$(\Delta_X - 1/4)A_j = 0, \quad j = 1, 2. \tag{8.33}$$

If $\psi \in \rho^\varepsilon L^2(X)$ then (8.32) implies that $\|A_2\psi\| < \infty$. Thus the range of A_2 consists of L^2 eigenfunctions of Δ_X with eigenvalue $1/4$. By Proposition 7.5 there are none; hence $A_2 = 0$.

Thus the singular part of $R_X(s)$ at $\frac{1}{2}$ is $(2s-1)^{-1}A_1$. The multiplicity of the resonance at $s = \frac{1}{2}$ is given by $m_{1/2} = \operatorname{rank} A_1$ by (8.3). Consider the decomposition of $R_X(s)$ given in Theorem 6.3. The term $M_c(s)$ has a pole at $s = \frac{1}{2}$, coming from the zero mode in $R_{C_\infty}(s)$. In each cusp the singular part could be written

$$M_c(s; z, z')\Big|_{z,z' \in C_j} = (1 - \chi_0)\frac{(\rho_c\rho_c')^{s-1}}{2s-1}(1 - \chi_1) + (\text{holo}). \tag{8.34}$$

A pole could occur also in the term $Q(s; \cdot, \cdot) \in (\rho_f\rho_f')^s (\rho_c\rho_c')^{s-1} C^\infty(\overline{X} \times \overline{X})$. These facts, along with the symmetry $R_X(s) = R_X(s)^t$ from Lemma 6.5, allow us to represent the singular part of $R_X(s)$ near $s = \frac{1}{2}$ as $(2s-1)^{-1}A(s)$, where

$$A(s) = \sum_{k=1}^{m_{1/2}} \phi_k(s) \otimes \phi_k(s),$$

for some independent functions $\phi_k(s) \in \rho_f^s \rho_c^{s-1} C^\infty(\overline{X})$. Because $A(\frac{1}{2}) = A_1$, the fact that $(\Delta_X - 1/4)\phi_k(\frac{1}{2}) = 0$ follows from (8.33). $\qquad\square$

A resonance at the point $\frac{1}{2}$ can still be detected in the scattering matrix, as the following lemma demonstrates.

Lemma 8.5. *The scattering matrix at $\frac{1}{2}$ has the form*

$$S_X(\tfrac{1}{2}) = -I + 2P,$$

where P is an orthogonal projection of rank $m_{1/2}$. Hence

$$m_{1/2} = \frac{1}{2} \operatorname{tr}\left[S_X(\tfrac{1}{2}) + I\right].$$

Proof. Because $S_X(\frac{1}{2})$ is self-adjoint, P must be also. And $S_X(\frac{1}{2})^2 = I$ implies that $P^2 = P$. Hence P is an orthogonal projection.

To compute its rank, recall that the scattering matrix is defined through the equation

$$(2s-1)E_X(s)\psi \sim \rho_f^{1-s}\rho_c^{-s}\psi + \rho_f^s\rho_c^{s-1}S_X(s)\psi.$$

With $\phi_k(s)$ as defined in Lemma 8.4, let

$$\phi_k^\sharp(s) := (\rho_f^{-s}\rho_c^{1-s}\phi_k(s))|_{\partial\overline{X}}.$$

Note that the functions $\phi_k^\sharp := \phi_k^\sharp(\tfrac{1}{2})$ must be independent for $k = 1, \dots, m_{1/2}$. Otherwise, we could combine them to form a solution of $(\Delta_X - 1/4)\phi = 0$, with $\phi \in \rho_f^{3/2} \rho_c^{1/2} C^\infty(\overline{X})$, which would be an L^2 eigenfunction with eigenvalue $1/4$.

Applying the definition (7.14) of the F_j component of the Poisson kernel, we derive a decomposition

$$E_j^f(s) = \frac{1}{2s-1} \sum_{k=1}^{m_{1/2}} \phi_k(s) \otimes \phi_k^\sharp(s)|_{F_j} + (1-\chi_0)E_{F_j}(s) + G_j^f(s), \qquad (8.35)$$

where $E_{F_j}(s)$ is the model funnel Poisson operator that appears as the boundary limit of $\rho'^{-s} M_f(s)$, and $G_j^f(s; \cdot, \cdot) \in \rho_{n_f}^s \rho_{n_c}^{s-1} C^\infty(\overline{X} \times \partial_0 \overline{F}_j)$ is holomorphic. Using Lemma 7.12 to take the boundary limit of $E_{F_j}(s)$, we obtain, for $f_j \in C^\infty(\partial_0 \overline{F}_j)$,

$$(2s-1)E_j^f(s)f_j \sim \sum_{k=1}^{m_{1/2}} \rho_f^s \rho_c^{s-1} [\phi_k^\sharp \otimes \phi_k^\sharp|_{F_j}]f_j$$
$$+ (\rho|_{F_j})^{1-s} f_j + (\rho|_{F_j})^s S_{F_j}(s)f_j + O(2s-1), \qquad (8.36)$$

where ρ_{F_j} stands for ρ within F_j and zero in any other end. The eigenvalue formula (7.40) shows that

$$S_{F_j}(\tfrac{1}{2}) = -I,$$

so we conclude from (8.36) that

$$S_{ij}^{ff}(\tfrac{1}{2}) = -\delta_{ij}I + \sum_{k=1}^{m_{1/2}} \phi_k^\sharp|_{F_i} \otimes \phi_k^\sharp|_{F_j}.$$

We can also read off the funnel–cusp terms,

$$S_{ij}^{fc}(\tfrac{1}{2}) = \sum_{k=1}^{m_{1/2}} \phi_k^\sharp|_{F_i} \otimes \phi_k^\sharp|_{C_j}.$$

Analysis of the cusp–cusp term is a little different, because of the pole in $M_c(s)$. Let us write the singular part of the remainder term $Q(s)$ in the decomposition of $R_X(s)$ from Theorem 6.3 as $B(s)$, so that

$$Q(s) = (2s-1)^{-1}B(s) + (\text{holo}).$$

The cusp component of the Poisson operator, defined in (7.15), has singular part with contributions from $Q(s)$ and the model term $M_c(s)$,

$$E_j^c(s; z) = \frac{1}{2s-1} \lim_{\rho' \to 0} \rho'^{1-s} B(s; z, z')\big|_{z' \in C_j}$$
$$+ (1-\chi_0)E_{C_\infty}(s; z)\big|_{C_j} + (\text{holo}), \qquad (8.37)$$

where $E_{C_\infty}|_{C_j}$ denotes the model generalized eigenfunction pulled back from C_∞ to C_j by isometry. Since $E_{C_\infty}(s)$ contributes the ρ_c^{-s} term in the boundary expansion of $E_j^c(s)$, the cusp–cusp component of the scattering matrix comes entirely from $B(s)$,

$$S_{ij}^{cc}(s) = \lim_{\rho,\rho' \to 0} (\rho\rho')^{1-s} B(s; z, z')\big|_{z \in C_i, z' \in C_j} + O(2s - 1). \tag{8.38}$$

On the other hand, $A(\frac{1}{2})$ was defined to be the singular part of $R_X(s)$, and this includes a contribution from $M_c(s)$ as shown in (8.34),

$$A(s; z, z')|_{z \in C_i, z' \in C_j} = B(s; z, z')\big|_{z \in C_i, z' \in C_j} + (2s - 1) M_c(s; z, z')\big|_{z \in C_i, z' \in C_j} + O(2s - 1).$$

Now if we take the boundary limit using (8.38) and use the form of the singular part of $M_c(s)$ shown in (8.34), we find that

$$\sum_{k=1}^{m_{1/2}} \phi_k^\sharp|_{C_i} \otimes \phi_k^\sharp|_{C_j} = S_{ij}^{cc}(s) + \delta_{ij}.$$

These calculations show that the full scattering matrix at $s = \frac{1}{2}$ satisfies

$$S_X(\tfrac{1}{2}) = -I + \sum_{k=1}^{m_{1/2}} \phi_k^\sharp \otimes \phi_k^\sharp, \tag{8.39}$$

and the result follows by the independence of the $\phi_k^\sharp$'s.   □

At the points $\zeta \in \frac{1}{2} \pm \mathbb{N}$, the issue is of course the infinite-rank zeros and poles of $S_X(s)$ and the effect of the gamma-function factors used to define $\widetilde{S}_X(s)$. First note that since $\widetilde{S}_X(s)$ is holomorphic for all $\mathrm{Re}\, s > 1$, we have in particular that

$$N_{1/2-n}(\widetilde{S}_X) = 0, \tag{8.40}$$

for $n \in \mathbb{N}$.

At the positive half-integers, we need to connect the null-multiplicity of $\widetilde{S}_X(s)$ back to the original scattering matrix. The crucial fact is the following:

Lemma 8.6. *For $\zeta \in \frac{1}{2} + \mathbb{N}$,*

$$\ker \widetilde{S}^{ff}(\zeta) = 0. \tag{8.41}$$

Proof. Let $\zeta = \frac{1}{2} + n$. The decomposition formula (7.36) expresses $S_X(s)$ in terms of $S_F(s)$ and a remainder term $Q^\sharp(s)$, which is holomorphic near ζ, since there are no resonances for $\mathrm{Re}\, s > 1$. Thus we have

$$\widetilde{S}^{ff}(\zeta) = \lim_{s \to \zeta} \frac{\Gamma(s + \frac{1}{2})}{\Gamma(\frac{3}{2} - s)} S_F(s), \tag{8.42}$$

because the zero in $1/\Gamma(\frac{3}{2} - s)$ knocks out the contribution from $Q^{\sharp}(s)$. The eigenvalue formula (7.40) shows that the operator

$$\lim_{s \to \zeta} \frac{\Gamma(s + \frac{1}{2})}{\Gamma(\frac{3}{2} - s)} S_{F_\ell}(s)$$

has no zero eigenvalues and hence is invertible. It then follows from (8.42) that $\widetilde{S}^{\mathrm{ff}}(\zeta)$ is invertible.   □

Before continuing, let us note an interesting consequence of Lemma 8.6. If X has no cusps, then S^{ff} is the full scattering matrix, and (8.40) and (8.41) give us the following:

Corollary 8.7. *If X is a geometrically finite hyperbolic surface without cusps, then for all $n \in \mathbb{N}$,*

$$\nu_{1/2-n} = 0.$$

Even when X has cusps, Lemma 8.6 is still quite useful. Using the notation of (8.21), we will compare $\widetilde{S}_X(s)$ to the meromorphic family $\Lambda(s) S_X(s) \Lambda(s)$ of bounded operators on $L^2(\partial \overline{X})$. This family has poles of infinite rank at $s = \frac{1}{2} + n$, but the null-multiplicity is still well-defined at such points. In fact we have the following:

Lemma 8.8. *For $n \in \mathbb{N}$,*

$$N_{1/2+n}(\widetilde{S}_X) = N_{1/2+n}(\Lambda S_X \Lambda).$$

Proof. Let $\zeta = \frac{1}{2} + n$. By Lemma 8.6, $\widetilde{S}^{\mathrm{ff}}(s)$ is invertible in some neighborhood of ζ. Thus the n_c-dimensional matrix

$$T(s) := \widetilde{S}^{\mathrm{cc}}(s) - \widetilde{S}^{\mathrm{cf}}(s) \widetilde{S}^{\mathrm{ff}}(s)^{-1} \widetilde{S}^{\mathrm{fc}}(s)$$

is well-defined near ζ. We can then write

$$\widetilde{S}_X(s) = \begin{pmatrix} I & 0 \\ \widetilde{S}^{\mathrm{cf}}(s) \widetilde{S}^{\mathrm{ff}}(s)^{-1} & I \end{pmatrix} \begin{pmatrix} I & 0 \\ 0 & T(s) \end{pmatrix} \begin{pmatrix} \widetilde{S}^{\mathrm{ff}}(s) & \widetilde{S}^{\mathrm{fc}}(s) \\ 0 & I \end{pmatrix}. \tag{8.43}$$

Because the first and last factors on the right are invertible near ζ, this implies

$$N_\zeta(\widetilde{S}_X) = N_\zeta(T).$$

On the other hand, we have a very similar formula for $\Lambda S_X \Lambda$,

$$\Lambda(s) S_X(s) \Lambda(s) = \begin{pmatrix} I & 0 \\ \Gamma(s + \frac{1}{2}) \widetilde{S}^{\mathrm{cf}}(s) \widetilde{S}^{\mathrm{ff}}(s)^{-1} & I \end{pmatrix} \begin{pmatrix} \frac{\Gamma(\frac{3}{2}-s)}{\Gamma(s+\frac{1}{2})} I & 0 \\ 0 & T(s) \end{pmatrix}$$

$$\times \begin{pmatrix} \widetilde{S}^{\mathrm{ff}}(s) & \frac{1}{\Gamma(\frac{3}{2}-s)} \widetilde{S}^{\mathrm{fc}}(s) \\ 0 & I \end{pmatrix}. \tag{8.44}$$

Again, the first and last factors are invertible near ζ. The upper left block of the middle term,

$$\frac{\Gamma(\frac{3}{2} - s)}{\Gamma(s + \frac{1}{2})} I,$$

is singular at ζ and clearly has no root vectors. Hence,

$$N_\zeta(\Lambda S_X \Lambda) = N_\zeta(T),$$

which proves the result. $\square$

8.6 Coincidence of resonances and scattering poles

Recall the formulas from Proposition 7.15 exhibiting the scattering matrix kernel as a boundary limit of the resolvent kernel. These show immediately that a scattering pole at ζ implies a resonance at ζ also (at least for $\zeta \notin \frac{1}{2} + \mathbb{N}$). On the other hand, in Corollary 7.14 we found the relation

$$R_X(s) - R_X(1-s) = (2s-1)E_X(1-s) \, S_X(s) \, E_X(1-s)^{\mathrm{t}}. \qquad (8.45)$$

For $\mathrm{Re}\, s < \frac{1}{2}$, the terms $R_X(1-s)$ and $E_X(1-s)$ are holomorphic except when $s(1-s) \in \sigma_{\mathrm{d}}(\Delta_X)$. Thus (8.45) shows that a resonance implies a scattering pole, except at finitely many points corresponding to the discrete spectrum.

These facts hint at a more precise connection between the multiplicities of resonances and scattering poles, proven by Guillopé–Zworski in [87].

Theorem 8.9 (Multiplicity formula). *For* $\mathrm{Re}\, \zeta < 1$ *the multiplicities of resonances and scattering poles have the relationship*

$$\nu_\zeta = m_\zeta - m_{1-\zeta}.$$

In particular, $\nu_\zeta = m_\zeta$ *except at finitely many points where* $\zeta(1-\zeta) \in \sigma_{\mathrm{d}}(\Delta_X)$.

Most of the proof is devoted to counting the ranks properly for poles of order higher than 1, and to dealing with the exceptional points where $\zeta \in \frac{1}{2} + \mathbb{Z}$ or $\zeta(1-\zeta) \in \sigma_{\mathrm{d}}(\Delta_X)$. Equating multiplicities is not difficult if the poles are *simple*, meaning of order and multiplicity one. To illustrate the basic ideas, let us first give the proof in this simple case.

Assume that $m_\zeta = 1$, with $\zeta \notin \frac{1}{2} + \mathbb{Z}$ and $\zeta(1-\zeta) \notin \sigma_{\mathrm{d}}(\Delta_X)$. The Laurent expansion of $R_X(s)$ at ζ gives

$$R_X(s) = \frac{\phi \times \phi}{s(1-s) - \zeta(1-\zeta)} + O(s-\zeta),$$

where $\phi \in \rho_{\mathrm{f}}^\zeta \rho_{\mathrm{c}}^{\zeta-1} C^\infty(\overline{X})$ and $(\Delta_X - \zeta(1-\zeta))\phi = 0$. If we set

$$\phi^\sharp = \rho_{\mathrm{f}}^{-\zeta} \rho_{\mathrm{c}}^{-\zeta+1} \phi|_{\partial X},$$

then Proposition 7.4 shows that $\phi^\sharp \neq 0$. By Proposition 7.15, the singular part of the scattering matrix at $s = \zeta$ is given by

$$S_X(s) = (2s - 1)\frac{\phi^\sharp \times \phi^\sharp}{s(1 - s) - \zeta(1 - \zeta)} + O(s - \zeta). \qquad (8.46)$$

Using (8.46) in conjunction with the Gohberg–Sigal factorization (8.27) shows that $N_\zeta(S_X^{-1}) = 1$. And since $S_X(1 - s)$ is holomorphic in a neighborhood of ζ, we have $N_\zeta(S_X) = 0$. This yields $v_\zeta = 1$.

On the other hand, for ζ in the same range suppose that $v_\zeta = 1$. This means $N_\zeta(S_X^{-1}) = 1$ so $S_X(s)$ has a simple pole at ζ. Then Proposition 7.15 shows that $m_\zeta > 0$, and (8.45) shows that $m_\zeta \leq 1$. Hence $m_\zeta = 1$.

This short argument has shown that

$$m_\zeta = 1 \quad \Longleftrightarrow \quad v_\zeta = 1,$$

for $\zeta \notin \frac{1}{2} + \mathbb{Z}$ and $\zeta(1 - \zeta) \notin \sigma_d(\Delta_X)$. One way to proceed to a general proof would be to introduce a perturbation to make all the poles simple. This method was used in Borthwick–Perry [9], where it was shown that all resonances become simple under generic perturbations of Δ_X by compactly supported potentials.

Here we will follow the direct approach of Guillopé–Zworski [87] and Guillarmou [79]. To deal with higher multiplicities, one needs to arrange the singular parts of the resolvent and scattering matrix carefully to make sure that ranks coincide. The first step in the general proof is to a derive a decomposition of $S_X(s)$ somewhat analogous to the Gohberg–Sigal factorization, with orders related to the resonance multiplicity.

Lemma 8.10. *Let ζ be a pole of $S_X(s)$ with $\mathrm{Re}\,\zeta < 1$. Then ζ is also a resonance ($m_\zeta > 0$) and the scattering matrix admits a decomposition*

$$S_X(s) = \Psi^t F_1(s)\left(\sum_{j=1}^m \frac{P_j}{[s(1 - s) - \zeta(1 - \zeta)]^{k_j}}\right)F_2(s)\Psi + H(s), \qquad (8.47)$$

where $\Psi : L^2(\partial\overline{X}) \to \mathbb{C}^{m_\zeta}$, $F_1(s)$ and $F_2(s)$ are holomorphically invertible matrices of dimension m_ζ, and $H(s) : L^2(\partial\overline{X}) \to L^2(\partial\overline{X})$ is holomorphic near ζ. The P_j's are mutually orthogonal projectors on $\mathbb{C}^q$ with rank $P_j = 1$ for $j > 0$. Finally, $k_j > 0$ and

$$m_\zeta = \sum_{j=1}^m k_j,$$

the multiplicity of ζ as a resonance.

Proof. The first step is to take the Laurent series of $R_X(s)$, in the form (8.11), and plug it into the formulas for the scattering matrix from Proposition 7.15. If we denote the restriction of the Laurent coefficients $B_j(\zeta)$ to the boundary by

$$B_j^\sharp(\zeta) = B_j(\zeta; \cdot, \cdot)|_{\partial\overline{X} \times \partial\overline{X}},$$

then this yields

$$S_X(s) = (2s - 1) \sum_{j=1}^{p} \frac{B_j^\sharp(\zeta)}{[s(1-s) - \zeta(1-\zeta)]^j} + H_1(s), \qquad (8.48)$$

where $H_3(s) \in \mathcal{L}(L^2(\partial \overline{X}))$ is holomorphic near $s = \zeta$.

We want to show that the coefficients $B_j^\sharp(\zeta)$ have the structure derived for the coefficients $A_j(\zeta)$ in Proposition 8.1. Indeed, the difference between the two is simply a change of basis. Define the power series coefficients $h_l = h_l(\zeta; \rho_f, \rho_c)$ by

$$\rho_f^{-s} \rho_c^{1-s} = \sum_{l=0}^{\infty} [s(1-s) - \zeta(1-\zeta)]^l h_l.$$

Then we can invert the relationship (8.13) to give

$$B_j(\zeta) = \sum_{i=j}^{p} \sum_{\substack{k,l: \\ k+l=i-j}} h_k A_i(\zeta) h_l. \qquad (8.49)$$

Introducing the basis $\{\phi_k\}$ and nilpotent matrix D from Proposition 8.1, we have

$$B_j(\zeta) = \sum_{m,n=1}^{m_\zeta} \sum_{i=j}^{p} \sum_{k+l=i-j} \left[D^{i-1} \right]_{mn} (h_k \phi_m) \otimes (h_l \phi_n),$$

where $q = m_\zeta$. Because D is nilpotent and symmetric, we can rewrite this as

$$B_j(\zeta) = \sum_{m,n=1}^{m_\zeta} \sum_{k,l=0}^{p-1} \left[D^{j-1} \right]_{mn} h_k (D^k \phi)_m \otimes h_l (D^l \phi)_n,$$

where $(D^k \phi)_m := \sum_j \left[D^k \right]_{jm} \phi_j$. Thus, if we set

$$\psi_k := \sum_{j=0}^{p-1} h_j (D^j \phi)_k,$$

the expression (8.49) reduces to

$$B_k(\zeta) = \sum_{m,n=1}^{m_\zeta} \left[D^{i-1} \right]_{mn} \psi_m \otimes \psi_n.$$

Since $B_k(\zeta)$ has a smooth kernel, the coefficients satisfy

$$\psi_k \in C^\infty(\overline{X}).$$

(The ϕ_k's had log terms in their boundary expansions, as given in (8.14).)

Now the Laurent expansion for $S_X(s)$ can be obtained by restricting ψ_k to $\partial \overline{X}$. Let

$$\psi_k^\sharp := \psi_k|_{\partial \overline{X}} \in C^\infty(\partial \overline{X}).$$

(Note that $\psi_k = \phi_k$ up to terms with at least one power of $\log \rho$, so $\psi_k^\sharp$ could be obtained directly as the coefficient of $\rho_f^s \rho_c^{s-1}$ in the boundary asymptotic expansion of ϕ_k.) Then

$$B_j^\sharp(\zeta) = \sum_{l,m=1}^{m_\zeta} \left[D^{i-1}\right]_{lm} \psi_l^\sharp \otimes \psi_m^\sharp.$$

Defining $\Psi : L^2(\partial \overline{X}) \to \mathbb{C}^{m_\zeta}$ by

$$\Psi(f) = \left[\int_{\partial \overline{X}} \psi_k^\sharp f \, dh\right]_{k=1}^{m_\zeta}$$

allows us to rewrite (8.48) as

$$S_X(s) = (2s - 1)\Psi^t \left(\sum_{j=1}^p \frac{D^{j-1}}{[s(1-s) - \zeta(1-\zeta)]^j}\right) \Psi + H_1(s). \qquad (8.50)$$

The problem is thereby reduced to linear algebra in $\mathbb{C}^{m_\zeta}$.

The next step is to put the nilpotent matrix D in Jordan normal form. Define N_k to be the $k \times k$ nilpotent Jordan block:

$$N_k := \begin{pmatrix} 0 & 1 & \cdots & 0 \\ 0 & 0 & \cdots & 0 \\ \vdots & \vdots & \ddots & 1 \\ 0 & 0 & \ldots & 0 \end{pmatrix}.$$

Then for some integers $k_1 \geq \cdots \geq k_m > 0$ with $k_1 + \cdots + k_m = m_\zeta$, the Jordan normal form of D is the block matrix

$$FDF^{-1} := \begin{pmatrix} N_{k_1} & 0 & \cdots & 0 \\ 0 & N_{k_2} & \cdots & 0 \\ \vdots & \vdots & \ddots & \vdots \\ 0 & 0 & \ldots & N_{k_n} \end{pmatrix}. \qquad (8.51)$$

For notational convenience, set

$$a = a(s, \zeta) := s(1 - s) - \zeta(1 - \zeta).$$

When we substitute for D in (8.50) using (8.51), we obtain blocks on the diagonal of the form

$$\sum_{j=1}^p a^{-j} N_{k_l}^{j-1} = \begin{pmatrix} a^{-1} & a^{-2} & \cdots & a^{-k_l} \\ 0 & a^{-1} & \cdots & a^{1-k_l} \\ \vdots & \vdots & \ddots & a^{-2} \\ 0 & 0 & \ldots & a^{-1} \end{pmatrix},$$

for $l = 1, \ldots, m$. To put this in the desired form, observe that we can factor this block as

$$
\begin{pmatrix} 1 & 0 & \cdots & 0 \\ a & 1 & \cdots & 0 \\ \vdots & \vdots & \ddots & \vdots \\ a^{k_l} & a^{k_l-1} & \cdots & 1 \end{pmatrix}
\begin{pmatrix} a^{-k_l} & 0 & \cdots & 0 \\ 0 & 1 & \cdots & 0 \\ \vdots & \vdots & \ddots & \vdots \\ 0 & 0 & \cdots & 1 \end{pmatrix}
\begin{pmatrix} a^{k_l-1} & a^{k_l-2} & \cdots & 1 \\ -1 & 0 & \cdots & 0 \\ \vdots & \vdots & \ddots & \vdots \\ 0 & 0 & -1 & 0 \end{pmatrix},
$$

where the matrices left and right are invertible and holomorphic in s. The center matrix has the form $a^{-k_l} P_l + P_0$, where P_l is of rank one. Putting all of the blocks into this form and combining the resulting P_0 projectors with the remainder term $H_1(s)$ gives the formula (8.47). $\qquad \square$

Our next step will be to use Lemma 8.10 to show that

$$
N_{1-\zeta}(\widetilde{S}_X) = m_\zeta.
$$

Since $\Lambda S_X \Lambda$ differs from $\widetilde{S}_X$ by factors that are holomorphically invertible away from $\frac{1}{2} \pm \mathbb{N}$, Lemma 8.8 implies that

$$
N_{1-\zeta}(\widetilde{S}_X) = N_{1-\zeta}(\Lambda S_X \Lambda), \tag{8.52}
$$

for all $\operatorname{Re} \zeta < 1$. We will split the calculation into two separate lemmas, following the presentation of Guillarmou [79].

Lemma 8.11. *If ζ is a pole of $\widetilde{S}_X(s)$, then*

$$
m_\zeta \geq N_{1-\zeta}(\widetilde{S}_X).
$$

Proof. The regularized scattering matrix $\widetilde{S}_X(s)$ is holomorphic for $\operatorname{Re} s > 1$ and unitary at $s = \frac{1}{2}$, so by assumption we have $\operatorname{Re} \zeta < 1$ and $\zeta \neq \frac{1}{2}$. By (8.52) it suffices to consider $\Lambda S_X \Lambda$ in place of $\widetilde{S}_X$. Lemma 8.10 gives the decomposition

$$
\Lambda(s) S_X(s) \Lambda(s) = \Psi_1(s)^{\mathrm{t}} \left(\sum_{j=1}^m \frac{P_j}{[s(1-s) - \zeta(1-\zeta)]^{k_j}} \right) \Psi_2(s) + \widetilde{H}(s), \tag{8.53}
$$

where $\Psi_1(s), \Psi_2(s) : L^2(\partial \overline{X}) \to \mathbb{C}^{m_\zeta}$ and $\widetilde{H}(s) : L^2(\partial \overline{X}) \to L^2(\partial \overline{X})$, all holomorphic near $s = \zeta$.

We may assume that the k_j's are ordered $k_1 \geq \cdots \geq k_m$. Let $\{\psi_j\}_{j=1}^n$ be a set of root vectors for $\Lambda S_X \Lambda$ at $1 - \zeta$, with ranks $r_1 \geq \cdots \geq r_n$. By definition,

$$
N_{1-\zeta}(\Lambda S_X \Lambda) = \sum_{j=1}^n r_j,
$$

where $n = \dim \ker_{1-\zeta}(\Lambda S_X \Lambda)$. Our goal is to show that $n \leq m$ and $r_j \leq k_j$ for each $j = 1, \ldots, n$. This will complete the proof, since $m_\zeta = \sum_{j=1}^m k_j$.

For each root vector ψ_j there is a root function $\psi_j(s)$ such that $\psi_j = \psi_j(\zeta)$ and

$$\widetilde{S}_X(1-s)\psi_j(s) = [s(1-s) - \zeta(1-\zeta)]^{r_j} \, \phi_j(s),$$

for some $\phi_j(s)$ such that $\phi_j(\zeta) \neq 0$. Applying $\widetilde{S}_X(s)$ in the form (8.53) then gives

$$\psi_j(s) = \sum_{l=0}^{m} [s(1-s) - \zeta(1-\zeta)]^{r_j - k_l} \, \Psi_1(s)^{\mathrm{t}} P_l \Psi_2(s) \phi_j(s) + O(s - \zeta),$$

where we set $k_0 = 0$. Letting $s \to \zeta$ shows that the root vector ψ_j lies in the vector subspace

$$V_j := \sum_{l: r_j \leq k_l} \mathrm{range}\big[\Psi_1(\zeta)^{\mathrm{t}} P_l \Psi_2(\zeta)\big].$$

Because of the ordering $r_j \geq r_{j+1}$, these spaces are nested,

$$V_1 \subset V_2 \subset \cdots \subset V_n.$$

And since the ψ_j's are independent, we have $\dim V_j \geq j$. On the other hand, the P_j's have rank one, so

$$\dim V_j \leq \#\{l = 1, \ldots, m : k_l \geq r_j\}.$$

We conclude that

$$\#\{l = 1, \ldots, m : k_l \geq r_j\} \geq j \quad \text{for } j = 1, \ldots, n,$$

and it follows immediately that $m \geq n$ and $k_j \geq r_j$ for each j. $\square$

The primary ingredient for the inequality in the other direction is the formula (8.45). We also need the Gohberg–Sigal factorization (8.27) of $\widetilde{S}_X(s)$ at $s = \zeta$, which takes the form

$$\widetilde{S}_X(s) = F_1(s)\left(P_0 + \sum_{j=1}^{n}(s - \zeta)^{k_j} P_j\right)F_2(s), \tag{8.54}$$

where $F_1(s)$, $F_2(s)$ are holomorphically invertible operators on $L^2(\partial \overline{X})$ near $s = \zeta$. The P_j are mutually orthogonal projections on $L^2(\partial \overline{X})$, having rank 1 for $j > 0$, with

$$I = \sum_{j=0}^{n} P_j.$$

The k_j are integers with $k_1 \leq \cdots \leq k_n$, in terms of which the null-multiplicity we are interested in is

$$N_{1-\zeta}(\widetilde{S}_X) = N_\zeta(\widetilde{S}_X^{-1}) = -\sum_{k_j < 0} k_j.$$

In order to apply the Gohberg–Sigal factorization in (8.57), we need to write it in terms of $\Lambda S_X \Lambda$ also. For $\mathrm{Re}\,\zeta < 1$ and $\zeta \notin \frac{1}{2} - \mathbb{N}$, this is connected to $\widetilde{S}_X(s)$ by holomorphically invertible factors, so the decomposition (8.54) implies

$$\Lambda(s)S_X(s)\Lambda(s) = G_1(s)\left(P_0 + \sum_{j=1}^{n}(s - \zeta)^{k_j} P_j\right)G_2(s), \qquad (8.55)$$

with $G_1(s)$, $G_2(s)$ holomorphically invertible near ζ.

If $\zeta \in \frac{1}{2} - \mathbb{N}$, then $N_\zeta(\widetilde{S}_X) = 0$, so all of the k_j's in (8.54) are negative in this case. The two factorization formulas (8.43) and (8.44) show that for $\zeta \in \frac{1}{2} - \mathbb{N}$, (8.54) yields the decomposition

$$\Lambda(s)S_X(s)\Lambda(s) = G_1(s)\left((s - \zeta)P_0 + \sum_{j=1}^{n}(s - \zeta)^{k_j} P_j\right)G_2(s). \qquad (8.56)$$

(Note that the shift in exponent occurs only in the P_0 component.)

Lemma 8.12. *Let ζ be a resonance with $\zeta \neq \frac{1}{2}$. Then*

$$m_\zeta \leq N_{1-\zeta}(\widetilde{S}_X).$$

Proof. A resonance could occur only for $\mathrm{Re}\,\zeta < 1$. First consider the case $\zeta(1-\zeta) \notin \sigma_d(\Delta_X)$. Then $\widetilde{S}_X(s)$ is holomorphic near $1-\zeta$, implying that $N_\zeta(\widetilde{S}_X) = 0$ and hence $k_j < 0$ for $j = 1, \ldots, n$. For $\zeta \notin \frac{1}{2} - \mathbb{N}$, we substitute (8.55) into (8.45) to obtain the formula

$$R_X(s) = R_X(1 - s) + (2s - 1)E_X(1 - s)\Lambda(s)^{-1}G_1(s)$$

$$\times \left(P_0 + \sum_{j=1}^{n}(s - \zeta)^{k_j} P_j\right)G_2(s)\Lambda(s)^{-1}E_X(1 - s)^{\mathrm{t}}. \qquad (8.57)$$

Aside from $(s - \zeta)^{k_j}$, the factors on the right-hand side are all holomorphic near ζ. Let us introduce the Laurent expansions at ζ,

$$(2s - 1)E_X(1 - s)\Lambda(s)^{-1}G_1(s) = \sum_{l=0}^{\infty}(s - \zeta)^l E_l \qquad (8.58)$$

and

$$G_2(s)\Lambda(s)^{-1}E_X(1 - s)^{\mathrm{t}} = \sum_{l=0}^{\infty}(s - \zeta)^l F_l. \qquad (8.59)$$

Then, from (8.57) we read off the residue of $R_X(s)$ at ζ:

$$\mathrm{res}_\zeta R_X(s) = \sum_{j=1}^{n}\sum_{l=0}^{-k_j-1} E_l P_j F_{-k_j-1-l}. \qquad (8.60)$$

Since the rank of each P_j is one, we have

$$m_\zeta = \mathrm{rank}\,A \leq \sum_{j=1}^{n}(-k_j) = N_{1-\zeta}(\widetilde{S}_X).$$

If $\zeta \in \frac{1}{2} - \mathbb{N}$, then we substitute (8.56) into (8.45), obtaining a formula identical to (8.57) except for an extra factor of $(s - \zeta)$ in front of P_0. The rest of the argument is unchanged.

The remaining possibility is that $\zeta(1 - \zeta) \in \sigma_d(\Delta_X)$. If $\operatorname{Re}\zeta > \frac{1}{2}$ then m_ζ is the dimension of the $\zeta(1 - \zeta)$-eigenspace, and the resonance at ζ must have order 1 by Proposition 8.1. Let $\{\phi_i\}_{i=1}^{m_\zeta}$ be an orthonormal basis for the eigenspace, and set

$$\phi_i^\sharp := (\rho_{\mathrm{f}}^{-s} \rho_{\mathrm{c}}^{1-s} \phi_i)|_{\partial \overline{X}}. \tag{8.61}$$

The pole in $S_X(s)$ at ζ has order one, and the decomposition (8.48) takes the form

$$S_X(s) = -\frac{1}{s - \zeta} \sum_{i=1}^{m_\zeta} \phi_i^\sharp \otimes \phi_i^\sharp + H_1(s).$$

The functions $\phi_i^\sharp$ are independent, by Proposition 7.13. Since $\widetilde{S}_X(s)$ is related to $S_X(s)$ by holomorphically invertible factors near ζ, it is then clear that $\widetilde{S}_X(s)^{-1} = \widetilde{S}_X(1-s)$ has m_ζ independent root vectors of rank one at $s = \zeta$. Hence $N_{1-\zeta}(\widetilde{S}_X) = m_\zeta$ for $\operatorname{Re}\zeta > \frac{1}{2}$.

The argument is slightly trickier when $\zeta(1 - \zeta) \in \sigma_d(\Delta_X)$ and $\operatorname{Re}\zeta < \frac{1}{2}$. The formula (8.57) is still valid near ζ, but the factors $R_X(1 - s)$ and $E_X(1 - s)$ now have poles of order 1 at $s = \zeta$. The appropriate Laurent expansions are written

$$R_X(1 - s) = \sum_{l=-1}^{\infty} (s - \zeta)^l R_l,$$

$$(2s - 1)E_X(1 - s)\Lambda(s)^{-1}G_1(s) = \sum_{l=-1}^{\infty} (s - \zeta)^l E_l,$$

$$G_2(s)\Lambda(s)^{-1}E_X(1 - s)^{\mathrm{t}} = \sum_{l=-1}^{\infty} (s - \zeta)^l F_l.$$

Now we need to analyze

$$A := \operatorname*{res}_{\zeta} R_X(s) = R_{-1} + \sum_{\substack{j,l,m: \\ k_j + l + m = -1}} E_l P_j F_m. \tag{8.62}$$

Note that the terms in the sum in (8.62) that involve P_0 or P_j for $k_j > 0$ must include a factor of either E_{-1} or F_{-1}. These terms, and R_{-1} as well, can be expressed in terms of eigenfunctions. Using the eigenfunction basis $\{\phi_i\}_{i=1}^q$ as above, we have

$$R_{-1} = \frac{1}{2\zeta - 1} \sum_{i=1}^{q} \phi_i \otimes \phi_i.$$

Using the notation (8.61), the singular part of $E_X(1 - s)$ can be written

$$E_X(1-s) = \frac{1}{2\zeta - 1} \sum_{i=1}^{q} (s-\zeta)^{-1} \phi_i \otimes \phi_i^\sharp + \text{(holo)}.$$

This shows that

$$E_{-1} = \sum_{i=1}^{q} \phi_i \otimes f_i,$$

$$F_{-1} = \sum_{i=1}^{q} g_i \otimes \phi_i,$$

for some functions $f_i, g_i \in C^\infty(\partial \overline{X})$.

Let V be the span of the ϕ_i's, which is contained in $\rho_f^{1-\zeta} \rho_c^{-\zeta} C^\infty(\overline{X})$. The range of A consists of solutions of $(\Delta_X - \zeta(1-\zeta))^p u = 0$, with boundary behavior

$$\text{range } A \subset \sum_{k=0}^{p-1} \rho_f^\zeta \rho_c^{\zeta-1} (\log \rho)^k C^\infty(\overline{X})$$

by (8.14). This implies that

$$\text{range } A \cap V \subset \left\{ u \in \rho^\infty C^\infty(\overline{X}) : (\Delta_X - \zeta(1-\zeta))^p u = 0 \right\} = \emptyset$$

by Proposition 7.4. Therefore, we can find a decomposition $\rho^{-1} L^2(X) = V \oplus W$ such that range $A \subset W$. Let Π be a projection of $\rho^{-1} L^2(X)$ onto W parallel to V. In particular, $\Pi \phi_i = 0$ for each i and $\Pi A = A$. Applying Π on the left side of (8.62) knocks out the R_{-1} and E_{-1} terms, and applying Π^t on the right knocks out the term with G_{-1}. Since $A = A^t$, by the symmetry of $R_X(s)$ we have

$$A = \Pi A \Pi^t = \sum_{j:\, k_j < 0} \sum_{l=0}^{-k_j - 1} \Pi E_l P_j F_{-k_l - 1 - l} \Pi^t.$$

This now gives

$$m_\zeta = \text{rank } A \le \sum_{j:\, k_j < 0} (-k_j) = N_{1-\zeta}(\widetilde{S}_X). \qquad \square$$

Proof of Theorem 8.9. Lemmas 8.11 and 8.12 have together established that

$$N_{1-\zeta}(\widetilde{S}_X) = m_\zeta,$$

for $\zeta \ne \frac{1}{2}$. We can then simply evaluate

$$\nu_\zeta = N_{1-\zeta}(\widetilde{S}_X) - N_\zeta(\widetilde{S}_X) = m_\zeta - m_{1-\zeta}.$$

As for the case $\zeta = \frac{1}{2}$, we have $\nu_{1/2} = 0$ since $\widetilde{S}_X(\frac{1}{2})$ is unitary. On the resonance side we simply have the cancellation $m_{1/2} - m_{1/2} = 0$. $\qquad \square$

Notes

The multiplicity formula $\nu_\zeta = m_\zeta - m_{1-\zeta}$ was extended to the asymptotically hyperbolic manifolds in higher dimensions by Borthwick–Perry [26], except at the half-integer points $\frac{1}{2} - \mathbb{Z}$. Their method involved use of Agmon's perturbation theory for resonances [3] to show that the addition of a generic compactly supported potential makes all the resonances simple. Guillarmou [79] provided a complete multiplicity formula in the asymptotically hyperbolic case, adapting results of Graham–Zworski [74] to compute the proper correction terms at half-integer points.

The multiplicity formula is also known in the context of "black-box" scattering, which deals with very general compactly supported perturbations of $\mathbb{R}^n$; see Petkov–Zworski [172] and Nedelec [146].

Christiansen–Zworski [45] established resonance wave expansions for the hyperbolic cylinder C_ℓ as well as the modular surface $\mathrm{PSL}(2, \mathbb{Z}) \backslash \mathbb{H}$, showing that solutions of the wave equation admit expansions in terms of the resonance eigenfunctions.

9

Upper Bound for Resonances

We introduced the resonance set $\mathcal{R}_X$ for a geometrically hyperbolic surface X in Chapter 8. Henceforth, we will adopt the convention that resonances are repeated in $\mathcal{R}_X$ according to the multiplicity m_ζ defined in (8.3).

Definition. The *resonance counting function* $N_X(r)$ is defined by

$$N_X(r) := \#\{\zeta \in \mathcal{R}_X : |\zeta| \leq r\}. \tag{9.1}$$

For the elementary cases, we worked out the multiplicities explicitly in Section 8.2. In (8.16) we saw that the resonances of $\mathbb{H}$ occur at $-\mathbb{N}_0$, with multiplicity of $m_{-k} = 2k + 1$. A simple estimate gives the asymptotics of the counting function:

$$N_\mathbb{H}(r) \sim r^2.$$

For the hyperbolic cylinder C_ℓ, Proposition 5.3 showed the resonance set to be $-\mathbb{N}_0 + (2\pi i/\ell)\mathbb{Z}$, and by (8.17) each resonance has multiplicity two. This gives the asymptotic

$$N_{C_\ell}(r) \sim \frac{\ell}{2}r^2.$$

Finally, the parabolic cylinder had only a single resonance at $s = \frac{1}{2}$, of multiplicity one.

When X has infinite area, no exact asymptotics are known for $N_X(r)$ outside of these elementary cases, but we can at least prove that r^2 is the correct growth rate. This chapter is entirely devoted to the proof of the upper bound; the global lower bound will be proven in Section 12.1.

Theorem 9.1 (Guillopé–Zworski). *For any geometrically finite hyperbolic surface X of infinite area,*

$$N_X(r) = O(r^2).$$

Guillopé–Zworski [86] proved this result by adapting the "Fredholm determinant" method introduced by Melrose [132, 133], with refinements by Vodev

[212, 213]. The proof ultimately hinges on the application of Jensen's formula (Theorem A.1) for estimating the zeros of an entire function.

To make use of Jensen's formula, we must produce an entire function whose zeros correspond to the resonance set. The starting point is the parametrix $M(s)$ constructed in Section 6.3, which is related to the resolvent by

$$M(s) = R_X(s)(I - L(s)).$$

The poles of $M(s)$ are known explicitly from the model cylinders. And resonances in $R_X(s)$ that do not coincide with poles from the model cases must be canceled by the factor $I - L(s)$. Hence if one could define the determinant of $I - L(s)$, we'd expect its zeros to correspond to resonances. Now, the Fredholm determinant $\det(I - T)$ is defined when T is a trace class operator, which $L(s)$ is not. Indeed, we had to weight the Hilbert spaces just to make $L(s)$ compact. Observe, however, that a power of $L(s)$ would work just as well: the kernel of $I - L(s)^k$, for $k > 1$, includes the kernel of $I - L(s)$. Applying a cutoff $\chi \in C_0^\infty(X)$, the operator $(L(s)\chi)^3$ will be trace class and we can use $\det(I - (L(s)\chi)^3)$ as the basis for the argument (although this is meromorphic rather than entire). Details of the definition and correspondence between zeros and resonances will be worked out in Section 9.1.

The proof of the upper bound then rests on bounding the growth of this determinant. Since we have an explicit expression for $L(s)$ in terms of model resolvents, the growth estimate ultimately relies on estimates for the model cases. We will show how these lead to the proof of Theorem 9.1 in Section 9.3, and then prove the estimates in Section 9.4.

9.1 Resonances and zeros of determinants

Let χ_a be a cutoff function as in (6.10). Because the kernel $L(s; z, z')$ is compactly supported within $\{r \leq 2\}$ in the z variable, $\chi_3 L(s) = L(s)$. We can exploit this to write

$$M(s)\chi_3 = R_X(s)\chi_3(I - L(s)\chi_3). \tag{9.2}$$

Let us define

$$
\begin{aligned}
L_3(s) &:= L(s)\chi_3 \\
&= L_i(s) + [\Delta_X, \chi_0] R_F(s)(\chi_3 - \chi_1) + [\Delta_X, \chi_0] R_C(s)(\chi_3 - \chi_1). \tag{9.3}
\end{aligned}
$$

By its definition in (6.14), $L_i(s)$ is a compactly supported pseudodifferential operator of order -1. The terms in (9.3) that involve $R_F(s)$ and $R_C(s)$ are compactly supported smoothing operators, because of the nonoverlapping cutoffs. Thus $L_3(s)$ is also a compactly supported pseudodifferential operator of order -1. In two dimensions, a compactly supported pseudodifferential operator is trace class when the order is less than -2 (see Proposition A.26). Hence $L_3(s)^3$ is the lowest power guaranteed to be trace class. Using the Fredholm determinant we define the meromorphic function

$$D(s) := \det(1 - L_3(s)^3).$$ (9.4)

To connect the zeros of $D(s)$ to resonances, we appeal to the following result from Vodev [213, Appendix], relating the multiplicity of a zero of $\det A(s)$ (where $A(s)$ is holomorphic) to the residue of $A(s)^{-1}$ at the zero.

Proposition 9.2. *Let $A(s)$ be a holomorphic family of determinant class operators on a Hilbert space $\mathcal{H}$, defined for s in some neighborhood of ζ. And suppose that $\det A(s)$ has a zero of order m at $s = \zeta$ and no other zeros in this neighborhood. Then for any families $B(s)$, $C(s)$ of bounded holomorphic operators on $\mathcal{H}$,*

$$\operatorname{rank} \int_{\gamma_\zeta} B(s) A(s)^{-1} C(s) \, ds \le m,$$

where γ_ζ is a sufficiently small circle centered at ζ.

Proof. It suffices to prove the case $\zeta = 0$. The first stage in the proof is reduction to a finite-dimensional problem, a standard technique in the perturbation theory of eigenvalues (see, e.g., Kato [107]). Because $I - A(s)$ is trace class, $\ker A(0)$ will be finite-dimensional, and the obvious goal is to restrict our attention to this subspace. The obstruction is that $A(s)$ does not preserve $\ker A(0)$ when $s \ne 0$. So we consider the holomorphic family of projectors $\Pi(s)$ onto $\ker A(s)$, defined by

$$\Pi(s) := \frac{1}{2\pi i} \int_{|w|=\varepsilon} (A(s) + w)^{-1} \, dw.$$

To see that this is a projector, we use the resolvent identity

$$(A(s) + w')^{-1}(A(s) + w)^{-1} = \frac{1}{w' - w}\Big[(A(s) + w')^{-1} - (A(s) + w)^{-1}\Big]$$

to write $\Pi(s)^2$ as

$$\frac{1}{(2\pi i)^2} \int_{|w'|=2\varepsilon} \int_{|w|=\varepsilon} \frac{1}{w' - w}\Big[(A(s) + w')^{-1} - (A(s) + w)^{-1}\Big] dw \, dw'.$$

The integral of the first term with respect to w gives zero, because w' lies outside the circle $|w| = \varepsilon$. The integral with respect to w' in the second term gives a factor of $2\pi i$, showing that $\Pi(s)^2 = \Pi(s)$.

Each $\Pi(s)$ restricts to a finite-dimensional space, but these spaces vary with s. To make the problem truly finite-dimensional we want to "line up" the vector spaces, by introducing a holomorphically invertible family of operators $U(s)$ such that

$$U(s)\Pi(0)U(s)^{-1} = \Pi(s).$$

The existence of $U(s)$ in some neighborhood of $s = 0$, a result of Kato, can easily be verified in our case using the explicit formula

$$U(s) = \Big[I - (\Pi(s) - \Pi(0))^2\Big]^{-1/2}\Big[\Pi(s)\Pi(0) + (I - \Pi(s))(I - \Pi(0))\Big].$$

(See, e.g., [107, Remark II.4.4] or [181, Section XII].)

If we set

$$\widetilde{A}(s) := U(s)^{-1} A(s) U(s),$$

then

$$\int_{|w|=\varepsilon} (\widetilde{A}(s) + w)^{-1} \, dw = U(s)^{-1} \Pi(s) U(s) = \Pi(0).$$

Hence $\widetilde{A}(s)$ commutes with $\Pi(0)$. Under the decomposition

$$\mathcal{H} = \ker A(0) \oplus \ker \Pi(0),$$

we have

$$\widetilde{A}(s) = \begin{pmatrix} A_1(s) & 0 \\ 0 & A_2(s) \end{pmatrix},$$

where $A_1(s)$ is a finite-dimensional matrix and $A_2(s)$ is determinant class. Since $\Pi(0)$ projects onto $\ker \widetilde{A}(s)$, $A_2(s)$ is invertible. The original determinant in the problem now factors,

$$\det A(s) = \det A_1(s) \, \det A_2(s),$$

where $\det A_1(s)$ is an ordinary matrix determinant, and $\det A_2(0) \neq 0$. By assumption $\det A_1(s)$ has a zero of order m at $s = 0$ and so the problem is now finite-dimensional.

Note that $A_1(0) = 0$, but $A_1(s)$ is otherwise invertible for s in some neighborhood of 0. We can thus use the normal form given in (8.31) to write

$$A_1(s) = E(s) \left(\sum_{j=1}^q s^{r_j} P_j \right) F(s),$$

where $E(s)$ and $F(s)$ are holomorphically invertible, the P_j's are mutually orthogonal rank-one projections, and $r_j > 0$ for each j. Taking the determinant, we see that m, the order of the zero, is given by

$$m = \sum_{j=1}^q r_j.$$

Let us now return to the original problem. Because $A_2(s)$ is invertible, we can reduce the rank calculation to

$$\mathrm{rank} \int_{\gamma_0} B(s) A(s)^{-1} C(s) \, ds = \mathrm{rank} \int_{\gamma_0} B_1(s) A_1(s)^{-1} C_1(s) \, ds$$

$$= \mathrm{rank} \int_{\gamma_0} B_2(s) \left(\sum_{j=1}^q s^{-r_j} P_j \right) C_2(s) \, ds,$$

where $C_2(s) : \mathcal{H} \to \mathbb{C}^n$ and $B_2(s) : \mathbb{C}^n \to \mathcal{H}$ are holomorphic. Denote the power-series expansions of these operators by

$$B_2(s) = \sum_{k=0}^{\infty} s^k B^{(k)}, \qquad C_2(s) = \sum_{k=0}^{\infty} s^k C^{(k)}.$$

Since rank $P_j = 1$,

$$\mathrm{rank} \int_{\gamma_0} B(s) A(s)^{-1} C(s)\, ds \le \sum_{j=1}^{q} \sum_{i=0}^{r_j-1} \mathrm{rank}\left(B^{(i)} P_j C^{(r_j-1-i)} \right)$$

$$\le \sum_{j=1}^{q} r_j = m. \qquad \square$$

Recalling that F denotes the disjoint union of funnel ends of X, we denote by $\mathcal{R}_F$ the union of the sets of resonances for each model resolvent $R_{F_j}(s)$,

$$\mathcal{R}_F := \bigcup_{j=1}^{n_f} \left(-\mathbb{N}_0 + \frac{2\pi i}{\ell_j} \mathbb{Z} \right),$$

with each point counted with multiplicity 2.

Corollary 9.3. *The resonance set $\mathcal{R}_X$ (with multiplicities) is contained within the union of the set of zeros of $D(s)$ (with multiplicities) and three copies of the set $\mathcal{R}_F \cup \{\frac{1}{2}\}$.*

Proof. Note that (9.2) implies the factorization

$$R_X(s)\chi_3 = M(s)\chi_3(I + L_3(s) + L_3(s)^2)(I - L_3(s)^3)^{-1}. \qquad (9.5)$$

Suppose ζ is a resonance. Then by (8.9) the range of the residue of $R_X(s)$ at ζ consists of solutions of $(\Delta_X - s(1-s))^k \phi = 0$ for some k. Such functions ϕ cannot vanish on any open set in the interior, by the standard uniqueness theory for second-order elliptic operators [7]. In particular, $\phi\chi_3$ is not identically zero. This implies that the rank of the residue of $R_X(s)\chi_3$ is the same as that of $R_X(s)$, namely m_ζ.

In (9.5), the terms $M(s)$ and $L_3(s)$ have poles at $\mathcal{R}_F$ coming from the funnel model terms, and a pole at $\frac{1}{2}$ from the cusp model term. Each pole in a model term possibly contributes to a resonance of $R_X(s)$. The highest possible multiplicity comes from the $M(s)\chi_3 L_3(s)^2$ term, so these terms are accounted for by the three copies of $\mathcal{R}_F \cup \{\frac{1}{2}\}$. Proposition 9.2 shows that the multiplicity of a pole at ζ coming from $(I - L_3(s)^3)^{-1}$ is bounded above by the multiplicity of ζ as a zero of $D(s)$. So with these zeros we account for all possible poles of $R_X(s)\chi_3$. $\qquad \square$

9.2 Singular value estimates

Because of Jensen's formula (Theorem A.1), Corollary 9.3 reduces the problem of bounding the resonance counting function to growth estimates on $D(s)$. The fundamental tool for these estimates is singular values, which have nice behavior under sums and products. They will allow us to break up the estimate on $D(s)$ into separate estimates on the three components of $L_3(s)$ given in (9.3).

For a compact operator A on a Hilbert space $\mathcal{H}$, the singular values $\mu_j(A)$ are the nonzero eigenvalues of the self-adjoint operator $|A|$, arranged in decreasing order. In other words, they are the square roots of the nonzero eigenvalues of $A^* A$. The basic theory of singular values is reviewed in Section A.4. The application to the estimation of determinants is based on Weyl's inequality (Theorem A.22), which says, assuming that A is trace class,

$$|\det(I + A)| \le \prod_{j=1}^{\infty} (1 + \mu_j(A)) = \det(1 + |A|) \le e^{\operatorname{tr}|A|}. \tag{9.6}$$

For convenience, we'll note a few other basic inequalities here. If A is bounded and B compact, then

$$\mu_j(AB) \le \|A\| \, \mu_j(B), \tag{9.7}$$

for all j. (This follows from the min-max principle, Theorem A.16.) We also need the Fan inequalities (Theorem A.18): if both A and B are compact, then

$$\mu_{i+j-1}(A + B) \le \mu_i(A) + \mu_j(B) \tag{9.8}$$

and

$$\mu_{i+j-1}(AB) \le \mu_i(A)\mu_j(B). \tag{9.9}$$

We conclude the section with two more specialized inequalities, taken from Guillopé–Zworski [86].

Lemma 9.4. *Suppose A, B are operators on a Hilbert space $\mathcal{H}$ such that A^p and B^p are trace class for some $p \in \mathbb{N}$. Then*

$$\left|\det(1 + (A + B)^p)\right| \le \det\left(1 + 2^{p-1}|A|^p\right)^{2p} \det\left(1 + 2^{p-1}|B|^p\right)^{2p}.$$

Proof. From the Weyl inequality (9.6),

$$|\det(I + (A + B)^p)| \le \prod_{j=1}^{\infty} (1 + \mu_j((A + B)^p)).$$

By iterating (9.9), we obtain an estimate for powers:

$$\mu_{p(k-1)+1}(A^p) \le \mu_k(A)^p,$$

for any $p, k \ge 1$. To apply this, we can break up the product:

$$\prod_{j=1}^{\infty} (1 + \mu_j((A + B)^p)) = \prod_{k=1}^{\infty} \prod_{l=1}^{p} (1 + \mu_{p(k-1)+l}((A + B)^p))$$

$$\le \left[\prod_{k=1}^{\infty} (1 + \mu_{p(k-1)+1}((A + B)^p))\right]^p$$

$$\le \left[\prod_{k=1}^{\infty} (1 + \mu_k(A + B)^p)\right]^p.$$

Next we need to break up the sum. Observe that

$$\prod_{k=1}^{\infty}\big(1 + \mu_k(A + B)^p\big) = \prod_{j=1}^{\infty}\big(1 + \mu_{2j-1}(A + B)^p\big)\big(1 + \mu_{2j}(A + B)^p\big)$$

$$\le \prod_{j=1}^{\infty}\big(1 + \mu_{2j-1}(A + B)^p\big)^2.$$

By (9.8), $\mu_{2j-1}(A + B) \le \mu_j(A) + \mu_j(B)$, so our main estimate now becomes

$$\prod_{j=1}^{\infty}\big(1 + \mu_j((A + B)^p)\big) \le \left[\prod_{j=1}^{\infty}\big(1 + [\mu_j(A) + \mu_j(B)]^p\big)\right]^{2p}.$$

Since x^p is a convex function for $x > 0$,

$$\left(\frac{a + b}{2}\right)^p \le \frac{a^p + b^p}{2},$$

for $a, b > 0$. From this we can immediately derive

$$1 + (a + b)^p \le (1 + 2^{p-1}a^p)(1 + 2^{p-1}b^p),$$

and it follows that

$$\prod_{j=1}^{\infty}\big(1 + \mu_j((A + B)^p)\big) \le \left[\prod_{j=1}^{\infty}\big(1 + 2^{p-1}\mu_j(A)^p\big)\big(1 + 2^{p-1}\mu_j(B)^p\big)\right]^{2p}$$

$$= \det(1 + 2^{p-1}|A|^p)^{2p} \det(1 + 2^{p-1}|B|^p)^{2p}. \qquad \square$$

Lemma 9.5. *For trace class operators A, B,*

$$|\det(1 + AB)| \le \det(I + |A|)^2 \det(I + |B|)^2.$$

Proof. The starting point is of course the Weyl inequality,

$$|\det(1 + AB)| \le \prod_{j=1}^{\infty}(1 + \mu_j(AB)).$$

Then we apply (9.9) by the same trick as in the proof of Lemma 9.4,

$$\prod_{j=1}^{\infty}(1 + \mu_j(AB)) \le \prod_{k=1}^{\infty}(1 + \mu_{2k-1}(AB))^2$$

$$\le \prod_{k=1}^{\infty}(1 + \mu_k(A)\mu_k(B))^2.$$

The remaining step is now obvious, since

$$1 + \mu_k(A)\mu_k(B) \le (1 + \mu_k(A))(1 + \mu_k(B)). \qquad \square$$

9.3 Upper bound

The singular value estimates, Lemmas 9.4 and 9.5, allow us to break down the estimate of $D(s) = \det(I - L_3(s)^3)$ into estimates of determinants of the constituent pieces of $L_3(s)$. Define

$$T(s) := [\Delta_X, \chi_0] R_F(s)(\chi_3 - \chi_1) + [\Delta_X, \chi_0] R_C(s)(\chi_3 - \chi_1), \qquad (9.10)$$

so that $L_3(s) = L_i(s) + T(s)$. By Lemma 9.4,

$$|D(s)| \le \det\left(1 + 4|L_i(s)|^3\right)^6 \det\left(1 + 4|T(s)|^3\right)^6. \qquad (9.11)$$

The kernel of $T(s)$ is smooth and compactly supported, so $T(s)$ is trace class in particular. Lemma 9.5 allows us to reduce the power in the determinant:

$$\det(1 + 4|T(s)|^3) \le \det(1 + 2|T(s)|)^2 \det(1 + 2|T(s)|^2)^2$$
$$\le \det(1 + 2|T(s)|)^2 \det(1 + \sqrt{2}|T(s)|)^8. \qquad (9.12)$$

Substituting (9.12) in (9.11) gives the bound

$$|D(s)| \le \det(1 + 4|L_i(s)|^3)^6 \det(1 + 2|T(s)|)^{60}. \qquad (9.13)$$

Estimating the interior term is relatively easy. For the estimates we set

$$\langle s \rangle := \sqrt{1 + |s|^2}.$$

Lemma 9.6. *There is a constant C such that for all $s \in \mathbb{C}$,*

$$\det(1 + 4|L_i(s)|^3) \le e^{C\langle s \rangle^2}.$$

Proof. Recall that the interior remainder term has two parts,

$$L_i(s) = [\Delta, \chi_2] R_X(s_0)\chi_1 + (s_0(1 - s_0) - s(1 - s))M_i.$$

The $R_X(s_0)$ term is pseudodifferential of order -1, while M_i is of order -2. Using Lemma 9.4, we have

$$\det(1 + 4|L_i(s)|^3) \le \det\left(1 + 16\big|[\Delta, \chi_2] R_X(s_0)\chi_1\big|^3\right)^6$$
$$\times \det\left(1 + 16\big|(s_0(1 - s_0) - s(1 - s))M_i\big|^3\right)^6.$$

The first determinant on the right is constant, and by (A.35) we can bound the second determinant by

$$\det\left(1 + 16\big|(s_0(1 - s_0) - s(1 - s))M_i\big|^3\right) \le \prod_{k=1}\left(1 + \frac{c^3\langle s \rangle^6}{k^3}\right).$$

This product can be evaluated explicitly, using

$$\prod_{k=1}^{\infty}\left(1+\frac{x^3}{k^3}\right) = \frac{1}{x^3\Gamma(x)\Gamma(e^{2\pi i/3}x)\Gamma(e^{4\pi i/3}x)}.$$

With Stirling's approximation (7.41), we can show that

$$\prod_{k=1}^{\infty}\left(1+\frac{x^3}{k^3}\right) \sim (2\pi x)^{-3/2}e^{2\pi x/\sqrt{3}},$$

which implies

$$\det\left(1 + 16\left|(s_0(1-s_0) - s(1-s))M_i\right|^3\right) \le e^{C\langle s\rangle^2}. \qquad \square$$

Let us define the Weierstrass product over the resonance set $\mathcal{R}_F$:

$$P_F(s) := \prod_{\zeta\in\mathcal{R}_F}\left(1 - \frac{s}{\zeta}\right)e^{s/\zeta + s^2/(2\zeta^2)},$$

where each ζ will actually occur twice because of the multiplicity. The product $P_F(s)$ defines an entire function of order 2, but that only implies a bound $|P_F(s)| \le \exp(C\langle s\rangle^{2+\varepsilon})$ for any $\varepsilon > 0$. If we can set $\varepsilon = 0$ in the bound, the entire function is said to be of finite type. By Lindelöf's theorem (see Theorem A.4), a Weierstrass product of order p over a zero set $\mathcal{Z}$ has finite type if and only if

$$\sum_{\substack{\zeta\in\mathcal{Z} \\ |\zeta|\le r}}\frac{1}{\zeta^p} = O(1)$$

as $r \to \infty$. This shows that $P_F(s)$ has infinite type. But $P_F(s)P_F(is)$ has finite type because of cancellation among the zeros. Thus we define

$$g(s) := (2s - 1)P_F(s)P_F(is),$$

to obtain a function whose zero set includes $\mathcal{R}_F \cup \{\frac{1}{2}\}$ and that satisfies the bound

$$|g(s)| \le e^{C\langle s\rangle^2}. \tag{9.14}$$

Lemma 9.7. *For sufficiently large N, we have a bound*

$$\left|g(s)^N \det(1 + 2|T(s)|)\right| \le e^{C\langle s\rangle^2}.$$

This estimate is much trickier than those in the lemmas above, because of the poles in $T(s)$; we'll defer it to the next section. Assuming this result, the main theorem follows easily.

Proof of Theorem 9.1. Applying Lemmas 9.6 and 9.7 to (9.13) shows that for N sufficiently large, $g(s)^N D(s)$ is entire and

$$\left| g(s)^N D(s) \right| \leq e^{C\langle s \rangle^2}. \tag{9.15}$$

Jensen's formula (Theorem A.1) then implies that the number of zeros of $g(s)^N D(s)$ with $|s| \leq r$ is $O(r^2)$. Corollary 9.3 shows that this furnishes an upper bound for $N_X(r)$.   □

9.4 Estimates on model terms

The remaining step is the proof of Lemma 9.7. Before launching into the estimates, which are somewhat technical, let us explain the overall strategy. To bound $\det(1 + 2|T(s)|)$ we would like to control the growth of the singular values of $T(s)$. The method, which comes from an argument of Melrose in [133], is based on comparison with the Dirichlet Laplacian Δ_Z for a compact region Z. With reference to the geodesic coordinate r in the funnels and cusps, let

$$Z = \{0 \leq r \leq 1\} \subset X,$$

a disjoint union of closed collar neighborhoods. In the definition (9.10) of $T(s)$ we note that $[\Delta_X, \chi_0]$ has coefficients supported within Z. Since the Dirichlet Laplacian on Z is invertible, we can write

$$T(s) = \Delta_Z^{-m}(\Delta_Z^m T(s)).$$

Moreover, $\Delta_Z^m T(s) = \Delta_X^m T(s)$. By Weyl's asymptotic law (Theorem A.15),

$$\mu_k(\Delta_Z) \sim C_Z k,$$

so that by (9.7) the singular values satisfy the bound

$$\mu_k(T(s)) \leq C k^{-m} \|\Delta_X^m T(s)\|, \tag{9.16}$$

for any $m \in \mathbb{N}_0$.

The goal is thus to produce operator norm estimates for $\Delta_X^m T(s)$, which reduces to a study of the component terms:

$$\Delta_X^m [\Delta_X, \chi_0] R_{F_\ell}(s)(\chi_3 - \chi_1)$$

and

$$\Delta_X^m [\Delta_X, \chi_0] R_{C_\infty}(s)(\chi_3 - \chi_1).$$

Before analyzing the model resolvents explicitly, we can establish a general principle that will simplify the estimates. The proof of this result works on any Riemannian manifold; the fact that X is a hyperbolic surface shows up only in the use of $s(1-s)$ for the spectral parameter.

Lemma 9.8. *Let X be any hyperbolic surface, including possibly F_ℓ with Dirichlet boundary conditions. Suppose that for $s \in \Omega \subset \mathbb{C}$, the resolvent $R_X(s)$ satisfies a bound of the following form: for any $\psi_0, \psi_1 \in C_0^\infty(X)$ with disjoint supports we can find a constant C_Ω such that*

$$\|\psi_0 R_X(s)\psi_1\| \le C_\Omega \langle s \rangle^\tau. \tag{9.17}$$

Then for $\psi_0, \psi_1 \in C_0^\infty(X)$ with disjoint supports we have bounds of the form

$$\left\| \Delta_X^m \psi_0 R_X(s)\psi_1 \right\| \le C_{m,\Omega} \langle s \rangle^{\tau+2m} \tag{9.18}$$

and

$$\left\| \Delta_X^m [\Delta_X, \psi_0] R_X(s)\psi_1 \right\| \le C_{m,\Omega} \langle s \rangle^{\tau+2m+1}. \tag{9.19}$$

Proof. The proof is by iteration using the resolvent equation

$$\Delta_X R_X(s) = I + s(1-s)R_X(s).$$

Let us first consider the $m = 0$ case of (9.19). Let $f \in L^2(X)$, and for convenience introduce

$$u_s := R_X(s)\psi_1 f.$$

We wish to bound

$$[\Delta_X, \psi_0] R_X(s)\psi_1 f = (\Delta_X \psi_0)u_s + 2g(\nabla \psi_0, \nabla u_s), \tag{9.20}$$

where g denotes the metric on X here. The first term on the right is $O(\langle s \rangle^\tau \|f\|)$ by the assumption (9.17), so we need only worry about the second. Introduce $\widetilde{\psi}_0 \in C_0^\infty(X)$, with support including that of ψ_0 but still disjoint from the support of ψ_1. Then

$$\|g(\nabla \psi_0, \nabla u_s)\|^2 = \int_X \left| g(\nabla \psi_0, \widetilde{\psi}_0 \nabla u_s) \right|^2 dg,$$
$$\le C \|\widetilde{\psi}_0 \nabla u_s\|^2, \tag{9.21}$$

by the Schwarz inequality. Using integration by parts,

$$\|\widetilde{\psi}_0 \nabla u_s\|^2 = \int_X \widetilde{\psi}_0^2 \, \overline{u}_s \, \Delta_X u_s \, dg - 2\int_X \widetilde{\psi}_0 \overline{u}_s \, g(\nabla \widetilde{\psi}_0, \nabla u_s) \, dg.$$

By the resolvent equation,

$$\Delta_X u_s = \psi_1 f + s(1-s)u_s.$$

Since $\widetilde{\psi}_0 \psi_1 = 0$, this gives us

$$\|\widetilde{\rho} \, \nabla u_s\|^2 = s(1-s)\|\widetilde{\psi}_0 u_s\|^2 - 2\int_X \widetilde{\psi}_0 \overline{u}_s \, g(\nabla \widetilde{\psi}_0, \nabla u_s) \, dg.$$

Using the Schwarz inequality again, we obtain

$$\left\| \widetilde{\psi}_0 \nabla u_s \right\|^2 \leq \langle s \rangle^2 \left\| \widetilde{\psi}_0 u_s \right\|^2 + 2 \int_X |\widetilde{\psi}_0 \overline{u_s}| \, |\nabla \widetilde{\psi}_0|_g \, |\nabla u_s|_g \, dg.$$

$$\leq \langle s \rangle^2 \left\| \widetilde{\psi}_0 u_s \right\|^2 + 2 \| u_s \nabla \widetilde{\psi}_0 \| \, \left\| \widetilde{\psi}_0 \nabla u_s \right\|$$

$$\leq \langle s \rangle^2 \left\| \widetilde{\psi}_0 u_s \right\|^2 + 2 \| u_s \nabla \widetilde{\psi}_0 \|^2 + \tfrac{1}{2} \left\| \widetilde{\psi}_0 \nabla u_s \right\|^2.$$

Moving the last term on the right to the other side and using (9.17) yields

$$\left\| \widetilde{\psi}_0 \nabla u_s \right\| \leq C \langle s \rangle^{\tau+1} \| f \|.$$

Together with (9.20) and (9.21), this proves (9.19) for $m = 0$.

It is then trivial to deduce (9.18) for $m = 1$ from the identity

$$\Delta_X \psi_0 R_X(s) \psi_1 = [\Delta_X, \psi_0] R_X(s) \psi_1 + s(1 - s) \psi_0 R_X(s) \psi_1,$$

which follows from the resolvent equation and $\psi_0 \psi_1 = 0$.

The rest of the proof is straightforward induction. From the resolvent formula we derive

$$\Delta_X^m \psi_0 R_X(s) \psi_1 = \Delta_X^{m-2} [\Delta_X, [\Delta_X, \psi_0]] R_X(s) \psi_1$$
$$+ s(1 - s) \Delta_X^{m-1} \psi_0 R_X(s) \psi_1$$
$$+ s(1 - s) \Delta_X^{m-2} [\Delta_X, \psi_0] R_X(s) \psi_1.$$

This reduces case m of (9.18) to the cases $0, \ldots, m - 2$ of (9.18), for the first two terms on the right, and the $m - 2$ case of (9.19) for the third term. Similarly,

$$\Delta_X^m [\Delta_X, \psi_0] R_X(s) \psi_1 = \Delta_X^{m-1} [\Delta_X, [\Delta_X, \psi_0]] R_X(s) \psi_1$$
$$+ s(1 - s) \Delta_X^{m-1} [\Delta_X, \psi_0] R_X(s) \psi_1$$

reduces (9.19) to previous cases.   □

We should remark that the separate proof of (9.19) was actually unnecessary. The estimate (9.18) is equivalent to a bound on the operator norm of $\psi_0 R_X(s) \psi_1$ as a map between Sobolev spaces, $H^0(X, dg) \to H^{2m}(X, dg)$. Complex interpolation between Sobolev spaces (see, e.g., [207, Section 4.2]) then implies that

$$\| \psi_0 R_X(s) \psi_1 \|_{\mathcal{L}(H^0, H^\sigma)} \leq C \langle s \rangle^{\tau+\sigma},$$

for $0 \leq \sigma \leq 2m$. Induction over m then serves to establish this more general result for all $\sigma \geq 0$.

Corollary 9.9. *For* $\mathrm{Re}\, s > \tfrac{1}{2} + \varepsilon$, *we have the estimates*

$$\Delta_X^m [\Delta_X, \chi_0] R_{F_\ell}(s)(\chi_3 - \chi_1) = O(\langle s \rangle^{2m}),$$
$$\Delta_X^m [\Delta_X, \chi_0] R_{C_\infty}(s)(\chi_3 - \chi_1) = O(\langle s \rangle^{2m}),$$

with constants depending on m *and* ε.

Proof. One immediate consequence of the spectral theorem (Theorem A.10) is a formula for the operator norm of the resolvent. If A is self-adjoint and $z \notin \sigma(A)$, then

$$\left\| (A - z)^{-1} \right\| = \frac{1}{d(z, \sigma(A))}.$$

The full spectrum of Δ_{F_ℓ} is $[1/4, \infty)$, by the same arguments used in Section 7.1. And for $\operatorname{Re} s > \frac{1}{2} + \varepsilon$, we have $d(s(1-s), [1/4, \infty)) > \max\{\varepsilon, 2\varepsilon \operatorname{Im}(s)\}$; hence the resolvent bound gives

$$\| R_{F_\ell}(s) \| \leq \frac{1}{\varepsilon \langle s \rangle}. \tag{9.22}$$

The same reasoning applies to $R_{C_\infty}(s)$. The claimed estimates then follow from (9.19). $\qquad\square$

The next step is to use the explicit expressions for the resolvents from Chapter 5 to establish weaker estimates for a vertical strip that includes $\{\operatorname{Re} s = \frac{1}{2}\}$.

Lemma 9.10. *Suppose* $\psi_0, \psi_1 \in C_0^\infty(F_\ell)$, *with disjoint supports. Then for* $\varepsilon < \operatorname{Re} s < 1$,

$$\| \psi_0 R_{F_\ell}(s) \psi_1 \| = O(1), \tag{9.23}$$

with a constant depending on ε.

Proof. It's convenient to switch our attention to the full cylinder C_ℓ. The method of images formula (5.20) shows that it suffices to prove the corresponding result for $\psi_0 R_{C_\ell}(s) \psi_1$. Let us identify $L^2(C_\ell, dg)$ with with $L^2(F_\ell, dg)$, where F_ℓ is the fundamental domain

$$\mathcal{F}_\ell = \{1 \leq |z| \leq e^\ell\} \subset \mathbb{H}.$$

For $\operatorname{Re} s > 0$, the resolvent kernel as a function on $\mathcal{F}_\ell \times \mathcal{F}_\ell$ is given explicitly as an average over Γ_ℓ:

$$R_{C_\ell}(s; z, z') = \sum_{k \in \mathbb{Z}} \frac{1}{4\pi} \int_0^1 \frac{(t(1-t))^{s-1}}{(\sigma(z, e^{k\ell} z') - t)^s} \, dt. \tag{9.24}$$

Because $d(z, e^{k\ell} z') \geq (k-1)\ell$ for $z, z' \in \mathcal{F}_\ell$, we can bound

$$|\sigma(z, e^{k\ell} z')| \geq C e^{k\ell \operatorname{Re} s}.$$

For the $k = 0$ term, we note that $d(z, z')$ is bounded away from 0 within the support of $\psi_0(z) \psi_1(z')$; hence $\sigma(z, z')$ is bounded away from 1. Thus for $\varepsilon < \operatorname{Re} s < 1$, we have

$$\left| \psi_0(z) \big(\sigma(z, e^{k\ell} z') - t \big)^{-s} \psi_1(z') \right| \leq C e^{-k\ell \varepsilon}.$$

Using this with (9.24) gives

$$\left| \psi_0(z) R_{C_\ell}(s; z, z') \psi_1(z') \right| \leq C \sum_{k \in \mathbb{Z}} e^{-k\ell \varepsilon} \frac{\Gamma(\operatorname{Re} s)^2}{\Gamma(2 \operatorname{Re} s)}$$

$$\leq \frac{C}{\varepsilon(1 - e^{-\ell \varepsilon})}. \qquad\square$$

We need a similar result for the cusp terms.

Lemma 9.11. *Suppose* $\psi_0, \psi_1 \in C_0^\infty(\mathbb{R}_+)$ *have disjoint supports. If we regard these as functions of the coordinate* y *on* C_∞, *then for* $\varepsilon < \operatorname{Re} s < 1$ *and* $|s - \frac{1}{2}| > \varepsilon$ *we have*

$$\|\psi_0 R_{C_\infty}(s)\psi_1\| = O(\langle s \rangle). \tag{9.25}$$

Proof. Since the analogue of (9.24) for the parabolic cylinder converges only for $\operatorname{Re} s > \frac{1}{2}$, we'll instead draw on the explicit formulas from Section 5.3. For $y < y'$, the Fourier decomposition of the resolvent kernel is

$$R_{C_\infty}(s; z, z') = \frac{y^s y'^{1-s}}{2s - 1} + \sum_{k \neq 0} \sqrt{yy'} I_{s-\frac{1}{2}}(2\pi |k| y) K_{s-\frac{1}{2}}(2\pi |k| y') e^{2\pi i k(x-x')}.$$

The bound on the contribution to $\psi_0 R_{C_\infty}(s)\psi_1$ from the zero mode is trivial:

$$\left| \psi_0(y) \frac{y^s y'^{1-s}}{2s - 1} \psi_1(y') \right| \leq \frac{C}{|2s - 1|}. \tag{9.26}$$

To control the contribution from the kth Fourier mode, we'll make use of standard integral representations for the Bessel functions. First, for $\operatorname{Re} s > 0$,

$$I_{s-\frac{1}{2}}(w) = \frac{(\frac{w}{2})^{s-\frac{1}{2}}}{\sqrt{\pi} \Gamma(s)} \int_0^\pi e^{-w \cos t} \sin^{2s-1} t \, dt.$$

This gives a straightforward estimate

$$|I_{s-\frac{1}{2}}(w)| \leq C_\varepsilon \frac{(\frac{w}{2})^{\operatorname{Re} s - \frac{1}{2}}}{|\Gamma(s)|} e^w, \tag{9.27}$$

for $\varepsilon < \operatorname{Re} s < 1$. For the K-Bessel function we use a corresponding identity for $\operatorname{Re} s > 0$,

$$K_{s-\frac{1}{2}}(w) = \frac{\Gamma(s)}{2\sqrt{\pi} (\frac{w}{2})^{s-\frac{1}{2}}} \int_{-\infty}^\infty e^{iwt} (t^2 + 1)^{-s} \, dt.$$

This integral is estimated by shifting the contour by $t \to t + ia$, for some $a < 1$, yielding

$$K_{s-\frac{1}{2}}(w) = \frac{\Gamma(s)}{2\sqrt{\pi} (\frac{w}{2})^{s-\frac{1}{2}}} e^{-aw} \int_{-\infty}^\infty e^{iwt} (t^2 + 2iat + 1 - a^2)^{-s} \, dt.$$

The remaining integral doesn't converge absolutely unless $\operatorname{Re} s > \frac{1}{2}$. But a simple integration by parts takes care of this and gives the estimate

$$|K_{s-\frac{1}{2}}(w)| \leq C_{\varepsilon, a} \frac{|\Gamma(s)|}{(\frac{w}{2})^{\operatorname{Re} s - \frac{1}{2}}} e^{-aw} \left| \frac{s}{w} \right|, \tag{9.28}$$

for $\operatorname{Re} s > \varepsilon$. Combining the two Bessel function bounds, we have

$$\left| \sqrt{yy'} I_{s-\frac{1}{2}}(2\pi |k|y) K_{s-\frac{1}{2}}(2\pi |k|y') \right| \le C_{\varepsilon,a} e^{2\pi |k|(y-ay')} \left| \frac{s}{ky'} \right|. \tag{9.29}$$

Since the compact supports of ψ_0, ψ_1 are disjoint, we can choose $a < 1$ so that $y - ay' \le 0$ within the support of $\psi_0(y)\psi_1(y')$. The result then follows from (9.29) and (9.26). □

Before dealing with the issue of extending estimates to the half-plane $\operatorname{Re} s < \frac{1}{2}$, let us demonstrate the application of the results already proven.

Lemma 9.12. *For $T(s)$ given by (9.10), we have for $\operatorname{Re} s > \frac{1}{2} + \varepsilon$ the estimate*

$$\det(1 + 2|T(s)|) \le e^{C\langle s \rangle^2}.$$

For $\varepsilon < \operatorname{Re} s < 1$, we can bound

$$\det(1 + 2|T(s)|) \le e^{C\langle s \rangle^4}.$$

Proof. By Corollary 9.9, for $\operatorname{Re} s > \frac{1}{2} + \varepsilon$ we have

$$\|\Delta_X^m T(s)\| \le C_m \langle s \rangle^{2m}.$$

Then (9.16) implies the bounds

$$\mu_k(T(s)) \le C_m k^{-m} \langle s \rangle^{2m}, \tag{9.30}$$

for any $m \in \mathbb{N}_0$.

Combining the estimates (9.30) for $m = 0$ and $m = 2$ gives

$$\mu_k(T(s)) \le C \min\{k^{-2}\langle s \rangle^4, 1\}.$$

This leads to the following bound on the trace: for $\operatorname{Re} s > \varepsilon$,

$$\operatorname{tr}|T(s)| = \sum_{k=1}^{\infty} \mu_k(T(s))$$

$$\le C\left(\sum_{k \le \langle s \rangle^2} 1 + \sum_{k > \langle s \rangle^2} \frac{\langle s \rangle^4}{k^2} \right)$$

$$\le C\langle s \rangle^2.$$

The first claim then follows immediately from the Weyl inequality (9.6).
For $\varepsilon < \operatorname{Re} s < 1$, Lemmas 9.10 and 9.11 imply

$$\mu_k(T(s)) \le C \min\{k^{-2}\langle s \rangle^6, \langle s \rangle^2\}.$$

This proves the second claim by the same argument. □

Of course, the ultimate goal is to show that the bound given for $\operatorname{Re} s > \frac{1}{2} + \varepsilon$ in Lemma 9.12 is valid for all s. Our preliminary bound for the strip $\varepsilon < \operatorname{Re} s < 1$ could perhaps be improved, since we didn't fully exploit the exponential decay of the Fourier modes as $|k| \to \infty$. But there is no need to do so. The easiest way to improve the bound in the vertical strip is via the Phragmén–Lindelöf theorem. Any preliminary bound of the form $e^{C\langle s \rangle^m}$ would suffice for this purpose.

But before we get to that, we must produce estimates on $\det(1 + 2|T(s)|)$ in the left half-plane $\operatorname{Re} s < \frac{1}{2} - \varepsilon$. The key for this is the identity

$$R_{F_\ell}(s) - R_{F_\ell}(1-s) = (2s-1)E_{F_\ell}(1-s)S_{F_\ell}(s)E_{F_\ell}(1-s)^{\mathrm{t}}. \qquad (9.31)$$

This holds because the proofs of Proposition 7.9 and Corollary 7.14 are valid for a single funnel as well. Note that the poles occur only in $S_{F_\ell}(s)$, whose eigenvalues are known explicitly.

Before getting into these complications, let's deal with the cusp case, which is comparatively simple.

Lemma 9.13. *There exist constants such that for all $|s - \frac{1}{2}| > \varepsilon$,*

$$\mu_k \left(\left[\Delta_{C_\infty}, \chi_0 \right] (R_{C_\infty}(s) - R_{C_\infty}(1-s)) (\chi_3 - \chi_1) \right)$$
$$\leq \exp \left[C\langle s \rangle + \langle s \rangle \log \frac{\langle s \rangle}{k} - ck \right].$$

Proof. Although the proof of Corollary 7.14 does not apply to C_∞, we can easily deduce the identify corresponding to (9.31) directly from Proposition 5.8. The result is a formula,

$$R_{C_\infty}(s) - R_{C_\infty}(1-s) = A_0(s) + \frac{2\cos(\pi s)}{\pi} F(s)F(s)^{\mathrm{t}},$$

where $A_0(s)$ is the zero-mode contribution

$$A_0(s; z, z') := \frac{1}{2s-1} \left(y^s y'^{1-s} + y^{1-s} y'^s \right),$$

and $F(s) = F(1-s)$ is given by a sum over the nonzero modes,

$$F(s; z, t') := \sum_{k \neq 0} \sqrt{y} K_{s-\frac{1}{2}} (2\pi |k| y) e^{2\pi i k (x - t')}.$$

For convenience, set

$$P := \left[\Delta_{C_\infty}, \chi_0 \right], \qquad \psi_1(y) := \chi_3(\log y) - \chi_1(\log y),$$

so that

$$\left[\Delta_{C_\infty}, \chi_0 \right] (R_{C_\infty}(s) - R_{C_\infty}(1-s))(\chi_3 - \chi_1)$$
$$= P A_0(s)\psi_1 + \frac{2\cos(\pi s)}{\pi} P F(s)F(s)^{\mathrm{t}}\psi_1, \qquad (9.32)$$

The kernel of $P A_0(s)\psi_1$ is easily estimated to give

$$\|P A_0(s)\psi_1\| \leq \frac{e^{C\langle s\rangle}}{|2s - 1|}.$$

Note, however, that $A_0(s)$ is an operator of rank 2, so for $|s - \frac{1}{2}| > \varepsilon$ this norm bound implies

$$\mu_j(P A_0(s)\psi_1) \leq \begin{cases} e^{C_\varepsilon\langle s\rangle}, & j = 1, 2, \\ 0, & j > 2. \end{cases} \tag{9.33}$$

To handle $F(s)^t\psi_1$, note that $(\psi_1 F(s))^*(\psi_1 F(s))$ is a diagonal operator with eigenvalues

$$\sigma_k := \int_0^\infty \psi_1(y)^2 y \left| K_{s-\frac{1}{2}}(2\pi|k|y)\right|^2 \frac{dy}{y^2},$$

for $k \in \pm\mathbb{N}$. By (9.28), and using Stirling's approximation (7.41), these are bounded for $\mathrm{Re}\, s > \varepsilon$ by

$$\sigma_k \leq \exp\left[C\langle s\rangle + 2\,\mathrm{Re}\, s \log\frac{\mathrm{Re}\, s}{k} - ck\right].$$

Using $\mu_j(A) = \sqrt{\mu_j(A^* A)}$, we deduce from the bound on σ_k the estimate

$$\mu_k(F(s)^t\psi_1) \leq \exp\left[C\langle s\rangle + \mathrm{Re}\, s \log\frac{\mathrm{Re}\, s}{k} - ck\right], \tag{9.34}$$

for $\mathrm{Re}\, s > \varepsilon$. By a similar argument, the same type of bound applies to $\mu_j(P F(s))$, and then (9.9) and the symmetry $F(1 - s) = F(s)$ imply

$$\mu_j(P F(s) F(s)^t\psi_1) \leq \exp\left[C\langle s\rangle + \langle s\rangle \log\frac{\langle s\rangle}{k} - ck\right]. \tag{9.35}$$

Finally, we use (9.8) and the obvious bound $|\cos \pi s| \leq \exp\{\pi|\mathrm{Im}(s)|\}$ to apply (9.33) and (9.35) to (9.32), yielding the result.   □

The final step is to use (9.31) to prove the estimate corresponding to Lemma 9.13 for F_ℓ. As noted above, the poles that complicate this estimate occur only in $S_{F_\ell}(s)$, and our first job is to deal with estimating around the poles. The key observation is that if s approaches $\zeta \in \mathcal{R}_{F_\ell}$, only the first two singular values blow up (since the multiplicity of the pole in the scattering matrix is 2).

Lemma 9.14. *For* $\mathrm{Re}\, s < \frac{1}{2} - \varepsilon$,

$$\mu_j(S_{F_\ell}(s)) \leq \begin{cases} d(s, \mathcal{R}_{F_\ell})^{-2} e^{C\langle s\rangle \log\langle s\rangle}, & j = 1, 2, \\ \exp\left[C\langle s\rangle + 2\langle s\rangle \log\frac{\langle s\rangle}{j}\right], & j > 2. \end{cases} \tag{9.36}$$

Proof. We will follow Guillopé–Zworski's original argument [86, Lemma 4.2] closely here. With $\omega = 2\pi/\ell$, the eigenvalues of $S_{F_\ell}(s)$ given in (7.40) are

$$\gamma_k(s) = \frac{\Gamma(\frac{1}{2} - s)\Gamma(\frac{1+s+ik\omega}{2})\Gamma(\frac{1+s-ik\omega}{2})}{\Gamma(s - \frac{1}{2})\Gamma(\frac{2-s+ik\omega}{2})\Gamma(\frac{2-s-ik\omega}{2})},$$

for $k \in \mathbb{Z}$. The singular values of $S_{F_\ell}(s)$ are equal to $|\gamma_k(s)|$.

Using gamma function identities, we can write

$$\frac{\Gamma(\frac{1}{2} - s)}{\Gamma(s - \frac{1}{2})} = \frac{1}{\pi}(s - \frac{1}{2})\cos(\pi s)\Gamma(\frac{1}{2} - s)^2$$

and

$$\frac{\Gamma(\frac{1+s+ik\omega}{2})}{\Gamma(\frac{2-s-ik\omega}{2})} = \frac{\sqrt{\pi}\,2^{-s-ik\omega}}{\cos\frac{\pi}{2}(s + ik\omega)\Gamma(1 - s - ik\omega)}.$$

These transform the eigenvalue expression to

$$\gamma_k(s) = \frac{4^{-s}(s - \frac{1}{2})\cos(\pi s)\Gamma(\frac{1}{2} - s)^2}{\cos\frac{\pi}{2}(s + ik\omega)\cos\frac{\pi}{2}(s - ik\omega)\Gamma(1 - s - ik\omega)\Gamma(1 - s + ik\omega)}. \quad (9.37)$$

The numerator is now holomorphic for $\mathrm{Re}\,s < \frac{1}{2} - \varepsilon$ and easily estimated using Stirling's formula (7.41),

$$\left|4^{-s}(s - \tfrac{1}{2})\cos(\pi s)\Gamma(\tfrac{1}{2} - s)^2\right| \leq e^{C\langle s\rangle + (1 - 2\,\mathrm{Re}\,s)\log\langle s\rangle}. \quad (9.38)$$

To estimate the denominator of (9.37), we consider first the case $|\mathrm{Im}\,s \pm k\omega| > \omega/2$, for which the denominator has no zeros. Writing the cosine as a sum of exponentials leads to a simple bound,

$$|\cos\tfrac{\pi}{2}(s \pm ik\omega)|^{-1} \leq Ce^{-\frac{\pi}{2}|\mathrm{Im}\,s\pm k\omega|}. \quad (9.39)$$

For the gamma function we use Stirling's formula (7.41), and the fact that $|\arg(1 - (s \pm ik\omega))| < \frac{\pi}{2}$ for $\mathrm{Re}\,s < \frac{1}{2} - \varepsilon$, to estimate

$$|\Gamma(1 - (s \pm ik\omega))|^{-1} \leq Ce^{-\mathrm{Re}\,s}e^{(\mathrm{Re}\,s - \frac{1}{2})\log|1 - (s\pm ik\omega)|}e^{\frac{\pi}{2}|\mathrm{Im}\,s\pm k\omega|}.$$

This leads to the combined estimate

$$\left|\cos\tfrac{\pi}{2}(s \pm ik\omega)\Gamma(1 - (s \pm ik\omega))\right|^{-1} \leq Ce^{-\mathrm{Re}\,s}e^{(\mathrm{Re}\,s - \frac{1}{2})\log|1 - (s\pm ik\omega)|}. \quad (9.40)$$

Since $\mathrm{Re}\,s - \frac{1}{2} < -\varepsilon$ by assumption, we can use $|1 - (s \pm ik\omega)| \geq |\mathrm{Im}\,s \pm k\omega|$ to simplify the exponent. Then, using (9.39) and (9.40), we find that

$$|\gamma_k(s)| \leq e^{C\langle s\rangle}\exp\left[(\tfrac{1}{2} - \mathrm{Re}\,s)\log\frac{\langle s\rangle^2}{|(\mathrm{Im}\,s)^2 - k^2\omega^2|}\right], \quad (9.41)$$

for $\operatorname{Re} s < \frac{1}{2} - \varepsilon$ and $|\operatorname{Im} s \pm k\omega| > \omega/2$.

Now consider the case $|\operatorname{Im} s \pm k\omega| \leq \omega/2$. This restricts s to a horizontal strip containing the poles of $\cos\frac{\pi}{2}(s \pm ik\omega)$, so the estimate is

$$|\cos\tfrac{\pi}{2}(s \pm ik\omega)|^{-1} \leq Cd(s, \mathcal{R}_{F_\ell})^{-1}.$$

Thus (9.40) is replaced by the bound

$$\left|\cos\tfrac{\pi}{2}(s \pm ik\omega)\Gamma(1-(s \pm ik\omega))\right|^{-1} \leq Cd(s, \mathcal{R}_{F_\ell})^{-1}e^{(\operatorname{Re} s - \frac{1}{2})\log|1 - \operatorname{Re} s|}. \quad (9.42)$$

Then from (9.39), (9.40), and (9.42), we obtain

$$|\gamma_k(s)| \leq d(s, \mathcal{R}_{F_\ell})^{-1}e^{C\langle s\rangle \log\langle s\rangle}, \quad (9.43)$$

if $k \neq 0$ and $|\operatorname{Im} s \pm k\omega| \leq \omega/2$. On the other hand, if $k = 0$ and $|\operatorname{Im} s| \leq \omega/2$, then the zeros of the two cosine terms coincide. From (9.39) and (9.42) we have

$$|\gamma_0(s)| \leq d(s, \mathcal{R}_{F_\ell})^{-2}e^{C\langle s\rangle}, \quad (9.44)$$

for $|\operatorname{Im} s| \leq \omega/2$.

To finish the argument, for a given s the singular values of the scattering matrix are given by arranging the $|\gamma_k(s)|$ into a decreasing sequence. The condition $|\operatorname{Im} s \pm ik\omega| \leq \omega/2$ is satisfied for at most two values of k. The claimed bound (9.36) on the first two singular values thus follows from taking the maximum among the bounds in the cases (9.41), (9.43), and (9.44).

For $j > 2$, we can bound $\mu_j(S_{F_\ell}(s))$ by reordering the right-hand side of (9.41) as a decreasing sequence. This amounts to rearranging the sequence $|(\operatorname{Im} s)^2 - k^2\omega^2|$, for $k \in \mathbb{Z}$ such that $|\operatorname{Im} s \pm ik\omega| > \omega/2$, in increasing order. The terms in this increasing rearrangement can be bounded below by ck^2, where c is independent of s, giving the estimate

$$\mu_j(S_{F_\ell}(s)) \leq e^{C\langle s\rangle}\exp\left[(\tfrac{1}{2} - \operatorname{Re} s)\log\frac{\langle s\rangle^2}{cj^2}\right],$$

for $j > 2$. This gives the second bound in (9.36). $\qquad\square$

To exploit (9.31), we also need an explicit estimate on the Poisson kernel.

Lemma 9.15. *If* $\psi \in C_0^\infty(F_\ell)$ *then for* $\operatorname{Re} s > \varepsilon$,

$$\mu_j(\psi E_{F_\ell}(s)) \leq e^{C\langle s\rangle - cj}.$$

Proof. Using the coordinates $(z, x') \in \mathcal{F}_\ell \times [1, e^\ell]$, we found an explicit expression for the Poisson kernel in (7.26) and (7.27),

$$E_{F_\ell}(s; z, x') := \frac{1}{4\pi}\frac{\Gamma(s)^2}{\Gamma(2s)}(4x'y)^s\left(\sum_{k\in\mathbb{Z}}\left[e^{-k\ell}(x - e^{k\ell}x')^2 + e^{-k\ell}y^2\right]^{-s}\right.$$

$$\left. + \sum_{k\in\mathbb{Z}}\left[e^{-k\ell}(x + e^{k\ell/2}x')^2 + e^{-k\ell}y^2\right]^{-s}\right). \quad (9.45)$$

Stirling's formula (7.41) gives

$$\left| \frac{\Gamma(s)^2}{\Gamma(2s)} \right| \le e^{C\langle s \rangle},$$

so we concentrate on the two sums over k. Using the substitution $x' = e^\eta$, let's define

$$h_s(\eta) = e^{s\eta} \sum_{k \in \mathbb{Z}} \left[e^{-k\ell}(x \pm e^{k\ell+\eta})^2 + e^{-k\ell} y^2 \right]^{-s}, \tag{9.46}$$

which is periodic under $\eta \mapsto \eta + \ell$. With (x, y) restricted to the support of ψ, and assuming $\operatorname{Re} s > \varepsilon$, the argument used in Lemma 5.1 shows that $h_s(\eta)$ extends to an analytic function of $\eta \in \mathbb{C}$. We can apply Cauchy's integral formula on the boundary of a rectangle $K = \{|\operatorname{Re} \eta| \le \ell, |\operatorname{Im} \eta| < \delta\}$ to estimate derivatives:

$$\left| \frac{d^p}{d\eta^p} h_s(\eta) \right| \le p! \, \delta^{-p} \sup_{\eta \in K} |h_s(\eta)|.$$

From the explicit formula (9.46), we can produce a bound,

$$\sup_K |h_s(\eta)| \le e^{C\delta\langle s \rangle}.$$

These estimates on derivatives give us bounds of the form

$$|\partial_{t'}^p \psi(z) E_{F_\ell}(s; z, t')| \le C^p p! \, e^{C\langle s \rangle}, \tag{9.47}$$

for $\operatorname{Re} s > \varepsilon$ and $p \ge 1$, where C depends on ψ and ε but not on p.

With $\Delta_{\partial_0 \overline{F}_\ell}$ denoting the Laplacian on $\mathbb{R}/(\ell/\mathbb{Z})$, we have

$$\mu_j(\psi E_{F_\ell}(s)) \le \mu_j\left((\Delta_{\partial_0 \overline{F}_\ell} + 1)^{-m} \right) \left\| (\Delta_{\partial_0 \overline{F}_\ell} + 1)^m E_{F_\ell}(s)^{\mathrm{t}} \psi \right\|.$$

Since $\Delta_{\partial_0 \overline{F}_\ell}$ has eigenvalues $(2\pi k/\ell)^2$ for $k \in \mathbb{Z}$,

$$\mu_j\left((\Delta_{\partial_0 \overline{F}_\ell} + 1)^{-m} \right) \le C^m j^{-2m},$$

and by (9.47),

$$\left\| (\Delta_{\partial_0 \overline{F}_\ell} + 1)^m E_{F_\ell}(s)^{\mathrm{t}} \psi \right\| \le C^m (2m)! \, e^{C\langle s \rangle}.$$

Thus for $m \ge 1$, we have

$$\mu_j(\psi E_{F_\ell}(s)) \le C^{2m} j^{-2m} (2m)! \, e^{C\langle s \rangle},$$

with C independent of m. The final step is optimization in m. If we set $m = \frac{1}{2}\left[C^{-1} j - 1 \right]$, then the result follows from Stirling's formula (7.41). $\square$

With these explicit estimates of the funnel scattering matrix and Poisson kernel, we are now ready to prove the funnel analogue of Lemma 9.13.

Lemma 9.16. *There exist constants such that for all* $\operatorname{Re} s < \frac{1}{2} - \varepsilon$,

$$\mu_k \left(\left[\Delta_{F_\ell}, \chi_0 \right] (R_{F_\ell}(s) - R_{F_\ell}(1-s))(\chi_3 - \chi_1) \right)$$
$$\leq \begin{cases} d(s, \mathcal{R}_{F_\ell})^{-2} e^{C\langle s \rangle \log \langle s \rangle}, & k = 1, 2, \\ \exp\left[C\langle s \rangle + \langle s \rangle \log \frac{\langle s \rangle}{k} - ck \right], & k > 2. \end{cases}$$

Proof. With $P = \left[\Delta_{F_\ell}, \chi_0 \right]$ and $\psi = \chi_3 - \chi_1$, we start by using (9.31) to write

$$\left[\Delta_{F_\ell}, \chi_0 \right] (R_{F_\ell}(s) - R_{F_\ell}(1-s))(\chi_3 - \chi_1)$$
$$= (2s - 1) P E_{F_\ell}(1-s) S_{F_\ell}(s) E_{F_\ell}(1-s)^{\mathrm{t}} \psi.$$

Then by (9.9),

$$\mu_k \left(\left[\Delta_{F_\ell}, \chi_0 \right] (R_{F_\ell}(s) - R_{F_\ell}(1-s))(\chi_3 - \chi_1) \right)$$
$$\leq |2s - 1| \, \| P E_{F_\ell}(1-s) \| \, \mu_{j_1}(S_{F_\ell}(s)) \, \mu_{j_2}(\psi E_{F_\ell}(1-s)),$$

where $j_1 + j_2 = k - 1$. We easily see that

$$\| P E_{F_\ell}(1-s) \| \leq e^{C\langle s \rangle},$$

using (9.45). For $k < \langle s \rangle$, we set $j_1 = k$ and $j_2 = 1$, and the estimate follows from Lemma 9.14. For $k > \langle s \rangle$ we set $j_1 = [\langle s \rangle + 1]$ and $j_2 = [k - \langle s \rangle - 1]$ and apply Lemma 9.15.   □

At last we can extend the result of Lemma 9.12 to the left half-plane $\operatorname{Re} s < \frac{1}{2} - \varepsilon$.

Lemma 9.17. *With* $T(s)$ *given by (9.10), for sufficiently large* N *and* $\operatorname{Re} s < \frac{1}{2} - \varepsilon$, *we have*

$$\left| g(s)^N \det(1 + 2|T(s)|) \right| \leq e^{C\langle s \rangle^2}.$$

Proof. According to Lemma 9.4,

$$\det(1 + 2|T(s)|) \leq \det(1 + 2|T(1-s)|)^2 \det(1 + 2|T(s) - T(1-s)|)^2.$$

For $\operatorname{Re} s < \frac{1}{2} - \varepsilon$, Lemma 9.12 takes care of the first term. To handle the second term, Lemma 9.4 shows that we can consider separately the determinants corresponding to the terms in the sum,

$$T(s) - T(1-s) = \sum_{i=1}^{n_c} \left[\Delta_{C_i}, \chi_0 \right] (R_{C_i}(s) - R_{C_i}(1-s))(\chi_3 - \chi_1)$$
$$+ \sum_{j=1}^{n_f} \left[\Delta_{F_j}, \chi_0 \right] (R_{F_j}(s) - R_{F_j}(1-s))(\chi_3 - \chi_1).$$

Set

$$H_{C_i}(s) := \det\left(1 + \left\| \left[\Delta_{C_i}, \chi_0 \right] (R_{C_i}(s) - R_{C_i}(1-s))(\chi_3 - \chi_1) \right\| \right).$$

By Lemma 9.13 there is some threshold, of the form $a\langle s\rangle$ with $a > 0$ independent of s, above which the singular values of the operator appearing in $H_{C_i}(s)$ lie below 1 and decay exponentially. Splitting the product at this point yields

$$\left|H_{C_i}(s)\right| \le \prod_{k < a\langle s\rangle} \exp\left[C\langle s\rangle + \langle s\rangle \log \frac{\langle s\rangle}{k}\right] \prod_{l=0}^{\infty}(1 + e^{-cl})$$

$$\le C \exp\left[Ca\langle s\rangle^2 + a\langle s\rangle^2 \log\langle s\rangle - \langle s\rangle \log \Gamma(a\langle s\rangle)\right].$$

By Stirling's formula (7.41), the $\langle s\rangle^2 \log\langle s\rangle$ terms cancel, leaving

$$\left|H_{C_i}(s)\right| \le e^{C\langle s\rangle^2},$$

for $\mathrm{Re}\, s < \frac{1}{2} - \varepsilon$. Analysis of the funnel term,

$$H_{F_j}(s) := \det\left(1 + \left|\left[\Delta_{F_j}, \chi_0\right](R_{F_j}(s) - R_{F_j}(1-s))(\chi_3 - \chi_1)\right|\right),$$

using Lemma 9.16 is much the same. The only change is that we must split off the first two singular values separately, resulting in the estimate

$$\left|H_{F_j}(s)\right| \le d(s, \mathcal{R}_{F_{\ell_j}})^{-4} e^{C\langle s\rangle^2},$$

for $s \notin \mathcal{F}_{\ell_j}$ and $\mathrm{Re}\, s < \frac{1}{2} - \varepsilon$. The result follows, because

$$\left|d(s, \mathcal{R}_F)^{-1} g(s)\right| \le e^{C\langle s\rangle^2}. \qquad \square$$

Finally, we can fill in the missing step from the proof of Theorem 9.1.

Proof of Lemma 9.7. For sufficiently large N, $g(s)^N \det(1 + 2T(s))$ is an entire function. Lemmas 9.12 and 9.17, together with (9.14), give estimates

$$\left|g(s)^N \det(1 + 2T(s))\right| \le e^{C\langle s\rangle^2},$$

for $|\mathrm{Re}\, s - \frac{1}{2}| > \varepsilon$, and

$$\left|g(s)^N \det(1 + 2T(s))\right| \le e^{C\langle s\rangle^4},$$

in the strip $\varepsilon < \mathrm{Re}\, s < 1$. The Phragmén–Lindelöf theorem (Theorem A.5) then implies that the stronger bound holds for all $s \in \mathbb{C}$. $\qquad \square$

Notes

The Fredholm determinant method for resonance counting was initiated by Melrose [132, 133] in the context of odd-dimensional potential scattering. The method was subsequently extended and refined in work of Intissar [99], Zworski [223, 222], Sjöstrand–Zworski [197], and Vodev [211, 212, 213, 214].

For an n-dimensional hyperbolic manifold, the optimal upper bound on the resonance counting function would be $O(r^n)$. Perry [164] used a version of the Selberg trace formula to prove an $O(r^{n+\varepsilon})$ bound on the number of scattering poles. Guillopé–Zworski [85] proved an $O(r^{n+1})$ bound for metrics required to be hyperbolic only near infinity. The $O(r^n)$ bound was finally extended to this case in work of Froese–Hislop [67] and Cuevas–Vodev [47].

The distribution of resonances or scattering poles has also been studied in various regions near the continuous spectrum. We will present such a result in Section 15.4 and give further references there.

10

Selberg Zeta Function

The heat trace is perhaps the most commonly used spectral invariant in geometric spectral theory for compact manifolds, and we observed its application to compact hyperbolic surfaces in Section 3.2. However, the heat trace is not very helpful in the theory of resonances for an infinite-area surface. Of course, the heat kernel is still well-defined in this case, and to produce a regularized version of the trace is not so difficult (by methods that we'll discuss in Section 10.2). The problem is that we have no realization of the regularized heat trace as a sum over the resonances—it is not necessarily a spectral invariant. (Such a realization is possible in the finite-area case because the resonances are confined to a vertical strip, see, e.g., Müller [139].)

Fortunately, other invariants are available to fill the role played by the heat trace. The first we'll consider is the Selberg zeta function $Z_X(s)$ introduced in Section 2.6. By definition, the zeta function is associated to the length spectrum of X (or to traces of conjugacy classes of Γ). But we'll show in this chapter that it deserves to be thought of as a spectral invariant as well, by virtue of a beautiful correspondence between resonances and the zeros of $Z_X(s)$.

The other spectral invariant we'll make use of is a regularization of the wave trace. Like the heat trace, the wave trace is a more general object with a long history in spectral theory. We analyze the Selberg zeta function first for reasons of technical simplicity; we'll use the zeta function to develop the regularized wave trace in Chapter 11.

Recall the definition, for $\operatorname{Re} s > \delta$,

$$Z_X(s) := \prod_{\ell \in \mathcal{L}_X} \prod_{k=0}^{\infty} \left(1 - e^{-(s+k)\ell}\right), \tag{10.1}$$

where $\mathcal{L}_X$ is the primitive length spectrum (repeated according to multiplicity). In this chapter we will show that $Z_X(s)$ extends meromorphically to $s \in \mathbb{C}$, and establish a Hadamard-type factorization formula.

The central feature of the factorization will be a product over the resonance set $\mathcal{R}_X$ (repeated according to multiplicity). In Theorem 9.1 we saw that the resonance counting function $N_X(r)$ is bounded $O(r^2)$. By the Weierstrass factorization theory, this implies that the product

$$P_X(s) := s^{m_0} \prod_{\substack{\zeta \in \mathcal{R}_X \\ \zeta \neq 0}} \left(1 - \frac{s}{\zeta}\right) e^{s/\zeta + s^2/2\zeta^2} \tag{10.2}$$

converges uniformly on compact sets and defines an entire function with zero set given by $\mathcal{R}_X$.

The factorization also includes a "topological" term expressed in terms of the function

$$G_\infty(s) = (2\pi)^{-s} \Gamma(s) G(s)^2,$$

introduced in (3.8), where $G(s)$ is the Barnes G-function. This is an entire function with zeros at $s = -n$ of multiplicity $2n + 1$, for $n \in \mathbb{N}_0$. (The zeros of $G_\infty(s)$ correspond, with multiplicities, to the resonances of $\mathbb{H}$.)

Theorem 10.1 (Borthwick–Judge–Perry). *Suppose X is a geometrically finite hyperbolic surface of infinite area. The Selberg zeta function $Z_X(s)$ extends to a meromorphic function of $s \in \mathbb{C}$ and admits the factorization*

$$Z_X(s) = e^{q(s)} G_\infty(s)^{-\chi(X)} \Gamma(s - \tfrac{1}{2})^{n_c} P_X(s), \tag{10.3}$$

where $q(s)$ is a polynomial of degree ≤ 2 and $P_X(s)$ is the Hadamard product over resonances (10.2).

The structure of the divisor of $Z_X(s)$ implied by (10.3) is illustrated in Figure 10.1. Note that the resonance zeros and the topological singularities could overlap. If $n_c = 0$ then $Z_X(s)$ is entire.

For the context of Theorem 10.1, meromorphic continuation of the zeta function was proven by Guillopé [83]. For geometrically finite hyperbolic surfaces without cusps, Theorem 10.1 is a special case of the results of Patterson–Perry [160] (which

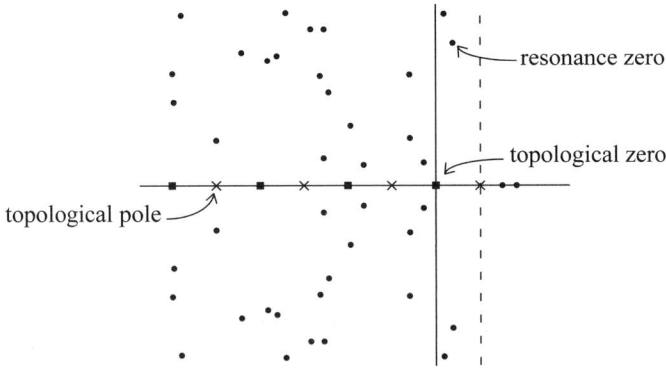

Fig. 10.1. Divisor of the zeta function.

also cover higher dimension). The full two-dimensional version stated here comes from Borthwick–Judge–Perry [23].

This chapter is primarily devoted to the proof of Theorem 10.1, but the scattering determinant and resolvent traces introduced along the way will also play important roles later on. The characterization of the divisor of the zeta function has obvious implications for the connection between the resonance set and the length spectrum, which we'll take up in Chapter 12.

10.1 Relative scattering determinant

The standard Fredholm theory for determinants requires an operator that differs from the identity by a trace class operator. The scattering matrix $S_X(s)$ is not of this form (in the funnel ends), and one way around this is to define the scattering matrix relative to some background operators. The natural choices in our context are the model scattering matrices $S_{F_j}(s)$ for the funnels with Dirichlet boundary conditions, first encountered in Section 7.4.

The funnel scattering matrix $S_F(s)$ was defined in (7.35) as the direct sum of the model terms $S_{F_j}(s)$, acting on $C^\infty(\partial_0 \overline{F})$. No regularization is required in the cusps, so we define a background operator acting on $C^\infty(\partial \overline{X})$ by

$$S_0(s) := \begin{pmatrix} S_F(s) & 0 \\ 0 & I \end{pmatrix},$$

where I here denotes the identity map on $\mathbb{C}^{n_c}$. The *relative scattering matrix* is defined by

$$S_{\mathrm{rel}}(s) := S_0(s)^{-1} S_X(s).$$

By Proposition 7.15, $S_{\mathrm{rel}}(s) - I$ is a smoothing operator, so that the Fredholm determinant of $S_{\mathrm{rel}}(s)$ is well-defined.

Definition. The *relative scattering determinant* is the meromorphic function

$$\tau_X(s) := \det S_{\mathrm{rel}}(s). \tag{10.4}$$

Our first task is to work out the divisor of the relative scattering determinant. This will ultimately lead to the crucial link between resonances and regularized traces of the resolvent. We have already introduced the collective resonance set for the funnels in Section 9.1,

$$\mathcal{R}_F := \bigcup_{j=1}^{n_f} \left(-\mathbb{N}_0 + \frac{2\pi i}{\ell_j} \mathbb{Z} \right)$$

(all of multiplicity two), with corresponding Weierstrass product

$$P_F(s) := \prod_{\zeta \in \mathcal{R}_F} \left(1 - \frac{s}{\zeta} \right) e^{s/\zeta + s^2/2\zeta^2}.$$

The following result comes from Guillopé–Zworski [87, Proposition 3.4].

Lemma 10.2. *The function τ_X extends to a meromorphic function in the complex plane having the form*

$$\tau_X(s) = e^{h(s)} \frac{P_X(1-s)}{P_X(s)} \frac{P_F(s)}{P_F(1-s)}, \tag{10.5}$$

for some entire function $h(s)$.

Proof. Since a nonzero entire function can always be written in the form $e^{h(s)}$, it suffices to show that the divisors of the two sides of (10.5) are equal. There are no resonances on $\operatorname{Re} s = \frac{1}{2}$, $s \neq \frac{1}{2}$ by Corollary 7.8. The functional relation $S_X(s)^{-1} = S_X(1-s)$ from Corollary 7.14, which holds for F as well, implies that

$$\tau_X(s)\tau_X(1-s) = 1, \tag{10.6}$$

with $|\tau_X(\frac{1}{2})| = 1$ in particular. On the right-hand side of (10.5), any singularities in P_X or P_F at $s = \frac{1}{2}$ will cancel. Thus both sides of (10.5) are analytic in some neighborhood of the critical line $\operatorname{Re} s = \frac{1}{2}$.

Since both sides of (10.5) have the same symmetry under $s \to 1 - s$, we need only compute the divisor of τ in the half-plane $\operatorname{Re} s < \frac{1}{2}$. Suppose ζ is a zero or pole of τ_X with $\operatorname{Re} \zeta < \frac{1}{2}$, and let γ_ζ be a circle around ζ enclosing no other pole or zero of either S_X or S_F. The order of ζ as a zero of τ_X (or pole if the order is negative) is computed by

$$\operatorname*{ord}_{s=\zeta} \tau_X(s) = \frac{1}{2\pi i} \int_{\gamma_\zeta} \frac{\tau_X'}{\tau_X}(s)\, ds,$$

where γ_ζ is a positively oriented circle containing no other zero or pole. The logarithmic derivative of a Fredholm determinant behaves just as in the finite-dimensional case (see, e.g., [72, Section IV.1]), so that

$$\frac{\tau_X'}{\tau_X}(s) = \frac{d}{ds} \log \det S_{\mathrm{rel}}(s) = \operatorname{tr}\!\left[S_{\mathrm{rel}}(s)^{-1} S_{\mathrm{rel}}'(s) \right],$$

$$\operatorname*{ord}_{s=\zeta} \tau_X(s) = \frac{1}{2\pi i} \int_{\gamma_\zeta} \operatorname{tr}\!\left[S_{\mathrm{rel}}(s)^{-1} S_{\mathrm{rel}}'(s) \right] ds$$

$$= M(S_{\mathrm{rel}}; \zeta),$$

where $M(S_{\mathrm{rel}}; \zeta)$ is the notation for the trace of the logarithmic residue introduced in (8.25).

Recall that the multiplicity ν_ζ of a scattering pole was defined by

$$\nu_\zeta := -M(\widetilde{S}_X; \zeta),$$

with $\widetilde{S}_X$ defined by (8.21). We could paraphrase the definition as

$$\widetilde{S}_X(s) := G(s)\Lambda(s)S_X(s)\Lambda(s)G(1-s)^{-1},$$

where

$$G(s) := \begin{pmatrix} \Gamma(s + \tfrac{1}{2}) & 0 \\ 0 & I \end{pmatrix},$$

and $\Lambda(s)$ is the holomorphically invertible family of operators defined in (8.20). By analogy, let us define the regularized reference operator

$$\widetilde{S}_0(s) := G(s)\Lambda(s)S_0(s)\Lambda(s)G(1-s)^{-1}.$$

With these definitions, the relative scattering matrix is

$$S_{\mathrm{rel}}(s) = G(1-s)^{-1}\Lambda(s)\widetilde{S}_0(s)^{-1}\widetilde{S}_X(s)\Lambda(s)^{-1}G(1-s). \tag{10.7}$$

Note that $G(1-s)$ is holomorphically invertible in $\operatorname{Re} s < \tfrac{1}{2}$. Applying Lemma 8.3 to (10.7) therefore yields

$$\operatorname*{ord}_{s=\zeta} \tau_X(s) = M(\widetilde{S}_X; \zeta) - M(\widetilde{S}_0; \zeta).$$

By Theorem 8.9, we have $\nu_\zeta = m_\zeta - m_{1-\zeta}$ for $\operatorname{Re} \zeta < \tfrac{1}{2}$, so

$$M(\widetilde{S}_X; \zeta) = m_{1-\zeta} - m_\zeta = \operatorname*{ord}_{s=\zeta} \frac{P_X(1-s)}{P_X(s)}.$$

The eigenvalues of $\widetilde{S}_0(s)$ can be written explicitly using (7.40), from which it is clear that $M(\widetilde{S}_0; \zeta) = -2$ for $\zeta \in \mathcal{R}_F$, and 0 for all other $\operatorname{Re} s < \tfrac{1}{2}$. Hence, for $\zeta < \tfrac{1}{2}$,

$$M(\widetilde{S}_0; \zeta) = \operatorname*{ord}_{s=\zeta} \frac{P_F(1-s)}{P_F(s)}.$$

This verifies that the two sides of (10.5) have the same divisor and completes the proof. $\qquad\square$

We will see in Section 10.5 that the function $h(s)$ appearing in Lemma 10.2 is polynomial. And in Section 12.3 we will study the asymptotics of τ_X on the critical strip $\operatorname{Re} s = \tfrac{1}{2}$.

10.2 Regularized traces

The next step in the proof of Theorem 10.1 involves the study of certain formal traces of the resolvent, defined by restricting a continuous kernel to the diagonal and then integrating. We cannot do this directly to the resolvent kernel, because it is singular on the diagonal. There are two options for removing the diagonal singularity that will be significant for us: subtracting the model resolvent $R_{\mathbb{H}}(s)$ from $R_X(s)$ and taking the difference $R_X(s) - R_X(1-s)$.

Even after these subtractions, the resulting formal traces will be divergent as $\rho \to 0$. To regularize the integrals, we recall a technique from the basic theory of distributions known as the Hadamard finite part. Suppose that $A(\varepsilon)$ diverges as $\varepsilon \to 0$, but has an asymptotic expansion of the form

$$A(\varepsilon) = \sum_{k=1}^{n} a_k \varepsilon^{-k} + a_0' \log \varepsilon + a_0 + o(1).$$

Then the *finite part* is defined by

$$\operatorname*{FP}_{\varepsilon \to 0} A(\varepsilon) = a_0.$$

Definition. If $f \in C^\infty(X)$ is polyhomogeneous in ρ, meaning that f has an asymptotic expansion as $\rho \to 0$ in powers of ρ and $\log \rho$, then the 0-*integral* of f over X is defined by

$$\int_X^0 f \, dA = \operatorname*{FP}_{\varepsilon \to 0} \int_{\{\rho \geq \varepsilon\}} f \, dA.$$

The term 0-integral is used here (as opposed to finite part) as a reminder that this regularization is not intrinsically defined on X but rather depends on the compactification $\overline{X}$, and more specifically on the 2-jet of ρ on $\partial \overline{X}$. Our definition assumes that ρ is chosen according to prescription given in Section 6.1. (As noted before, dependence on ρ could be avoided entirely by introducing density bundles, at the cost of increased notational complexity.)

The 0-*volume* of a hyperbolic surface is naturally the 0-integral of 1.

Lemma 10.3. *For X a geometrically finite hyperbolic surface,*

$$0\text{-vol}(X) = -2\pi \chi(X).$$

Proof. Recall from Section 2.4 the Nielsen decomposition $X = N \cup F_1 \cup \cdots \cup F_{n_{\mathrm{f}}}$, where N is the (finite-area) convex core. By (6.6), we have

$$\begin{aligned}
0\text{-vol}(F_j) &= \operatorname*{FP}_{\varepsilon \to 0} \int_0^{\ell_j} \int_\varepsilon^2 (1 + \rho^2/4) \frac{d\rho \, dt}{\rho^2} \\
&= \operatorname*{FP}_{\varepsilon \to 0} (\varepsilon^{-1} - \varepsilon/4)\ell_j \\
&= 0.
\end{aligned}$$

Hence $0\text{-vol}(X) = \text{area}(N)$. The result then follows by Gauss–Bonnet (Theorem 2.15). $\square$

It is worth noting here that a simple rescaling of ρ, for example dropping the factor of 2 from the funnel defining function in the definition (6.5), would not affect the zero volume.

For a smoothing operator A defined on $X \times X$, with continuous kernel $A(x, y)$ with respect to Riemannian measure on X, the 0-*trace* of A is defined by

$$0\text{-tr } A := \int_X^0 A(z, z) \, dA(z),$$

under the restriction that $A(z, z)$ is polyhomogeneous as $\rho \to 0$.

The 0-trace was introduced by Guillopé–Zworski [87]. The terminology refers to the "0-calculus" associated by Mazzeo–Melrose [128] to the Lie algebra of vector fields on $\overline{X}$ that vanish at the boundary.

Now we turn to the definition of regularized traces of $R_X(s)$. Using $X \cong \Gamma \backslash \mathbb{H}$, we can write the lift of the resolvent kernel to $\mathbb{H} \times \mathbb{H}$ as a sum

$$R_X(s; z, w) = \sum_{T \in \Gamma} R_{\mathbb{H}}(s; Tz, w),$$

valid for $\operatorname{Re} s > \delta$, where δ is the exponent of convergence of Γ. Since the $T = I$ term is the only term with a singularity on the diagonal, we obtain a smooth function by writing

$$\varphi_X(s; z) := \sum_{T \in \Gamma - \{I\}} R_{\mathbb{H}}(s; Tz, z), \qquad (10.8)$$

for $\operatorname{Re} s > \delta$. It's easy to check that $\varphi(s)$ is invariant under Γ; for $R \in \Gamma$, we have

$$\varphi_X(s; Rz) = \sum_{T \in \Gamma - \{I\}} R_{\mathbb{H}}(s; T Rz, Rz)$$

$$= \sum_{T \in \Gamma - \{I\}} R_{\mathbb{H}}(s; R^{-1} T Rz, z)$$

$$= \varphi_X(s; z).$$

Hence $\varphi_X(s)$ descends to a smooth function on X. The lift of $\varphi_X(s)$ to $\mathbb{H}$ could also be written

$$\varphi_X(s; z) = \Big[R_X(s; z, w) - R_{\mathbb{H}}(s; z, w) \Big]_{w=z},$$

from which it is clear that $\varphi_X(s)$ extends to a meromorphic function of $s \in \mathbb{C}$.

By Theorem 6.3, $\varphi_X(s)$ can be decomposed into model terms equal to φ_{F_ℓ} or φ_{C_∞} (with appropriate cutoffs) plus a function in $\rho_f^{2s} \rho_c^{2s-2} C^\infty(\overline{X})$. It is possible to deduce directly from Propositions 5.2 and 5.7 that

$$\varphi_{F_\ell}(s; z) \in \rho^{2s} C^\infty(\partial_0 \overline{F}_\ell) \qquad (10.9)$$

and

$$\varphi_{C_\infty}(s; z) \sim \sum_{k=0}^{\infty} a_k \rho^{-1+k}, \qquad (10.10)$$

as $\rho \to 0$. (We will see these facts confirmed explicitly below, in Propositions 10.6 and 10.7.) We conclude that

$$\varphi_X(s) \in \rho_f^{2s} \rho_c^{2s-2} C^\infty(\overline{X}) + \rho_c^{-1} C^\infty(\overline{X}). \qquad (10.11)$$

The integral over $\rho \geq \varepsilon$ therefore has an asymptotic expansion of the form

$$\int_{\{\rho \geq \varepsilon\}} \varphi_X(s) \, dA \sim a_0(s) + b_0(s) \log \varepsilon + \sum_{k=1}^{\infty} b_k(s) \varepsilon^k + \sum_{l=0}^{\infty} c_l(s) \varepsilon^{2s-1+l}, \quad (10.12)$$

where all of the coefficients are meromorphic in s. For $s \notin \mathcal{R}_X \cup (\frac{1}{2} - \mathbb{N}_0/2)$, we define

$$\Phi_X(s) := (2s - 1) \int_X^0 \varphi_X(s) \, dA = (2s - 1) a_0(s), \qquad (10.13)$$

where $a_0(s)$ is defined by (10.12). The restriction to $s \notin \frac{1}{2} - \mathbb{N}_0/2$ is necessary to avoid a contribution to the 0-integral from the coefficient c_l in (10.12) when $2s - 1 + l = 0$.

We'll conclude this section with explicit computations of $\Phi_X(s)$ for the model cases from Chapter 5. These will all be useful later on, and they serve to demonstrate the existence of the expansions (10.9) and (10.10). The basis of our computations is the following calculation from Patterson [159]:

Lemma 10.4. *If we write* $R_{\mathbb{H}}(s; z, w) = g_s(\sigma(z, w))$, *with* g_s *defined in (4.12), then for any* $\omega \in \mathbb{R}$,

$$(2s - 1) \int_{-\infty}^{\infty} g_s\left((1 + u^2)\cosh^2 \omega\right) du = \frac{e^{(1-2s)|\omega|}}{2\cosh \omega}.$$

Proof. Observe that

$$(\Delta_{\mathbb{H}} - s(1 - s))y^t = (t(1 - t) - s(1 - s))y^t.$$

Formally, this implies a simple formula for $R_{\mathbb{H}}(s)y^t$. To make this precise we introduce a cutoff function in the form $\psi(z/\lambda)$, where $\psi \in C_0^\infty(\mathbb{R}^2)$ is equal to 1 in the unit disk. With the cutoff we obtain

$$(\Delta_{\mathbb{H}} - s(1 - s))y^t \psi(y/\lambda) = (t(1 - t) - s(1 - s))y^t \psi(y/\lambda)$$
$$- \lambda^{-1}y^{t+1}\partial_y \psi(y/\lambda) - \lambda^{-2}y^{t+2}\Delta_{\mathbb{H}}\psi(y\lambda).$$

Since $y^t \psi(y/\lambda)$ and the error terms on the right-hand side are in L^2 for $\mathrm{Re}\, t > \frac{1}{2}$, we can apply $R_{\mathbb{H}}(s)$ to both sides for $\mathrm{Re}\, s > \frac{1}{2}$. The kernel $R_{\mathbb{H}}(s; z, z')$ is locally integrable, because, as we saw explicitly in Section 4.1, the singularity on the diagonal is only logarithmic. This makes it easy to remove the cutoff by taking the limit $\lambda \to \infty$, via the dominated convergence theorem. The error terms vanish under the assumption $\mathrm{Re}\, s > 1 + |\mathrm{Re}\, t|$, and we obtain

$$\int_{\mathbb{H}} R_{\mathbb{H}}(s; z, z')\, y'^t \, dA(z') = \frac{y^t}{t(1 - t) - s(1 - s)}. \tag{10.14}$$

Rewriting (10.14) in terms of $g_s(\sigma)$ gives

$$\int_0^\infty \int_{-\infty}^\infty g_s\left(\frac{x'^2 + (y + y')^2}{4yy'}\right)y'^{t-2} \, dx' \, dy' = \frac{y^t}{t(1 - t) - s(1 - s)}.$$

Next we change variables to (u, v) defined by

$$x' = y(1 + v)u,$$
$$y' = yv,$$

to obtain

$$\int_0^\infty \int_{-\infty}^\infty g_s\left((u^2 + 1)\frac{(1 + v)^2}{4v}\right)v^{t-2}(1 + v) \, du \, dv = \frac{1}{t(1 - t) - s(1 - s)}.$$

This is a Mellin transform, of the form

$$\int_0^\infty v^{t-1} f(v)\, dv = \left(\frac{1}{s-t} + \frac{1}{s+t-1}\right),$$

where

$$f(v) := (2s-1) \int_{-\infty}^\infty g_s\left((u^2+1)\frac{(1+v)^2}{4v}\right) v^{-1}(1+v)\, du.$$

The inverse transform is easily calculated by a contour integral in the strip $|\operatorname{Re} t| < \operatorname{Re} s - 1$,

$$f(v) = \begin{cases} v^{s-1}, & v \le 1, \\ v^{-s}, & v \ge 1. \end{cases}$$

From the g_s integral we thus obtain

$$(2s-1) \int_{-\infty}^\infty g_s\left((1+u^2)\frac{(1+v)^2}{4v}\right) du = \begin{cases} v^s/(1+v), & v \le 1, \\ v^{1-s}/(1+v), & v \ge 1. \end{cases}$$

The result follows immediately by setting $v = e^{2\omega}$. $\qquad\qquad\square$

Now we turn our attention to the calculation of $\Phi_X(s)$ in the model cases, starting with the hyperbolic cylinder $C_\ell = \Gamma_\ell \backslash \mathbb{H}$. Our computation is essentially a trace formula relating $\Phi_{C_\ell}(s)$ to the Selberg zeta function of C_ℓ. Since there is exactly one primitive closed geodesic (with two possible orientations), this is the entire function

$$Z_{C_\ell}(s) = \prod_{k \ge 0}\left(1 - e^{-(s+k)\ell}\right)^2. \tag{10.15}$$

The following formula comes from Patterson [159]:

Proposition 10.5. *For the hyperbolic cylinder, we have*

$$\Phi_{C_\ell}(s) = \frac{Z'_{C_\ell}}{Z_{C_\ell}}(s).$$

Proof. Consider the integrand of $\Phi_{C_\ell}(s)$, whose lift to $\mathbb{H}$ can be written as the sum

$$\varphi_{C_\ell}(s; z) = \sum_{m \ne 0} g_s\left(\frac{(x - e^{m\ell}x)^2 + (y + e^{m\ell}y)^2}{4e^{m\ell}y^2}\right)$$
$$= 2\sum_{m=1}^\infty g_s\left(\frac{x^2}{y^2}\sinh^2(m\ell/2) + \cosh^2(m\ell/2)\right). \tag{10.16}$$

This sum converges uniformly on compact sets for $\operatorname{Re} s > \frac{1}{2}$. In this range there is no need for the finite part regularization, and we can compute $\Phi_{C_\ell}(s)$ simply by integrating (10.16) term by term.

For this purpose we'll use the (non-Dirichlet) fundamental domain $\mathcal{F} := \mathbb{R} \times [1, e^\ell]$. The substitution

$$u = \frac{x}{y} \tanh(m\ell/2)$$

allows us to bring in the identity from Lemma 10.4,

$$
\begin{aligned}
(2s - 1) \int_{\mathcal{F}} g_s &\left(\frac{x^2}{y^2} \sinh^2(m\ell/2) + \cosh^2(m\ell/2) \right) \frac{dx\,dy}{y^2} \\
&= \frac{(2s - 1)\ell}{\tanh(m\ell/2)} \int_{-\infty}^{\infty} g_s \left((1 + u^2) \cosh^2(m\ell/2) \right) du \\
&= \frac{\ell e^{(1/2 - s)m\ell}}{2 \sinh(m\ell/2)} \\
&= \frac{\ell e^{-sm\ell}}{1 - e^{-m\ell}},
\end{aligned}
$$

for $m > 0$. Thus, integrating (10.16) term by term gives

$$\Phi_{C_\ell}(s) = 2\ell \sum_{m=1}^{\infty} \frac{e^{-sm\ell}}{1 - e^{-m\ell}},$$

for $\operatorname{Re} s > \frac{1}{2}$. We complete the proof by rearranging this sum,

$$
\begin{aligned}
\Phi_{C_\ell}(s) &= 2\ell \sum_{m=1}^{\infty} \sum_{k=0}^{\infty} e^{-m(s+k)\ell} \\
&= 2\ell \sum_{k=0}^{\infty} \frac{e^{-(s+k)\ell}}{1 - e^{-(s+k)\ell}} \\
&= \frac{d}{ds} \sum_{k=0}^{\infty} \log \left(1 - e^{-(s+k)\ell} \right)^2,
\end{aligned}
$$

and noting that both sides are meromorphic in s.   □

For the funnel F_ℓ, we can define $\Phi_{F_\ell}(s)$ by (10.13), with no regularization needed at the geodesic boundary. The standard definition of the zeta function does not apply to F_ℓ, but to make an analogy with the formula in Proposition 10.5 we will set

$$Z_{F_\ell}(s) := e^{-s\ell/4} \prod_{k \geq 0} \left(1 - e^{-(s+2k+1)\ell} \right)^2. \tag{10.17}$$

This definition is justified by the following calculation from Borthwick–Judge–Perry [24]:

Proposition 10.6. *If the funnel zeta function is defined as above, then*

$$\Phi_{F_\ell}(s) = \frac{Z'_{F_\ell}}{Z_{F_\ell}}(s). \tag{10.18}$$

Proof. Using the method of images formula, (5.20), the integrand in the definition on $\Phi_F(s)$ can be written

$$\varphi_{F_\ell}(s;z) = \sum_{m \neq 0} g_s\left(\frac{(x - e^{m\ell}x)^2 + (y + e^{m\ell}y)^2}{4e^{m\ell}y^2}\right)$$
$$- \sum_{m \in \mathbb{Z}} g_s\left(\frac{(x + e^{m\ell}x)^2 + (y + e^{m\ell}y)^2}{4e^{m\ell}y^2}\right). \tag{10.19}$$

Convergence is uniform on compact sets for $\operatorname{Re} s > \frac{1}{2}$. In this range the claim (10.9) can be deduced immediately from this sum, because we can identify ρ with x/y as in (7.24).

The contributions to $\Phi_{F_\ell}(s)$ from the $m \neq 0$ terms are computed as in the proof of Proposition 10.5, using Lemma 10.4 to compute integrals over $\mathcal{F}^+ := \mathbb{R}_+ \times [1, e^\ell]$. The result is

$$(2s - 1)\int_{\mathcal{F}^+} g_s\left(\frac{(x \mp e^{m\ell}x)^2 + (y + e^{m\ell}y)^2}{4e^{m\ell}y^2}\right)\frac{dx\,dy}{y^2} = \frac{\ell}{2}\frac{e^{-s|m|\ell}}{1 \mp e^{-|m|\ell}}.$$

Summing this expression over the $m \neq 0$ terms gives

$$\frac{\ell}{2}\sum_{m \neq 0}\left(\frac{e^{-s|m|\ell}}{1 - e^{-|m|\ell}} - \frac{e^{-s|m|\ell}}{1 + e^{-|m|\ell}}\right) = 2\ell\sum_{m \geq 1}\left(\frac{e^{-(s+1)m\ell}}{1 - e^{-2m\ell}}\right)$$
$$= 2\frac{d}{ds}\sum_{m \geq 1}\sum_{k=0}^{\infty}e^{-m(s+2k+1)\ell}$$
$$= \frac{d}{ds}\sum_{k=0}^{\infty}\log\left(1 - e^{-(s+2k+1)\ell}\right)^2.$$

When the $m = 0$ term in the second sum (10.19) is integrated over $\mathcal{F}^+$, we obtain

$$-(2s - 1)\int_1^{e^\ell}\int_0^{\infty} g_s\left(1 + \frac{x^2}{y^2}\right)\frac{dx\,dy}{y^2} = -\frac{2s - 1}{2}\int_1^{e^\ell}\int_{-\infty}^{\infty} g_s(1 + t^2)\frac{dt\,dy}{y}$$
$$= -\frac{\ell}{4},$$

by Lemma 10.4. Thus,

$$\Phi_{F_\ell}(s) = -\frac{\ell}{4} + \frac{d}{ds}\sum_{k=0}^{\infty}\log\left(1 - e^{-(s+2k+1)\ell}\right)^2. \qquad \square$$

A parabolic cylinder has no closed geodesics, and hence no sensible definition of a zeta function. But we still have an explicit formula for the resolvent trace, also from Borthwick–Judge–Perry [24].

Proposition 10.7. *For the parabolic cylinder C_∞ we have*

$$\Phi_{C_\infty}(s) = -\log 2 - \Psi(s + \tfrac{1}{2}) + \frac{1}{2s - 1}, \tag{10.20}$$

where Ψ is the digamma function Γ'/Γ.

Proof. For the proof it suffices to assume $\operatorname{Re} s > \frac{1}{2}$. The integrand in (10.20) can be written

$$\varphi_{C_\infty}(s; z) = \sum_{k \neq 0} R_{\mathbb{H}}(s, z, z + k) = 2 \sum_{k=1}^{\infty} g_s \left(1 + \frac{k^2}{4y^2}\right).$$

We will use the standard fundamental domain $\mathcal{F} = [0, 1] \times \mathbb{R}_+$ for C_∞. The integral in the definition of $\Phi_{C_\infty}(s)$ is well behaved at the funnel end if $\operatorname{Re} s$ is sufficiently large, but needs regularization at the cusp end:

$$\Phi_{C_\infty}(s) = 2(2s - 1) \operatorname*{FP}_{\varepsilon \to 0} \int_0^{1/\varepsilon} \int_0^1 \sum_{k=1}^{\infty} g_s \left(1 + \frac{k^2}{4y^2}\right) \frac{dx\, dy}{y^2}.$$

Substituting $u = \frac{k}{2y}$, this becomes

$$\Phi_{C_\infty}(s) = 4(2s - 1) \operatorname*{FP}_{\varepsilon \to 0} \sum_{k=1}^{\infty} \int_{k\varepsilon/2}^{\infty} g_s(1 + u^2) \frac{1}{k}\, du$$

$$= 4(2s - 1) \operatorname*{FP}_{\varepsilon \to 0} \int_0^{\infty} g_s(1 + u^2) \left[\sum_{1 \leq k \leq 2u/\varepsilon} \frac{1}{k}\right] du.$$

In order to extract the finite part, we use the classical estimate of the harmonic series,

$$\sum_{k=1}^{N} \frac{1}{k} = \log N + \gamma + O(N^{-1}),$$

where γ is Euler's constant. Under the assumption that $\operatorname{Re} s > \frac{1}{2}$, the error term can be controlled using the facts that $g_s(1 + u^2) = O(u^{-2s})$ as $u \to \infty$ and $g_s(1 + u^2) = O(\log u)$ as $u \to 0$. The result is

$$\int_0^{\infty} g_s(1 + u^2) \left[\sum_{1 \leq k \leq 2u/\varepsilon} \frac{1}{k}\right] du = \int_0^{\infty} g_s(1 + u^2) \left(\log \frac{2u}{\varepsilon} + \gamma\right) du + O(\varepsilon).$$

By including the full expansion of the harmonic series here, we could actually compute all of the constants a_k appearing in the expansion (10.10).

To compute the finite part of the integral, we simply drop the $\log \varepsilon$ term, leaving us with

$$\Phi_{C_\infty}(s) = 4(2s - 1) \int_0^{\infty} g_s(1 + u^2)(\log 2u + \gamma)\, du. \tag{10.21}$$

By Lemma 10.4, we have

$$4(2s - 1) \int_0^\infty g_s(1 + u^2)\, du = 1. \tag{10.22}$$

For the integral involving $\log u$ we return to (4.13) to write

$$\int_0^\infty g_s(1 + u^2) \log u\, du = \frac{1}{4\pi} \int_0^\infty \int_0^1 \frac{t^{s-1}(1 - t)^{s-1}}{(1 - t + u^2)^s} \log u\, du\, dt.$$

Substituting $u = \sqrt{(1 - t)(1 - w)/w}$ then yields

$$\int_0^\infty g_s(1 + u^2) \log u\, du$$
$$= \frac{1}{16\pi} \int_0^1 \int_0^1 \frac{t^{s-1}(1 - t)^{-1/2} w^{s-3/2}}{(1 - w)^{1/2}} \log \left[\frac{(1 - t)(1 - w)}{w} \right] dw\, dt.$$

This breaks down into a sum of beta-type integrals, easily evaluated using the identity

$$\int_0^1 t^{\alpha-1}(1 - t)^{\beta-1} \log t\, dt = \frac{\Gamma(\alpha)\Gamma(\beta)}{\Gamma(\alpha + \beta)} [\Psi(\alpha) - \Psi(\alpha + \beta)].$$

The result is

$$\int_0^\infty g_s(1 + u^2) \log u\, du = \frac{1}{4(2s - 1)} \left[\Psi(\tfrac{1}{2}) - \Psi(s + \tfrac{1}{2}) + \frac{1}{2s - 1} \right].$$

Since $\Psi(\tfrac{1}{2}) = -\gamma - \log 4$, we obtain

$$4(2s - 1) \int_0^\infty g_s(1 + u^2) \log u\, du = -\gamma - \log 4 - \Psi(s + \tfrac{1}{2}) + \frac{1}{2s - 1}.$$

Using this last equation together with (10.22) in (10.21) gives the stated result.   □

10.3 The resolvent 0-trace calculation

The structure of the diagonal singularity of $R_X(s; z, z')$ is independent of s, so we'd expect the kernel of $R_X(s) - R_X(1 - s)$ to be continuous on the diagonal. In fact, we can see quite explicitly from Proposition 7.9 that the function

$$\left[R_X(s; z, w) - R_X(1 - s; z, w) \right]_{w=z}$$

is meromorphic in s, smooth in z, and polyhomogeneous as $\rho \to 0$. For $s \notin \mathcal{R}_X \cup (1 - \mathcal{R}_X) \cup \mathbb{Z}/2$ we set

$$\Upsilon_X(s) := (2s - 1)\, 0\text{-tr}\left[R_X(s) - R_X(1 - s) \right]. \tag{10.23}$$

For $\mathrm{Re}\, s = \tfrac{1}{2}$, this is a regularization of the trace of the spectral resolution, as it appears in Stone's formula.

Proposition 10.8. *For any geometrically finite hyperbolic surface (including cylinders and funnels), the function Υ_X extends to a meromorphic function of $s \in \mathbb{C}$ satisfying*

$$\Phi_X(s) + \Phi_X(1 - s) = \Upsilon_X(s) + (2s - 1)\chi(X)\,\pi\cot(\pi s). \tag{10.24}$$

Proof. For $s \notin \mathcal{R}_X \cup (1 - \mathcal{R}_X) \cup \mathbb{Z}/2$, the left-hand side of (10.24) is

$$\Phi_X(s) + \Phi_X(1 - s) = (2s - 1)\int_X^0 [\varphi_X(s) - \varphi_X(1 - s)]\,dA, \tag{10.25}$$

where $\varphi_X(s)$ was defined in (10.8). The lift of the integrand to $\mathbb{H}$ is given for $\operatorname{Re} s > \delta$ by

$$\varphi_X(s; z) - \varphi_X(1 - s; z) = \sum_{T \in \Gamma - \{I\}} \left[R_{\mathbb{H}}(s; Tz, z) - R_{\mathbb{H}}(1 - s; Tz, z) \right].$$

In (4.18), we saw that

$$\left[R_{\mathbb{H}}(s, z, w) - R_{\mathbb{H}}(1 - s, z, w) \right]_{z=w} = \frac{1}{2}\cot \pi s.$$

By adding and subtracting this term in the sum we obtain

$$\varphi_X(s; z) - \varphi_X(1 - s; z) = \left[R_X(s; z, w) - R_X(1 - s; z, w) \right]_{w=z} - \frac{1}{2}\cot(\pi s).$$

Multiplying by $(2s - 1)$ and taking the 0-integral gives

$$\Phi_X(s) + \Phi_X(1 - s) = \Upsilon_X(s) - \frac{(2s - 1)}{2}\cot(\pi s)\ 0\text{-vol}(X).$$

The result then follows from the 0-volume calculation of Lemma 10.3. $\qquad\square$

Having computed the divisor of τ_X in Section 10.1, we are ready to forge the crucial link between the resonance set and the resolvent traces. This result is motivated by similar calculations in Guillopé–Zworski [87] and Patterson–Perry [160].

Proposition 10.9. *The relative scattering determinant is related to the resolvent trace Υ_X by*

$$\Upsilon_X(s) - \Upsilon_F(s) = -\frac{\tau_X'}{\tau_X}(s) \tag{10.26}$$

(meromorphically in s).

Proof. It suffices to assume that $s \notin \mathcal{R}_X \cup (1 - \mathcal{R}_X) \cup \mathbb{Z}/2$. Using the measure dh on $\partial \overline{X}$ as in (7.18), the identity from Proposition 7.9 allows us to write

$$\Upsilon_X(s) = \operatorname*{FP}_{\varepsilon \to 0} I_\varepsilon(s),$$

where

$$I_\varepsilon(s) := -(2s-1)^2 \int_{X_\varepsilon} \left[\int_{\partial \overline{X}} E_X(s; z, q) E_X(1-s; z, q) \, dh(q) \right] dA(z),$$

with $X_\varepsilon = \{\rho \geq \varepsilon\} \subset X$.

To derive the asymptotic expansion of $I_\varepsilon(s)$ as $\varepsilon \to 0$, we use a trick inspired by the Maass–Selberg relation from the finite-area case (see, e.g., [210, Chapter 4]). Because $E_X(s)$ satisfies the eigenvalue equation, we have

$$E_X(s+a)\big(\Delta_X E_X(1-s)\big)^{\mathrm{t}} - \big(\Delta_X E_X(s+a)\big)E_X(1-s)^{\mathrm{t}}$$
$$= \big[(2s-1)a + a^2\big]E_X(s+a)E_X(1-s)^{\mathrm{t}}.$$

This relation implies the formula

$$I_\varepsilon(s) = -\lim_{a\to 0^+} \frac{2s-1}{a} \int_{X_\varepsilon} \int_{\partial\overline{X}} \Big[E_X(s+a; z, q) \, \Delta_X E_X(1-s; z, q)$$
$$- \Delta_X E_X(s+a; z, q) \, E_X(1-s; z, q) \Big] dh(q) \, dA(z).$$

The expressions in the integrand are smooth, since z is kept away from the boundary, so the order of integration is not significant here. We can apply Green's formula to the integral over X_ε, just as in the proof of Proposition 7.9, to obtain

$$I_\varepsilon(s) = \lim_{a\to 0^+} \frac{2s-1}{a} \int_{\partial X_\varepsilon} \int_{\partial\overline{X}} \Big[E_X(s+a; z, q) \, \partial_r E_X(1-s; z, q)$$
$$- \partial_r E_X(s+a; z, q) \, E_X(1-s; z, q) \Big] dh(q) \, d\sigma_\varepsilon(z), \qquad (10.27)$$

where $d\sigma_\varepsilon$ is the measure induced on $\{\rho = \varepsilon\}$ by g, and ∂_r is the outward unit normal to $\{\rho = \varepsilon\}$ (acting on the z coordinate). By the same integration by parts, with a set equal to zero, we can see that

$$0 = (2s-1) \int_{\partial X_\varepsilon} \int_{\partial\overline{X}} \Big[E_X(s; z, q) \, \partial_r E_X(1-s; z, q)$$
$$- \partial_r E_X(s; z, q) \, E_X(1-s; z, q) \Big] dh(q) \, d\sigma_\varepsilon(z).$$

Subtracting this expression from (10.27) and taking the limit $a \to 0$ leaves us with our version of the Maass–Selberg relation,

$$\Upsilon_X(s) = (2s-1) \operatorname*{FP}_{\varepsilon\to 0} \int_{\partial X_\varepsilon} \int_{\partial\overline{X}} \Big[\partial_s E_X(s; z, q) \, \partial_r E_X(1-s; z, q)$$
$$- \partial_s \partial_r E_X(s; z, q) \, E_X(1-s; z, q) \Big] dh(q) \, d\sigma_\varepsilon(z). \qquad (10.28)$$

Because of the singular behavior of $E_X(s; z, q)$ as $z \to q$ in a funnel, we want to subtract off $\Upsilon_F(s)$ before evaluating the finite part. The Maass–Selberg relation for this term reads

$$\Upsilon_F(s) = (2s - 1) \operatorname*{FP}_{\varepsilon \to 0} \int_{\partial F_\varepsilon} \int_{\partial_0 \overline{F}} \Big[\partial_s E_F(s; z, q) \, \partial_r E_F(1 - s; z, q)$$

$$- \partial_s \partial_r E_F(s; z, q) \, E_F(1 - s; z, q) \Big] dh(q) \, d\sigma_\varepsilon(z). \qquad (10.29)$$

(Note that the Dirichlet boundary conditions on the geodesic boundary of F imply that there are no terms coming from this boundary in the integration by parts.) The boundaries ∂F_ε and $\partial_0 \overline{F}$ are naturally identified with the funnel components of ∂X_ε and $\partial \overline{X}$, respectively.

To keep the notation under control, let us break down the integrals (10.28) and (10.29) according which boundary components the variables z and q occupy. We set

$$\Upsilon_X(s) - \Upsilon_F(s)$$

$$= \operatorname*{FP}_{\varepsilon \to 0} \Bigg[\sum_{i,j=1}^{n_f} A_{ij}^{\mathrm{ff}}(s, \varepsilon) + \sum_{i=1}^{n_f} \sum_{k=1}^{n_c} \Big(A_{ik}^{\mathrm{fc}}(s, \varepsilon) + A_{ki}^{\mathrm{cf}}(s, \varepsilon) \Big) + \sum_{k,l=1}^{n_c} A_{kl}^{\mathrm{cc}}(s, \varepsilon) \Bigg],$$

$$(10.30)$$

where, for example, A_{ij}^{ff} is the integral over $z \in \partial F_{i,\varepsilon}$ and $q \in \partial_0 \overline{F}_j$, A_{ki}^{cf} is the integral over $z \in \partial C_{k,\varepsilon}$ and $q \in \partial_0 \overline{F}_i$, etc. Note that the subtraction of $\Upsilon_F(s)$ affects only the A_{jj}^{ff} terms.

To see what happens as $\varepsilon \to 0$, consider first the off-diagonal funnel–funnel case. For $i \neq j$,

$$A_{ij}^{\mathrm{ff}}(s, \varepsilon) := (2s - 1) \int_0^{\ell_i} \int_0^{\ell_j} \Big[\partial_s E_j^{\mathrm{f}}(s; \varepsilon, t, t') \, \partial_r E_j^{\mathrm{f}}(1 - s; \varepsilon, t, t')$$

$$- \partial_s \partial_r E_j^{\mathrm{f}}(s; \varepsilon, t, t') \, E_j^{\mathrm{f}}(1 - s; \varepsilon, t, t') \Big] dt' \, d\sigma_\varepsilon(t), \qquad (10.31)$$

where (ρ, t) are standard funnel coordinates for F_i, with ρ set equal to ε, and t' is the coordinate for $\partial_0 \overline{F}_j$.

To evaluate the finite part, we use the asymptotic

$$E_j^{\mathrm{f}}(s; \rho, t, t') = \frac{\rho^s}{2s - 1} S_{ij}^{\mathrm{ff}}(s; t, t') + \rho^{s+1} f(\rho, t), \qquad (10.32)$$

where $f \in C^\infty(\overline{F}_i)$. Noting that $\partial_r = -\rho \partial_\rho$, and $d\sigma_\varepsilon(t) = \varepsilon^{-1} dt + O(\varepsilon)$, we simply substitute (10.32) into (10.31) to obtain

$$A_{ij}^{\mathrm{ff}}(s, \varepsilon) = \int_0^{\ell_i} \int_0^{\ell_j} \Big[-\partial_s S_{ij}^{\mathrm{ff}}(s; t, t') \, S_{ij}^{\mathrm{ff}}(1 - s; t, t')$$

$$+ \frac{1}{2s - 1} S_{ij}^{\mathrm{ff}}(s; t, t') S_{ij}^{\mathrm{ff}}(1 - s; t, t') \Big] dt' \, dt$$

$$+ (\log \varepsilon \text{ terms}) + O(\varepsilon).$$

The $\log \varepsilon$ terms (which came from $\partial_s \rho^s$) are dropped in the finite part. For $i \neq j$, $S_{ij}^{ff}(s)$ is a smoothing operator, so the remaining integrals can be identified as traces over $L^2(\partial_0 \overline{F}_j)$. Using the fact that $S_X(s)^t = S_X(s)$, we can write the result as

$$\operatorname*{FP}_{\varepsilon \to 0} A_{ij}^{ff}(s, \varepsilon) = -\operatorname{tr}\left[S_{ji}^{ff}(1-s)\, \partial_s S_{ij}^{ff}(s) \right] + \frac{1}{2s-1}\operatorname{tr}\left[S_{ij}^{ff}(s) S_{ji}^{ff}(1-s) \right], \quad (10.33)$$

for $i \neq j$.

Now consider a funnel–funnel term with $i = j$,

$$A_{jj}^{ff}(s, \varepsilon) := (2s-1) \int_0^{\ell_j} \int_0^{\ell_j} \left[\partial_s E_j^f(s; \varepsilon, t, t')\, \partial_r E_j^f(1-s; \varepsilon, t, t') \right.$$
$$- \partial_s \partial_r E_j^f(s; \varepsilon, t, t')\, E_j^f(1-s; \varepsilon, t, t')$$
$$- \partial_s E_{F_j}(s; \varepsilon, t, t')\, \partial_r E_{F_j}(1-s; \varepsilon, t, t')$$
$$\left. + \partial_s \partial_r E_{F_j}(s; \varepsilon, t, t')\, E_{F_j}(1-s; \varepsilon, t, t') \right] dt'\, d\sigma_\varepsilon(t). \quad (10.34)$$

Propositions 7.11 and 7.12 give the asymptotics of $E_{F_j}(s; \cdot, \cdot)$ and $E_j^f(s; \cdot, \cdot)$ in a distributional sense. For example, from (7.20) and Proposition 7.15, we obtain

$$E_j^f(s; \rho, t, t') \sim \frac{1}{2s-1}\left[\rho^s \delta(t-t') + \rho^{1-s} S_j^f(s; t, t') \right]. \quad (10.35)$$

In the same way, (7.22) becomes

$$E_{F_j}(s; \rho, t, t') \sim \frac{1}{2s-1}\left[\rho^s \delta(t-t') + \rho^{1-s} S_{F_j}(s; t, t') \right]. \quad (10.36)$$

The easiest way to handle the computation of the $\varepsilon \to 0$ limit in (10.34) is to substitute (10.35) and (10.36) into the integral and manipulate the expressions formally. This is not hard to justify rigorously, using the asymptotics from Propositions 7.11 and 7.12 and the fact that $E_j^f(s; \rho, t, t') - E_{F_j}(s; \rho, t, t')$ is nonsingular. Assuming that $\operatorname{Re} s = \frac{1}{2}$, $s \neq \frac{1}{2}$, the resulting expression is

$$A_{jj}^{ff}(s, \varepsilon) := -\operatorname{tr}\left[S_{jj}^{ff}(1-s)\, \partial_s S_{jj}^{ff}(s) + S_{F_j}(1-s)\, \partial_s S_{F_j}(s) \right]$$
$$+ \frac{1}{2s-1}\operatorname{tr}\left[S_{jj}^{ff}(s) S_{jj}^{ff}(1-s) - S_{F_j}(s) S_{F_j}(1-s) \right]$$
$$+ \frac{\varepsilon^{1-2s}}{2s-1}\operatorname{tr}\left[S_{jj}^{ff}(1-s) - S_{F_j}(1-s) \right]$$
$$- \frac{\varepsilon^{2s-1}}{2s-1}\operatorname{tr}\left[S_{jj}^{ff}(s) - S_{F_j}(s) \right]$$
$$+ (\log \varepsilon \text{ terms}) + O(\varepsilon). \quad (10.37)$$

Note that $S_{F_j}(s)S_{F_j}(1-s) = I_{F_j}$. Assuming $s \neq \frac{1}{2}$, the final two traces drop from the finite part, leaving

$$A_{ij}^{\mathrm{ff}}(s, \varepsilon) := -\operatorname{tr}\left[S_{jj}^{\mathrm{ff}}(1-s)\,\partial_s S_{jj}^{\mathrm{ff}}(s) + S_{F_j}(1-s)\,\partial_s S_{F_j}(s)\right]$$

$$+ \frac{1}{2s-1}\operatorname{tr}\left[S_{jj}^{\mathrm{ff}}(s)S_{jj}^{\mathrm{ff}}(1-s) - I_{F_j}\right]. \tag{10.38}$$

For the funnel–cusp and cusp–funnel terms, the argument is very similar to the derivation of (10.31), since the components are nonsingular. The finite parts are

$$\operatorname*{FP}_{\varepsilon \to 0} A_{ik}^{\mathrm{fc}}(s, \varepsilon) = -S_{ki}^{\mathrm{cf}}(1-s)\,\partial_s S_{ik}^{\mathrm{fc}}(s)$$

$$+ \frac{1}{2s-1}S_{ik}^{\mathrm{fc}}(s)S_{ki}^{\mathrm{cf}}(1-s) \tag{10.39}$$

(no trace here; this is just a number) and

$$\operatorname*{FP}_{\varepsilon \to 0} A_{kj}^{\mathrm{cf}}(s, \varepsilon) = -\operatorname{tr}\left[S_{jk}^{\mathrm{fc}}(1-s)\,\partial_s S_{kj}^{\mathrm{cf}}(s)\right]$$

$$+ \frac{1}{2s-1}\operatorname{tr}\left[S_{kj}^{\mathrm{cf}}(s)S_{jk}^{\mathrm{fc}}(1-s)\right]. \tag{10.40}$$

Finally, we turn to the cusp–cusp terms,

$$A_{km}^{\mathrm{cc}}(s, \varepsilon) := (2s-1)\int_0^1 \left[\partial_s E_m^{\mathrm{c}}(s; \varepsilon, t)\,\partial_r E_m^{\mathrm{c}}(1-s; \varepsilon, t)\right.$$

$$\left. - \partial_s \partial_r E_m^{\mathrm{c}}(s; \varepsilon, t)\,E_m^{\mathrm{c}}(1-s; \varepsilon, t)\right]\varepsilon\,dt.$$

Here (ρ, t) denote standard cusp coordinates in C_k, with ρ set equal to ε. By (7.30) and Proposition 7.15, we can write the asymptotic

$$E_m^{\mathrm{c}}(s, \rho, t) = \frac{1}{2s-1}\left[\delta_{km}\rho^{-s} + \rho^{s-1}S_{km}^{\mathrm{cc}}(s)\right] + O(\rho^\infty).$$

(Note that $S_{km}^{\mathrm{cc}}(s)$ is just a complex number.) When we substitute this into A_{km}^{cc}, we find that

$$A_{km}^{\mathrm{cc}}(s, \varepsilon) = -S_{mk}^{\mathrm{cc}}(1-s)\,\partial_s S_{km}^{\mathrm{cc}}(s) + \frac{1}{2s-1}\left[S_{km}^{\mathrm{cc}}(s)S_{mk}^{\mathrm{cc}}(1-s) - \delta_{km}\right]$$

$$+ \frac{\delta_{km}\varepsilon^{1-2s}}{2s-1}S_{kk}^{\mathrm{cc}}(1-s) - \frac{\delta_{km}\varepsilon^{2s-1}}{2s-1}S_{kk}^{\mathrm{cc}}(s)$$

$$+ (\log \varepsilon \text{ terms}) + O(\varepsilon). \tag{10.41}$$

For $s \neq \frac{1}{2}$, the finite part is

$$A_{km}^{\mathrm{cc}}(s, \varepsilon) = -S_{mk}^{\mathrm{cc}}(1-s)\partial_s S_{km}^{\mathrm{cc}}(s) + \frac{1}{2s-1}\left[S_{mk}^{\mathrm{cc}}(1-s)S_{km}^{\mathrm{cc}}(s) - \delta_{km}\right]. \tag{10.42}$$

Once we combine these calculations together in (10.30), we can simplify by noting that

$$0 = \text{tr}\Big[S_X(1-s)S_X(s) - I \Big]$$
$$= \text{tr}\Big[S^{\text{ff}}(1-s)S^{\text{ff}}(s) + S^{\text{fc}}(1-s)S^{\text{cf}}(s) - I_F \Big]$$
$$+ \text{tr}\Big[S^{\text{cc}}(1-s)S^{\text{cc}}(s) + S^{\text{cf}}(1-s)S^{\text{fc}}(s) - I_C \Big].$$

With these terms removed, we find that

$$\Upsilon_X(s) - \Upsilon_F(s) = -\text{tr}\Big[S_X^{\text{ff}}(1-s)\partial_s S_X^{\text{ff}}(s) - S_F(1-s)\partial_s S_F(s) \Big]$$
$$- \text{tr}\Big[S_X^{\text{fc}}(1-s)\partial_s S^{\text{cf}}(s) \Big] - \text{tr}\Big[S_X^{\text{cf}}(1-s)\partial_s S^{\text{fc}}(s) \Big]$$
$$- \text{tr}\Big[S_X^{\text{cc}}(1-s)\partial_s S^{\text{cc}}(s) \Big]. \tag{10.43}$$

It is not hard to see that this is the answer we're looking for. Recalling the definition $S_{\text{rel}}(s) = S_0^{-1}(s)S_X(s)$, we have

$$\frac{\tau_X'}{\tau_X}(s) = \partial_s \log\det S_{\text{rel}}(s)$$
$$= \text{tr}\Big[S_{\text{rel}}(s)^{-1}\,\partial_s S_{\text{rel}}(s) \Big]$$
$$= \text{tr}\Big[S_X(s)^{-1} S_0(s)\,\partial_s\big(S_0(s)^{-1}S_X(s)\big) \Big]$$
$$= \text{tr}\Big[S_X(s)^{-1}\big(\partial_s S_X(s) - \partial_s S_0(s)\,S_0(s)^{-1}S_X(s)\big) \Big]$$
$$= \text{tr}\Big[S_X(1-s)\,\partial_s S_X(s) - S_0(1-s)\,\partial_s S_0(s) \Big].$$

(In the last step, we used the cyclicity of the trace and $S_X(s)^{-1} = S_X(1-s)$ from Corollary 7.14.) If we then decompose the trace into funnel and cusp parts, using the decomposition (7.34) and $S_0 = S_F(s) \otimes I$, this breaks this down into the sum of traces appearing in (10.43).   □

10.4 Structure of the zeta function

We next want to connect the zeta function to the regularized trace $\Phi_X(s)$, so that Proposition 10.9 can be used to understand the divisor of the zeta function. For a hyperbolic surface without cusps (convex cocompact group) Patterson [159] identified $\Phi_X(s)$ with the logarithmic derivative of the zeta function. A slight modification is required if there are cusps; this version was given in Borthwick–Judge–Perry [24].

Proposition 10.10. *The logarithmic derivative of the zeta function is given by*

$$\frac{Z_X'}{Z_X}(s) = \Phi_X(s) - n_c \Phi_{C_\infty}(s). \tag{10.44}$$

Proof. Recall the definition (10.8) of $\varphi_X(s)$, the integrand for $\Phi_X(s)$,

$$\varphi_X(s; z) = \sum_{S \in \Gamma - \{I\}} R_{\mathbb{H}}(s; z, Sz),$$

for $\mathrm{Re}\, s > \delta$. We can compute $\Phi_X(s)$ by applying the strategy of the length trace formula from the compact case (Proposition 3.2), which is based on a decomposition of Γ into conjugacy classes.

Let Π_h and Π_p be lists of representatives of conjugacy classes of generators of maximal hyperbolic and parabolic subgroups of Γ. By Lemma 2.18, any $S \in \Gamma - \{I\}$ can be written uniquely as

$$S = RT^k R^{-1},$$

where $T \in \Pi_h \cup \Pi_p$, $k \in \mathbb{Z} - \{0\}$, and $R \in \Gamma / \langle T \rangle$. Applying this decomposition in the sum for $\varphi_X(s)$ gives

$$\varphi_X(s; z) = \sum_{T \in \Pi_h \cup \Pi_p} \sum_{R \in \Gamma / \langle T \rangle} \sum_{k \neq 0} R_{\mathbb{H}}(s; z, R^{-1} T^k R z). \tag{10.45}$$

For some particular $T \in \Pi_h \cup \Pi_p$, consider the associated contribution to $\Phi_X(s)$,

$$(2s - 1) \sum_{R \in \Gamma / \langle T \rangle} \sum_{k \neq 0} {\int_X^0} R_{\mathbb{H}}(s; z, R^{-1} T^k R z) \, dA(z). \tag{10.46}$$

The trick for evaluating such expressions is to group the integrals in the sum over $\Gamma / \langle T \rangle$ into a single integral over the cylinder $\langle T \rangle \backslash \mathbb{H}$. The 0-integral over X can be realized as an integral over some fundamental domain $\mathcal{F} \subset \mathbb{H}$ for Γ, by lifting the boundary-defining function ρ to $\mathbb{H}$. Because the model resolvent depends only on hyperbolic distance (and the hyperbolic measure is invariant under the group action), a simple change of variables gives

$${\int_{\mathcal{F}}^0} R_{\mathbb{H}}(s; z, R^{-1} T^k R z) \, dA(z) = {\int_{R\mathcal{F}}^0} R_{\mathbb{H}}(s; z, T^k z) \, dA(z).$$

Thus the sum (10.46) can be rewritten as

$$(2s - 1) \sum_{k \neq 0} {\int_{\widetilde{\mathcal{F}}}^0} R_{\mathbb{H}}(s; z, T^k z) \, dA(z),$$

where

$$\widetilde{\mathcal{F}} := \bigcup_{R \in \Gamma / \langle T \rangle} R\mathcal{F}.$$

It is easily checked that $\widetilde{\mathcal{F}}$ is a fundamental domain for the cyclic group $\langle T \rangle$, and so (10.46) is equal to

$$(2s - 1) {\int_{\widetilde{\mathcal{F}}}^0} \sum_{k \neq 0} R_{\mathbb{H}}(s; z, T^k z) \, dA(z) = \Phi_{\langle T \rangle \backslash \mathbb{H}}(s).$$

(We're still assuming $\mathrm{Re}\, s > \delta$, in order to interchange summation and integration.)

Adding up the terms (10.46) for all $T \in \Pi_h \cup \Pi_p$, we conclude that

$$\Phi_X(s) = \sum_{T \in \Pi_h} \Phi_{\langle T \rangle \backslash \mathbb{H}}(s) + \sum_{S \in \Pi_p} \Phi_{\langle S \rangle \backslash \mathbb{H}}(s).$$

For T hyperbolic, $\langle T \rangle \backslash \mathbb{H}$ is isomorphic to the cylinder $C_{\ell(T)}$. Proposition 2.17 shows the correspondence between $T \in \Pi_h$ and $\ell(T) \in \mathcal{L}_X$, but we must remember that elements of $\mathcal{L}_X$ are repeated for both orientations, whereas Π_h is a list of generators and would include only one of T and T^{-1}. Thus,

$$\sum_{T \in \Pi_h} \Phi_{\langle T \rangle \backslash \mathbb{H}}(s) = \frac{1}{2} \sum_{\ell \in \mathcal{L}_X} \Phi_{C_\ell}(s).$$

By Proposition 10.5 this can be evaluated for $\mathrm{Re}\, s > \delta$,

$$\sum_{T \in \Pi_h} \Phi_{\langle T \rangle \backslash \mathbb{H}}(s) = \frac{1}{2} \sum_{\ell \in \mathcal{L}_X} \frac{d}{ds} \sum_{k=0}^{\infty} \log\left(1 - e^{-(s+k)\ell}\right)^2 = \frac{Z_X'}{Z_X}(s).$$

For S parabolic, we have $\langle S \rangle \backslash \mathbb{H} \cong C_\infty$. The set Π_p has n_c elements, since each cusp corresponds to an equivalence class of parabolic fixed points, so that

$$\sum_{S \in \Pi_p} \Phi_{\langle S \rangle \backslash \mathbb{H}}(s) = n_c \Phi_{C_\infty}(s). \qquad \square$$

Since the right-hand side of (10.44) is already known to be meromorphic, Proposition 10.10 shows that $Z_X'/Z_X(s)$ has a meromorphic extension from $\mathrm{Re}\, s > \delta$ to the whole complex plane. This does not yet establish the meromorphic continuation of $Z_X(s)$. We need also to show that $Z_X'/Z_X(s)$ has only simple poles with integer residues. For $\mathrm{Re}\, s \geq \frac{1}{2}$ we will calculate the poles of $\Phi_X(s)$ directly.

Proposition 10.11. *The meromorphic function $Z_X'(s)/Z_X(s)$ is analytic in $\mathrm{Re}\, s > \frac{1}{2}$ except for simple poles at points ζ where $\zeta(1 - \zeta) \in \sigma_d(\Delta_X)$, with*

$$\mathop{\mathrm{res}}_{s=\zeta} \frac{Z_X'}{Z_X}(s) = m_\zeta. \tag{10.47}$$

And $Z_X'(s)/Z_X(s)$ has no poles on the critical line $\mathrm{Re}\, s = \frac{1}{2}$ except possibly a simple pole at $s = \frac{1}{2}$ with

$$\mathop{\mathrm{res}}_{s=1/2} \frac{Z_X'}{Z_X}(s) = m_{1/2} - n_c.$$

Proof. Consider the decomposition from Theorem 6.3,

$$R_X(s) = M_i'(s) + M_f(s) + M_c(s) + Q(s). \tag{10.48}$$

For $\operatorname{Re} s > \frac{1}{2}$, the model terms are holomorphic, and the poles appear only in the remainder $Q(s)$. The subtraction of $R_{\mathbb{H}}(s)$ used to define $\varphi_X(s)$ affects the model terms, which are singular on the diagonal, but not $Q(s)$. Thus, in this case, the singular part of $(2s - 1)\varphi_X(s)$ is the restriction to the diagonal of the singular part of $R_X(s)$. Such poles occur only when $\zeta(1 - \zeta) \in \sigma_d(\Delta_X)$, and Proposition 8.1 shows that the residue of $(2s - 1)R_X(s)$ at such a ζ is the orthogonal projection onto the $\zeta(1 - \zeta)$-eigenspace.

By this reasoning, if $\{\phi_k\}$ is an orthonormal basis of the eigenspace, then $(2s - 1)\varphi_X(s)$ has the Laurent expansion

$$(2s - 1)\varphi_X(s) = \frac{1}{s - \zeta} \sum_{k=1}^{m_\zeta} |\phi_k|^2 + (2s - 1)h(s),$$

where $h(s) \in \rho_f^{2s} \rho_c^{2s-2} C^\infty(\overline{X}) + \rho_c^{-1} C^\infty(\overline{X})$ is holomorphic near $s = \zeta$. Taking the 0-integral gives

$$\Phi_X(s) = \frac{m_\zeta}{s - \zeta} + \int_X^0 h(s)\, dA,$$

which shows that $\Phi_X(s)$ has a simple pole of residue m_ζ at $s = \zeta$. Because $\Phi_{C_\infty}(s)$ is holomorphic for $\operatorname{Re} s > \frac{1}{2}$, Proposition 10.10 then gives (10.47).

Corollary 7.8 shows that the resolvent has no poles on the critical line, except possibly at $s = \frac{1}{2}$. By the same reasoning as above, $\Phi_X(s)$ has no poles for $\operatorname{Re} s = \frac{1}{2}$, $s \neq \frac{1}{2}$.

This leaves the point $s = \frac{1}{2}$, which must be treated with greater care. Since the pole of $R_X(s)$ at $s = 1$ has order one, by Lemma 8.4, $(2s - 1)\varphi_X(s)$ is holomorphic near $\frac{1}{2}$. Despite this, a pole can still occur in $\Phi_X(s)$, because of the 0-integration.

To compute the residue at this pole, we need to analyze the contributions from components of $\partial \overline{X}$ separately. Funnels are simplest, because the model term $M_f(s)$ has no pole at $s = \frac{1}{2}$. The argument used above can be applied, together with Lemma 8.4, to show that

$$(2s - 1)\varphi_X(s)\big|_{F_j} = \sum_{k=1}^{m_{1/2}} \phi_k(s)^2 \big|_{F_j} + (2s - 1)h(s),$$

where $h(s) \in \rho^{2s} C^\infty(\overline{F}_j)$ is analytic near $s = \frac{1}{2}$. Here the ϕ_k's are linearly independent functions, analytic in s, satisfying

$$(\Delta_X - \tfrac{1}{4})\phi_k(\tfrac{1}{2}) = 0, \qquad \phi_k(s) \in \rho_f^s \rho_c^{s-1} C^\infty(\partial \overline{X}).$$

Let $\phi_k^\sharp(s) = \rho_f^{-s} \rho_c^{1-s} \phi_k(s)|_{\partial \overline{X}}$. For s near but not equal to $\frac{1}{2}$, we have

$$\int_{F_j \cap \{\rho \geq \varepsilon\}} \phi_k(s)^2\, dA = \frac{\varepsilon^{2s-1}}{2s - 1} \int_{\partial_0 \overline{F}_j} \phi_k^\sharp(s)^2\, dt + \text{(holo)},$$

where the factor of $(2s - 1)^{-1}$ comes from integrating $\rho^{2s-2} d\rho$. The contribution to the residue of $\Phi_X(s)$ at $s = \frac{1}{2}$ is then

$$\frac{1}{2} \int_{\partial_0 \overline{F}_j} \phi_k^\sharp(s)^2 \, dt.$$

Using the formula (8.39) for $S_X(\frac{1}{2})$, we can identify the full contribution from the boundary of funnel F_j as

$$\sum_{k=1}^{m_{1/2}} \frac{1}{2} \int_{\partial_0 \overline{F}_j} \phi_k^\sharp(s)^2 \, dt = \frac{1}{2} \operatorname{tr}\left[S_{jj}^{\text{ff}}(\tfrac{1}{2}) + I\right].$$

Analysis of the cusp contributions is different; the pole in $M_i(s)$ implies that the singular part of $(2x - 1)\varphi_X(s)$ at $s = \frac{1}{2}$ is not the restriction to the diagonal of the singular part of $(2x - 1)R_X(s)$. Using the notation from the proof of Lemma 8.5, let $(2s - 1)^{-1}B(s)$ denote the singular part of the term $Q(s)$ in (10.48). What we can deduce from (10.48) is

$$(2s - 1)\varphi_X(s; z)\big|_{C_i} = (1 - \chi_0)\varphi_{C_\infty}(s; z) + B(s; z, z)\big|_{C_i} + O(2s - 1).$$

By (8.38), we have

$$B(s; z, z)\big|_{C_i} = \rho^{2s-2} S_{ii}^{\text{cc}}(\tfrac{1}{2}) + O(\rho^{2s-1}).$$

As in the funnel case, integrating $\rho^{2s-2} \, d\rho$ gives a factor of $(2s - 1)^{-1}$. The contribution to the residue from the $B(s; z, z)$ term is thus equal to $\frac{1}{2} S_{ii}^{\text{cc}}(\tfrac{1}{2})$.

We claim that

$$\operatorname*{res}_{s=1/2} \int_{C_j}^0 (2s - 1)(1 - \chi_0)\, \varphi_{C_\infty}(s) \, dA = 0, \tag{10.49}$$

so that the $\varphi_{C_\infty}(s; z)$ term does not contribute to the residue. To see this, note first that

$$\operatorname*{res}_{s=1/2} \Phi_{C_\infty}(s) = \frac{1}{2}.$$

To prove (10.49), it suffices to show that

$$\operatorname*{res}_{s=1/2}\left[(2s - 1) \int_{C_\infty}^0 \chi_0 \, \varphi_{C_\infty}(s) \, dA\right] = \frac{1}{2}, \tag{10.50}$$

where χ_0, which was defined for $y \geq 1$ (in terms of $r = \log y$), is extended to be equal to 1 for $y \leq 1$.

As in the proof of Proposition 10.7, we can write $\chi_0\varphi_{C_\infty}(s)$ as a series,

$$\chi_0(y)\, \varphi_{C_\infty}(s; y) = 2 \sum_{k=1}^\infty \chi_0(y)\, g_s\left(1 + \frac{k^2}{4y^2}\right).$$

The cutoff is useful here, because with it the series converges uniformly on compact sets for $\operatorname{Re} s > 0$. We can thus see directly that

$$(2s - 1)\chi_0(y)\,\varphi_{C_\infty}(s; y) = y^{2s} f_s(y),$$

where $f_s \in C_0^\infty(\mathbb{R}_+)$. Moreover, we can interchange the limit and sum to compute

$$f_s(0) = 2(2s - 1) \sum_{k=1}^{\infty} \lim_{y \to 0} y^{-2s} g_s\left(1 + \frac{k^2}{4y^2}\right)$$

$$= 2(2s - 1) \sum_{k=1}^{\infty} \frac{1}{4\pi} \frac{\Gamma(s)^2}{\Gamma(2s)} \left(\frac{4}{k^2}\right)^s$$

$$= (2s - 1) \frac{4^s}{2\pi} \frac{\Gamma(s)^2}{\Gamma(2s)} \zeta(2s),$$

where $\zeta(s)$ is the Riemann zeta function.

Integration by parts shows that for $s \neq \frac{1}{2}$,

$$\operatorname*{FP}_{\varepsilon \to 0} \int_\varepsilon^\infty y^{2s-2} f_s(y)\, dy = -\frac{1}{2s - 1} \int_0^\infty y^{2s-1} f_s'(y)\, dy.$$

Thus,

$$\operatorname*{res}_{s=1/2}\left[(2s - 1) \int_{C_\infty}^0 \chi_0\,\varphi_{C_\infty}(s)\, dA\right] = -\frac{1}{2} \int_0^\infty f_{1/2}'(y)\, dy = \frac{1}{2} f_{1/2}(0).$$

By the formula given above,

$$f_{1/2}(0) = \lim_{s \to 1/2}(2s - 1)\zeta(2s) = 1.$$

This completes the computation of (10.50), and (10.49) follows. Hence $\frac{1}{2} S_{ii}^{\mathrm{cc}}(\frac{1}{2})$ is the full contribution to the residue from the cusp C_i.

Putting the funnel and cusp residue computations together gives

$$\operatorname*{res}_{s=1/2} \Phi_X(s) = \frac{1}{2} \operatorname{tr}\left[S_X^{\mathrm{ff}}(s) + I\right] + \frac{1}{2} \operatorname{tr}\left[S_X^{\mathrm{cc}}(s)\right].$$

This trace was computed in Lemma 8.5, which gives

$$\operatorname*{res}_{s=1/2} \Phi_X(s) = m_{1/2} - n_{\mathrm{c}}/2.$$

By (10.44) and the fact that $\operatorname{res}_{s=1/2} \Phi_{C_\infty}(s) = 1/2$, the residue of $Z_X'/Z_X(s)$ at $s = \frac{1}{2}$ is $m_{1/2} - n_{\mathrm{c}}$. $\square$

Although we could also analyze the poles of $\Phi_X(s)$ for $\operatorname{Re} s < \frac{1}{2}$ directly, it proves to be unnecessary because of a functional equation connecting $Z_X(s)$ to $Z_X(1 - s)$ via the relative scattering determinant. This is the analogue of Selberg's functional equation (Corollary 3.10) from the compact case.

Proposition 10.12 (Functional equation). *The zeta function $Z_X(s)$ has a meromorphic extension to $s \in \mathbb{C}$. Moreover, the formula*

$$\tau_X(s) = \frac{c}{4^{sn_c}} \frac{Z_X(1-s)}{Z_X(s)} \left(\frac{G_\infty(s)}{G_\infty(1-s)} \right)^{-\chi(X)} \frac{Z_F(s)}{Z_F(1-s)} \left(\frac{\Gamma(s-\frac{1}{2})}{\Gamma(\frac{1}{2}-s)} \right)^{n_c} \quad (10.51)$$

holds for some constant c.

Proof. Combining Propositions 10.8 and 10.10 gives the formula

$$\Upsilon_X(s) = \Phi_X(s) + \Phi_X(1-s) - (2s-1)\chi(X)\,\pi\,\cot(\pi s)$$

$$= \frac{Z'_X}{Z_X}(s) + \frac{Z'_X}{Z_X}(1-s) - (2s-1)\chi(X)\,\pi\,\cot(\pi s) + n_c\Upsilon_{C_\infty}(s). \quad (10.52)$$

Recall from (3.10) that

$$\frac{d}{ds} \log \frac{G_\infty(s)}{G_\infty(1-s)} = -(2s-1)\pi\,\cot(\pi s).$$

This allows us to rewrite (10.52) as

$$\Upsilon_X(s) = \frac{Z'_X}{Z_X}(s) + \frac{Z'_X}{Z_X}(1-s) + \chi(X)\frac{d}{ds} \log \frac{G_\infty(s)}{G_\infty(1-s)} + n_c\Upsilon_{C_\infty}(s). \quad (10.53)$$

Breaking R_F up into a direct sum of the R_{F_j}'s and applying Propositions 10.6 and 10.8, we see that

$$\Upsilon_F(s) = \sum_{j=1}^{n_f} \Upsilon_{F_j}(s) = \frac{d}{ds} \log \left(\frac{Z_F(s)}{Z_F(1-s)} \right). \quad (10.54)$$

Proposition 10.9 now shows that

$$-\frac{\tau'_X}{\tau_X}(s) = \frac{Z'_X}{Z_X}(s) + \frac{Z'_X}{Z_X}(1-s) - \frac{Z'_F}{Z_F}(s) - \frac{Z'_F}{Z_F}(1-s)$$

$$+ \chi(X)\frac{d}{ds} \log \frac{G_\infty(s)}{G_\infty(1-s)} + n_c\Upsilon_{C_\infty}(s). \quad (10.55)$$

By Propositions 10.7 and 10.8, we have

$$\Upsilon_{C_\infty}(s) = -2\log 2 - \Psi(s+\tfrac{1}{2}) - \Psi(\tfrac{3}{2}-s)$$

$$= -\frac{d}{ds} \log \frac{\Gamma(s-\frac{1}{2})}{4^s\Gamma(\frac{1}{2}-s)}.$$

Thus, except possibly for $Z'_X/Z_X(s)$, all terms in (10.55) are logarithmic derivatives of meromorphic functions. Since Proposition 10.11 shows that $Z'_X/Z_X(s)$ has simple poles with integer residues in the half-plane $\operatorname{Re} s \geq \frac{1}{2}$, the relation (10.55) now extends this result to the reflected half-plane for $\operatorname{Re} s \leq \frac{1}{2}$. Hence $Z'_X/Z_X(s)$ can be integrated and so $Z_X(s)$ has a meromorphic extension to $\mathbb{C}$.

The relation (10.51) follows immediately by integrating (10.55). □

Combining Lemma 10.2 with Proposition 10.12 and the definition of Z_F yields

$$\frac{Z_X(1-s)}{Z_X(s)} = e^{h_1(s)}\frac{P_X(1-s)}{P_X(s)}\left(\frac{G_\infty(s)}{G_\infty(1-s)}\right)^{\chi(X)}\left(\frac{\Gamma(s-\frac{1}{2})}{\Gamma(\frac{1}{2}-s)}\right)^{-n_c}, \qquad (10.56)$$

for some entire function h_1. If we set

$$f(s) := \frac{Z_X(s)G_\infty(s)^{\chi(X)}}{P_X(s)\Gamma(s-\frac{1}{2})^{n_c}},$$

Proposition 10.11 shows that $f(s)$ has no zeros or poles in $\mathrm{Re}\, s \geq \frac{1}{2}$. And (10.56) gives the relation

$$f(1-s) = e^{h_1(s)}f(s),$$

implying that $f(s)$ has no zeros or poles in $\mathrm{Re}\, s \leq \frac{1}{2}$ either. Hence $f(s)$ is the exponential of an entire function $q(s)$, and we have

$$Z_X(s) = e^{q(s)}P_X(s)G_\infty(s)^{-\chi(X)}\Gamma(s-\tfrac{1}{2})^{n_c}. \qquad (10.57)$$

10.5 Order bound

With the structure of the zeta function established in (10.57), we have one final step in the proof of Theorem 10.1, which is to prove that $q(s)$ is polynomial of degree at most 2. In fact, the main task is to see that it's polynomial; the bound on the degree will be easy after that.

Before embarking on the (rather technical) proof, we should point out that if X has no cusps (Γ is convex cocompact), then the same dynamical methods that furnish a simple proof of meromorphic continuation of $Z_X(s)$ also show that $|Z_X(s)| \leq e^{C|s|^2}$. See Theorem 15.8 and Proposition 15.9.

From the definition (2.22) and the elementary estimate on the counting function for lengths of closed geodesics given in Proposition 2.19, we can easily show that $|Z_X(s)|$ is bounded by a constant in the half-plane $\mathrm{Re}\, s \geq 1 - \alpha$, for $0 < \alpha < 1 - \delta$ (where δ denotes the exponent of convergence for Γ). Since $P_X(s)$ and $1/G_\infty(s)$ are entire of order 2 and $1/\Gamma(s-\frac{1}{2})$ is of order 1, we conclude from (10.57) that

$$|q(s)| \leq C_\varepsilon \langle s\rangle^{2+\varepsilon} \quad \text{for } \mathrm{Re}\, s \geq 1 - \alpha. \qquad (10.58)$$

If we try to estimate $q(s)$ directly, in a region that includes resonances, then the best we can do is an exponential bound. In order to extend the polynomial bound, we will first show that the function $h_1(s)$ appearing in (10.56) is polynomial, because (10.56) implies a functional relation for $q(s)$ in terms of $h_1(s)$. The fact that $h_1(s)$ is polynomial will be a consequence of a bound on the order of the relative scattering determinant.

Lemma 10.13. *In the factorization formula from Lemma 10.2,*

$$\tau_X(s) = e^{h(s)} \frac{P_X(1-s)}{P_X(s)} \frac{P_F(s)}{P_F(1-s)},$$

$h(s)$ is a polynomial of degree at most four.

Proof. Controlling the growth of

$$h(s) = \log \left[\tau_X(s) \frac{P_X(1-s)}{P_X(s)} \frac{P_F(s)}{P_F(1-s)} \right]$$

is difficult because the individual terms inside the logarithm have zeros and poles. In many ways the problem of working around the singularities is analogous to the complications we encountered in Section 9.4, and similar techniques of estimation will be needed here.

To estimate τ_X we must isolate its poles. Recall the determinant $D(s)$ defined in (9.4). Let $\mathcal{Z}$ be the set of zeros of $D(s)$. For $\sigma > 0$, we can define a covering of the possible poles by a union of disks,

$$B := \bigcup_{\zeta \in \mathcal{Z} \cup \mathcal{R}_F \cup (1-\mathcal{R}_F)} B_{\mathbb{C}}(\zeta, \langle \zeta \rangle^{-2-\sigma}).$$

The decrease in radius for large $|\zeta|$ ensures that the disks are disjoint outside some compact set, since the counting function for $\mathcal{Z} \cup \mathcal{R}_F \cup (1 - \mathcal{R}_F)$ grows at most quadratically. By the Weyl inequality (9.6),

$$|\tau_X(s)| \le \prod_{k=1}^{\infty} (1 + \mu_k(S_{\text{rel}}(s) - I)).$$

We claim that for $\varepsilon > 0$,

$$\mu_k(S_{\text{rel}}(s) - I) \le e^{C\langle s \rangle^{2+\varepsilon} - ck}, \tag{10.59}$$

for $s \notin B$. This will complete the proof, because then for $s \notin B$,

$$|\tau_X(s)| \le C_1 \prod_{k < C_2 \langle s \rangle^{2+\varepsilon}} e^{C\langle s \rangle^{2+\varepsilon}} \le e^{C|s|^{4+2\varepsilon}}.$$

The Weierstrass products P_X and P_F are of order 2, so this gives the bound

$$|h(s)| \le C \langle s \rangle^{4+2\varepsilon}$$

for $s \notin B$. Since $h(s)$ is entire, the bound is easily extended to all of $\mathbb{C}$ by the maximum modulus principle, and this shows that $h(s)$ is a polynomial of degree at most 4.

It remains to prove (10.59). For this we must break down $S_{\text{rel}}(s) - I$ into pieces we can estimate. By the formula (7.36) for the scattering matrix we have

$$S_X^{\text{ff}}(s) = S_F(s) + (2s - 1) \lim_{\rho_f, \rho_f' \to 0} (\rho_f \rho_f')^{-s} Q(s; z, z'), \qquad (10.60)$$

where $Q(s)$ is the remainder term from Theorem 6.3. Using the notation from the proof of that theorem, $Q(s) = M(s)K(s)$, where $M(s)$ is the parametrix and $K(s) = (I - L(s))^{-1}L(s)$. Since $\chi_3 L(s) = L(s)$, we can introduce $L_3(s) := L(s)\chi_3$ as in Section 9.1, yielding the formula

$$Q(s) = M(s)(1 - L_3(s))^{-1}L(s).$$

The limits $\rho, \rho' \to 0$ can then be expressed directly in terms of model Poisson operators. In the funnels,

$$\lim_{\rho_f \to 0} \rho_f^{-s} M(s; z, z') = E_F(s; z', x)(1 - \chi_1(z'))$$

and

$$\lim_{\rho_f' \to 0} \rho_f'^{-s} L(s; z, z') = [\Delta_X, \chi_0] E_F(s; z, x').$$

Together, these limits reduce (10.60) to the formula

$$S_X^{\text{ff}}(s) = S_F(s) + (2s - 1)E_F(s)^{\text{t}}(1 - \chi_1)(I - L_3(s))^{-1}[\Delta_X, \chi_0] E_F(s).$$

Since $E_F(s)S_F(s)^{-1} = -E_F(1 - s)$, we have

$$S_{\text{rel}}^{\text{ff}}(s) - I^{\text{ff}} = -(2s - 1)E_F(1-s)^{\text{t}}(1 - \chi_1)(I - L_3(s))^{-1}[\Delta_X, \chi_0] E_F(s). \quad (10.61)$$

In the same way, we can use (7.36) to derive the formulas

$$S_{\text{rel}}^{\text{fc}}(s) = (2s - 1)E_F(1 - s)^{\text{t}}(1 - \chi_1)(I - L_3(s))^{-1}[\Delta_X, \chi_0] E_C(s),$$
$$S_{\text{rel}}^{\text{cf}}(s) = (2s - 1)E_C(s)^{\text{t}}(1 - \chi_1)(I - L_3(s))^{-1}[\Delta_X, \chi_0] E_F(s), \qquad (10.62)$$
$$S_{\text{rel}}^{\text{cc}}(s) = (2s - 1)E_C(s)^{\text{t}}(1 - \chi_1)(I - L_3(s))^{-1}[\Delta_X, \chi_0] E_C(s).$$

To produce bounds on $\mu_k(S_{\text{rel}} - I)$, it suffices to control terms of the form $(I - L_3(s))^{-1}$, $JE_F(s)$, and $JE_C(s)$, where J denotes either the cutoff $(1 - \chi_1)$ or the differential operator $[\Delta_X, \chi_0]$. We saw in Section 7.5 that

$$E_{C_\infty}(s; z) = \frac{y^s}{2s - 1},$$

in the standard fundamental domain $[0, 1] \times \mathbb{R}_+$. So away from $s = \frac{1}{2}$ we have the trivial bound

$$\|JE_{C_\infty}(s)\| \le e^{C\langle s \rangle}. \qquad (10.63)$$

The model operator $E_{F_\ell}(s)$ is more difficult, but the work has already been done in Section 9.4. Using Lemma 9.15 (for $\operatorname{Re} s > \varepsilon$) and the relation $E_{F_\ell}(s) = -E_{F_\ell}(1 - s)S_{F_\ell}(s)$ with Lemmas 9.14 (for $\operatorname{Re} s < \frac{1}{2} - \varepsilon$), we can see that

$$\mu_k(JE_{F_\ell}(s)) \leq \begin{cases} d(s, \mathcal{R}_{F_\ell})^{-2} e^{C\langle s \rangle \log \langle s \rangle}, & k = 1, 2, \\ \exp\left[C\langle s \rangle + 2\langle s \rangle \log \frac{\langle s \rangle}{k} - ck \right], & k > 2, \end{cases} \tag{10.64}$$

for $J = (1 - \chi_1)$. The same methods work for $J = [\Delta_X, \chi_0]$ (or any compactly supported differential operator, for that matter).

To estimate $(I - L_3(s))^{-1}$, we use

$$(I - L_3(s))^{-1} = (I + L_3(s) + L_3(s)^2)(I - L_3(s)^3)^{-1}. \tag{10.65}$$

Recall the decomposition $L_3(s) = L_i(s) + T(s)$ from Section 9.3. By definition, $\|L_i(s)\| \leq C\langle s \rangle^2$, and $\|T(s)\|$ can be bounded by $Ce^{\langle s \rangle^2}$ for $s \notin B$ using Lemma 9.7. So the first term on the right-hand side of (10.65) can be bounded:

$$\|I + L_3(s) + L_3(s)^2\| \leq Ce^{\langle s \rangle^2}.$$

The second term is handled using a resolvent estimate from Gohberg–Krein [72] (see Theorem A.23), which yields

$$\|(1 - L_3(s)^3)^{-1}\| \leq \frac{\det(I + |L_3(s)|^3)}{D(s)},$$

where $D(s) = \det(I - L_3(s)^3)$ as in (9.4). By (9.15) we know that $D(s)$ is a ratio of entire functions of order 2. And the estimate

$$\det(I + |L_3(s)|^3) \leq e^{C\langle s \rangle^{2+\varepsilon}}$$

for $s \notin B$ follows from Lemmas 9.6 and 9.7 together with (9.12). We conclude that for $s \notin B$

$$\|(I - L_3(s))^{-1}\| \leq e^{C\langle s \rangle^{2+\varepsilon}}. \tag{10.66}$$

With these estimates, we can prove (10.59) by estimating the singular values of the components of $S_{\mathrm{rel}}(s) - I$. Since $S_{\mathrm{rel}}^{\mathrm{fc}}(s)$, $S_{\mathrm{rel}}^{\mathrm{fc}}(s)$, and $S_{\mathrm{rel}}^{\mathrm{cc}}(s)$ have finite rank, it suffices to estimate the operator norms of these components. Using the estimates (10.63), (10.64), and (10.66) with the decompositions (10.62), we can bound these norms by $\exp(C\langle s \rangle^{2+\varepsilon})$. Applying the estimates to (10.61) gives the singular value estimate

$$\mu_k(S_{\mathrm{rel}}^{\mathrm{ff}}(s) - I^{\mathrm{ff}}) \leq e^{C\langle s \rangle^{2+\varepsilon} - ck}.$$

Combining these estimates using (9.8) establishes (10.59). $\square$

With 10.13, we will obtain a polynomial bound on $q(s)$ for $\mathrm{Re}\, s < \frac{1}{2}$. If $\delta < \frac{1}{2}$, then we can choose $\alpha > \frac{1}{2}$ and we will be done. If $\delta \geq \frac{1}{2}$, then we can extend the polynomial bound to the strip $\alpha < \mathrm{Re}\, s < 1 - \alpha$ by first proving an exponential bound and then applying the Phragmén–Lindelöf theorem.

Lemma 10.14. *Within the strip $\alpha \leq \mathrm{Re}\, s \leq 1 - \alpha$,*

$$|q(s)| \leq C_\varepsilon \exp(|s|^{2+\varepsilon})$$

for $\varepsilon > 0$.

Proof. From (10.57) and Proposition 10.10 we derive that

$$q'(s) = \Phi_X(s) - n_c \Phi_{C_\infty}(s) + \frac{P'_X}{P_X}(s) - \chi(X)\frac{G'_\infty}{G_\infty}(s) - n_c \Psi(s - \tfrac{1}{2}).$$

Other than $\Phi_X(s)$, the terms on the right are logarithmic derivatives of entire functions of order 2. Hence it suffices to prove the exponential bound for $\Phi_X(s)$. Since $\Phi_X(s)$ is meromorphic, we must work around its poles. Define (for some fixed $\sigma > 0$)

$$F_\alpha := \{\alpha \leq \mathrm{Re}\, s \leq 1 - \alpha\} - \bigcup_{\zeta \in \mathcal{R}_X \cup \{1/2\}} B_{\mathbb{C}}(\zeta; \langle\zeta\rangle^{-2-\sigma}),$$

where $B_{\mathbb{C}}(\zeta; r)$ denotes a Euclidean disk of radius r centered on ζ. We claim that

$$|\Phi_X(s)| \leq C_\varepsilon \exp(|s|^{2+\varepsilon}), \tag{10.67}$$

for $s \in F_\alpha$. Integrating this to obtain a bound for $q(s)$ and using the maximum modulus principle to fill in the gaps will then complete the proof.

To prove (10.67), we break $\Phi_X(s)$ up into pieces. Let s_0 be the fixed value used in the definition of M_i in the parametrix construction of Section 6.3. Then, using $R_X(s) = M(s)(I + K(s))$ from the parametrix construction, we have

$$\begin{aligned}
\Phi_X(s) - \Phi_X(s_0) = {}& 0\text{-tr}\,[M(s) - M(s_0)] \\
& + 0\text{-tr}\,[(M(s) - M(s_0))K(s)] \\
& + 0\text{-tr}\,[M(s_0)(K(s) - K(s_0))] \\
& + a(s, s_0)\, 0\text{-vol}(X),
\end{aligned} \tag{10.68}$$

where

$$a(s, s_0) := \Big[R_{\mathbb{H}}(s; z, z') - R_{\mathbb{H}}(s_0; z, z')\Big]_{z=z'},$$

which is independent of z.

We can evaluate $a(s, s_0)$ explicitly by (4.14),

$$a(s, s_0) = \frac{1}{2\pi}\,[\Psi(s_0) - \Psi(s)].$$

In particular, the growth of $|a(s, s_0)|$ in F_α is only logarithmic.

Consider the first line of (10.68), which we can reduce to a sum of traces of the model operators by writing

$$M(s) - M(s_0) = M_f(s) - M_f(s_0) + M_c(s) - M_c(s_0).$$

First we'll look at the contribution from the funnel terms. Although $M_f(s) - M_f(s_0)$ has a continuous kernel, it is not trace class and its 0-trace is difficult to estimate directly. Fortunately, we can reduce to an actual trace by noting that

$$0\text{-tr}\big[M_f(s) - M_f(s_0)\big] = \Phi_F(s) - \Phi_F(s_0) - \mathrm{tr}\big[\chi_1(R_F(s) - R_F(s_0))\chi_1\big],$$

where we have used $0\text{-vol}(F) = 0$ to drop the $R_{\mathbb{H}}(s) - R_{\mathbb{H}}(s_0)$ term. Note that the explicit formula for $\Phi_{F_\ell}(s)$ from Proposition 10.6 shows that $\Phi_F(s)$ is bounded by a constant in the strip $\alpha \leq \operatorname{Re} s \leq 1 - \alpha$.

By (4.13) and (5.2) we can write the hyperbolic cylinder resolvent as a sum (5.2),

$$R_{C_\ell}(s; z, z') = \frac{1}{4\pi} \sum_{k \in \mathbb{Z}} \int_0^1 \frac{(t(1-t))^{s-1}}{(\sigma(z, e^{k\ell}z') - t)^s} \, dt.$$

Differentiating with respect to s gives an integral convergent even on the diagonal $\{\sigma = 1\}$,

$$\frac{d}{ds} R_{C_\ell}(s; z, z') = \frac{1}{4\pi} \sum_{k \in \mathbb{Z}} \int_0^1 \frac{(t(1-t))^{s-1}}{(\sigma(z, e^{k\ell}z') - t)^s} \log\left[\frac{t(1-t)}{\sigma(z, e^{k\ell}z') - t}\right] dt.$$

In the strip $\alpha \leq \operatorname{Re} s \leq 1 - \alpha$, we have

$$\left|\frac{d}{ds} R_{C_\ell}(s; z, z')\right| = O(1).$$

This implies the same estimate for $(d/ds)R_{F_\ell}(s; z, z')$, which can then be integrated to

$$\left|R_{F_\ell}(s; z, z') - R_{F_\ell}(s_0; z, z')\right| = O(|s|), \tag{10.69}$$

for $\alpha \leq \operatorname{Re} s \leq 1 - \alpha$. This implies directly that

$$\left|\operatorname{tr}\left[\chi_1 (R_F(s) - R_F(s_0))\chi_1\right]\right| \leq O(|s|).$$

Similarly for $M_{\mathrm{c}}(s) - M_{\mathrm{c}}(s_0)$, we first write

$$0\text{-tr}\left[M_{\mathrm{c}}(s) - M_{\mathrm{c}}(s_0)\right] = \Phi_C(s) - \Phi_C(s_0) - \operatorname{tr}\left[\chi_1(R_C(s) - R_C(s_0))\chi_1\right],$$

and use the explicit formula for $\Phi_C(s)$ from Proposition 10.7 to control this term. The remainder can be estimated directly from the formulas for the Fourier decomposition of $R_{C_\infty}(s; z, z')$ from Section 5.3. The Fourier decomposition of $R_{C_\infty}(s; z, z') - R_{C_\infty}(s_0; z, z')$ converges even on the diagonal, so we can estimate the Fourier series term by term. The Bessel function estimates (9.27) and (9.28) show that

$$\left|R_{C_\infty}(s; z, z') - R_{C_\infty}(s_0; z, z')\right| = O(|s|), \tag{10.70}$$

for $\alpha \leq \operatorname{Re} s \leq 1 - \alpha$. This in turn gives

$$\left|\operatorname{tr}\left[\chi_1(R_C(s) - R_C(s_0))\chi_1\right]\right| = O(|s|),$$

for s in the same range and bounded away from $\frac{1}{2}$.

With these estimates we have now shown that

$$\left|0\text{-tr}\left[M(s) - M(s_0)\right]\right| = O(|s|), \tag{10.71}$$

for $s \in F_\alpha$.

Turning to the second line of (10.68), we note that

$$K(s) = \chi_3(I - L_3(s))^{-1} L(s).$$

We want to exploit the fact that $(M(s) - M(s_0))\chi_3$ is trace class, but the problem is that $L(s)$ is not a bounded operator unless $\mathrm{Re}\, s > \frac{1}{2}$. Since the problem lies in the boundary behavior in the z' variable, we can correct this by using $L(s)\rho$ instead. Thus we will write the second line of (10.68) as

$$0\text{-tr}\left[\rho^{-1}(M(s) - M(s_0))\chi_3(I - L_3(s))^{-1}L(s)\rho\right].$$

(Inserting the factors of ρ and ρ^{-1} is justified by the fact that for the 0-trace we restrict the integral kernel to the diagonal.) The expression in brackets is now trace class for $\mathrm{Re}\, s > -\frac{1}{2}$. Using (10.69) and (10.70), we can see that

$$\left\|\rho^{-1}(M(s) - M(s_0))\chi_3\right\|_1 \leq C\langle s\rangle,$$

for $s \in F_\alpha$, where $\|A\|_1 := \mathrm{tr}\,|A|$ is the trace norm. The operator norm of $(I - L_3(s))^{-1}$ was bounded by $\exp(C|s|^{2+\varepsilon})$, for $s \in F_\alpha$, in (10.66). And finally, we can bound

$$\|L(s)\rho\| = O(|s|^2),$$

for $\alpha \leq \mathrm{Re}\, s \leq 1 - \alpha$ and s bounded away from $\frac{1}{2}$, by extending the arguments of Lemmas 9.8, 9.10, and 9.11. Combining these estimates gives

$$\left|0\text{-tr}\left[(M(s) - M(s_0))K(s)\right]\right| \leq e^{C|s|^{2+\varepsilon}}, \tag{10.72}$$

for $s \in F_\alpha$.

Finally, we have the third line of (10.68), which we rewrite introducing a weight as above,

$$0\text{-tr}\left[\rho^{-1}M(s_0)\chi_3(K(s) - K(s_0))\rho\right].$$

The operator $\rho^{-1}M(s_0)\chi_3$ is of order -2 in the interior, implying a logarithmic singularity on the diagonal. Assuming that $\mathrm{Re}\, s_0 > \frac{3}{2}$, $\rho^{-1}M(s_0)\chi_3$ will be Hilbert–Schmidt (its kernel is square-integrable). The same holds for $K(s)\rho$, for $s \in F_\alpha$. Thus $\rho^{-1}M(s_0)\chi_3 K(s)\rho$ is trace class for $s \in F_\alpha$ even without the subtraction of $K(s_0)$. This yields the estimate,

$$\left|0\text{-tr}\left[\rho^{-1}M(s_0)\chi_3 K(s)\rho\right]\right| = \left|\mathrm{tr}\left[\rho^{-1}M(s_0)\chi_3 K(s)\rho\right]\right|$$
$$= \left\|\rho^{-1}M(s_0)\chi_3\right\|_2 \left\|(I - L_3(s))^{-1}\right\| \|L(s)\rho\|_2.$$

Obviously, the first Hilbert–Schmidt norm is constant, and we have an exponential bound on the second (operator) norm in (10.66). For the third norm, since Lemmas 9.10 and 9.11 were proven by direct estimation of the kernels, we can easily adapt the proofs to show that

$$\|L(s)\rho\|_2 = O(|s|^2),$$

for $\alpha \le \operatorname{Re} s \le 1 - \alpha$ and bounded away from $\frac{1}{2}$. Hence

$$\left|0\text{-tr}\big[\rho^{-1}M(s_0)\chi_3(K(s) - K(s_0))\rho\big]\right| \le e^{C|s|^{2+\varepsilon}}, \qquad (10.73)$$

for $s \in F_\alpha$. With (10.71), (10.72), and (10.73), we have completed the proof of (10.67).   □

Now let us assemble these ingredients to finish the proof of the main result of this chapter.

Proof of Theorem 10.1. The factorization formula was established in (10.57). According to (10.56), $q(s)$ and $h_1(s)$ are related by

$$q(1 - s) = q(s) + h_1(s) \qquad (10.74)$$

(up to an integer multiple of $2\pi i$). From (10.56) and (10.51), the relation between $h(s)$ and $h_1(s)$ is

$$e^{h(s)} = c4^{-sn_c}e^{h_1(s)}\frac{P_F(1 - s)}{P_F(s)}\frac{Z_F(s)}{Z_F(1 - s)}.$$

Since P_F and Z_F are entire functions of order 2, and $h(s)$ is a polynomial of degree at most 4 by Lemma 10.13, we conclude that $h_1(s)$ is also polynomial of degree at most 4. Thus, the relation (10.74) allows us to extend the bound (10.58) to

$$|q(s)| \le C\langle s\rangle^4,$$

for $\operatorname{Re} s \notin [\alpha, 1 - \alpha]$. With the exponential estimate in the strip from Lemma 10.14, we can extend the $C\langle s\rangle^4$ bound to all $s \in \mathbb{C}$ by Phragmén–Lindelöf (Theorem A.5). This implies that $q(s)$ is a polynomial of degree at most 4.

To see that the degree is actually at most 2, we take the log of (10.57) to get

$$q(s) = \log Z_X(s) - \log P_X(s) + \chi(X)\log G_\infty(s) - n_c \log \Gamma(s - \tfrac{1}{2}),$$

and then consider the asymptotics as $\operatorname{Re}(s) \to \infty$. In this limit, $\log Z_X(s)$ decays exponentially, by the defining formula (2.22). We also have $\log G_\infty(s) = O(s^2 \log s)$ and $\log \Gamma(s - \frac{1}{2}) = O(s \log s)$. Since $P_X(s)$ is entire of order 2, it follows that $q(s) = O(|s|^{2+\varepsilon})$ as $\operatorname{Re} s \to \infty$, for any $\varepsilon > 0$. A polynomial grows at the same rate in all directions, so $q(s)$ is at most quadratic.   □

10.6 Determinant of the Laplacian

For X an infinite-area geometrically finite hyperbolic surface, we can define (up to two constants of integration) a "determinant" of the Laplacian $D_X(s)$ formally equal to $\det(\Delta_X - s(1 - s))$ by integrating the expression

$$\left(\frac{1}{2s-1}\frac{d}{ds}\right)^2 \log D_X(s) = -\text{0-tr } R_X(s)^2. \tag{10.75}$$

In this short section, we observe that the calculations done above can be used to compute $D_X(s)$, up to two constants of integration not fixed by (10.75).

If we differentiate the definition of $\Phi_X(s)$, then the traces can be separated to obtain

$$\frac{1}{2s-1}\frac{d}{ds}\left(\frac{\Phi_X(s)}{2s-1}\right) = -\text{0-tr } R_X(s)^2 + \text{0-tr } R_{\mathbb{H}}(s)^2.$$

The first term on the right-hand side makes sense because $R_X(s)^2$ has a continuous kernel. For the second term on the right, we can see from (4.14) that

$$\frac{d}{ds}R_{\mathbb{H}}(s;z,w)\bigg|_{z=w} = -\frac{1}{2\pi}\Psi'(s).$$

Since $(d/ds)R_{\mathbb{H}}(s) = -(2s-1)R_{\mathbb{H}}(s)^2$, this gives

$$\text{0-tr } R_{\mathbb{H}}(s)^2 = \frac{-\chi(X)}{2s-1}\Psi'(s),$$

where $\text{0-vol}(X) = -2\pi\chi$ has been used to take the 0-integral. By reinterpreting this expression using (3.9), we now have that

$$\frac{1}{2s-1}\frac{d}{ds}\left(\frac{\Phi_X(s)}{2s-1}\right) = -\text{0-tr } R_X(s)^2 - \chi(X)\left(\frac{1}{2s-1}\frac{d}{ds}\right)^2 \log G_\infty(s).$$

Combining this formula with the differentiation of (10.44) and Proposition 10.7, we can deduce that

$$-\text{0-tr } R_X(s)^2 = \left(\frac{1}{2s-1}\frac{d}{ds}\right)^2 \log\left[\frac{Z_X(s)G_\infty(s)^{\chi(X)}}{2^{sn_c}(s-\frac{1}{2})^{n_c/2}\Gamma(s-\frac{1}{2})^{n_c}}\right]. \tag{10.76}$$

In view of Theorem 10.1, this proves the following:

Theorem 10.15. *If $D_X(s)$ is a function satisfying (10.75) for s in some region, then $(s-\frac{1}{2})^{n_c/2}D_X(s)$ extends to an entire function with zeros given exactly by $\mathcal{R}_X$ (with multiplicities). This function can be expressed as*

$$(s-\tfrac{1}{2})^{n_c/2}D_X(s) = e^{[Fs(1-s)+E+sn_c\log 2]}\frac{Z_X(s)G_\infty(s)^{\chi(X)}}{\Gamma(s-\frac{1}{2})^{n_c}}$$

$$= e^{q_1(s)}P_X(s),$$

where F, E are the constants of integration left unspecified by (10.75), and $q_1(s)$ is a polynomial of degree ≤ 2.

This is consistent with calculations of the determinant of the Laplacian by Sarnak [185], Efrat [54], and Borthwick–Judge–Perry [23].

Notes

A trace formula for the logarithmic derivative of the zeta function was proven for hyperbolic manifolds without cusps by Perry [163], refining the formula of Patterson [159].

Patterson–Perry [160] established the factorization formula of Theorem 10.1 for conformally compact hyperbolic manifolds in any dimension. Their proof of the meromorphic continuation of the zeta function, with a bound on its order, is dynamical, based on work of Ruelle [183] and Fried [65]. (In two dimensions, this approach can be formulated quite simply; see Chapter 15.) The dynamical methods do not apply to manifolds with cusps. For an overview see the review article by Perry [168].

To define the relative scattering determinant we need the model funnel scattering matrix as a "comparison operator." This is a serious obstacle to the definition of a relative scattering determinant in higher dimensions, because we do not have global models for the ends. However, for a natural class of asymptotically hyperbolic manifolds, Guillarmou [77] has recently developed a regularized scattering determinant based on the Kontsevich–Vishik trace for pseudodifferential operators. This is an intrinsic regularization with no need for a comparison operator. His results include a functional equation, analogous to that of Proposition 10.12, but without the need to include Z_F. This version generalizes the compact case (Corollary 3.10) more directly.

Bunke–Olbrich [32, 34, 35] introduced new tools for the study of the Selberg zeta function involving hyperfunctions supported on the limit set of a Kleinian group Γ. In particular, they proved Patterson's conjecture [159] that the singularities of the zeta function could be characterized in terms of group cohomology. For an overview see Juhl [105].

See Perry [170] for a survey of the spectral geometry of geometrically finite hyperbolic and asymptotically hyperbolic manifolds in higher dimensions.

11

Wave Trace and Poisson Formula

On a compact manifold, the wave trace is defined to be the distributional trace of the wave operator, $U(t) := e^{it\sqrt{\Delta}}$. This is a spectral invariant because we could write it directly as

$$\operatorname{tr} U(t) = \sum_{\lambda_j \in \sigma_d(\Delta)} e^{it\sqrt{\lambda_j}}.$$

(The sum doesn't converge, but is well-defined as a distribution.) In the early 1970s, work of Colin de Verdière [48, 49], Chazarain [41], and Duistermaat–Guillemin [53], showed that the wave trace has singularities only at values of t equal to lengths of closed geodesics. Such a connection had previously appeared in the physics literature, for example in the work of Balian–Bloch. The Selberg trace formula itself furnishes a prototype for these results. For compact hyperbolic surfaces one can use it to express the wave trace explicitly as a sum over the length spectrum.

For an infinite-area hyperbolic surface, the trace of the wave operator does not exist even in the distributional sense, but we can define a regularized wave trace using the 0-integral. We will prove in Theorem 11.2 an analogue of the Selberg trace formula expressing this as a sum over the length spectrum. It is far from obvious that the regularized wave trace could also be expressed as a sum over the resonance set $\mathcal{R}_X$, but fortunately this does turn out to be the case. This is the content of the Poisson formula, Theorem 11.3.

We can illustrate the Poisson formula explicitly in the case of a hyperbolic cylinder, $C_\ell = \Gamma_\ell \backslash \mathbb{H}$, for which

$$\mathcal{R}_{C_\ell} = \frac{2\pi i}{\ell} \mathbb{Z} - \mathbb{N}_0 \quad \text{(multiplicity 2)}.$$

Given a test function, $\phi \in C_0^\infty(\mathbb{R}_+)$, we observe that

$$\frac{1}{2} \sum_{\zeta \in \mathcal{R}_{C_\ell}} \int_0^\infty e^{(\zeta - 1/2)t} \phi(t)\, dt = \sum_{n=0}^\infty \sum_{k \in \mathbb{Z}} \int_0^\infty \exp\left(\frac{2\pi i k t}{\ell} - (n + \tfrac{1}{2})t \right) \phi(t)\, dt$$

$$= \sum_{k \in \mathbb{Z}} \int_0^\infty e^{2\pi i k t/\ell} \frac{\phi(t)}{2 \sinh t} \, dt$$

$$= \sum_{k \in \mathbb{Z}} \hat{f}(2\pi k/\ell),$$

where $f(t) := \phi(t)/(2 \sinh t/2)$. Noting that $f \in C_0^\infty(\mathbb{R})$, we can apply the classical Poisson summation formula (Theorem A.6) to obtain

$$\sum_{k \in \mathbb{Z}} \hat{f}(2\pi k/\ell) = \ell \sum_{m \in \mathbb{Z}} f(m\ell).$$

We have thus shown that

$$\frac{1}{2} \sum_{\zeta \in \mathcal{R}_{C_\ell}} \int_0^\infty e^{(\zeta - 1/2)t} \phi(t) \, dt = \sum_{m \in \mathbb{N}} \frac{\ell}{2 \sinh(m\ell/2)} \phi(m\ell).$$

We could write this more cleanly as a relation between distributions on $\mathbb{R}_+$,

$$\frac{1}{2} \sum_{\zeta \in \mathcal{R}_{C_\ell}} e^{(\zeta - 1/2)t} = \sum_{m=1}^\infty \frac{\ell}{2 \sinh(m\ell/2)} \delta(t - m\ell). \tag{11.1}$$

The right-hand side turns out to be precisely the regularized wave trace for the hyperbolic cylinder.

11.1 Regularized wave trace

Consider the wave equation on a hyperbolic surface X,

$$(\partial_t^2 + \Delta_X - \tfrac{1}{4})u = 0,$$

with initial conditions $u|_{t=0} = f$, $\partial_t u|_{t=0} = g$. The functional calculus allows us to express the general solution as

$$u = \partial_t W_X(t) f + W(t) g,$$

using the wave operator

$$W_X(t) := \frac{\sin\left(t\sqrt{\Delta_X - \tfrac{1}{4}}\right)}{\sqrt{\Delta_X - \tfrac{1}{4}}}$$

and its derivative

$$\partial_t W_X(t) := \cos\left(t\sqrt{\Delta_X - \tfrac{1}{4}}\right).$$

Even in the compact case, the wave trace, $\operatorname{tr} \partial_t W_X(t)$, is not well-defined as a function, but rather must be interpreted as a distribution on $\mathbb{R}$. In our infinite-area setting, the trace does not exist even in this sense, but Guillopé–Zworski [87] introduced the following substitute:

Definition. For a geometrically finite hyperbolic surface X of infinite area, the *wave 0-trace* is the distribution on $\mathbb{R}$ given by

$$\Theta_X(t) := 0\text{-tr}\left[\cos\left(t\sqrt{\Delta_X - \tfrac{1}{4}}\right)\right]. \tag{11.2}$$

To understand this more explicitly, let $\varphi \in C_0^\infty(\mathbb{R})$. The operator

$$\int_{-\infty}^\infty \varphi(t) \cos\left(t\sqrt{\Delta_X - \tfrac{1}{4}}\right) dt$$

has a smooth kernel, and its 0-trace over X is well-defined. The meaning of (11.2) is expressed by the pairing

$$\langle \Theta_X, \varphi \rangle := 0\text{-tr}\left(\int_{-\infty}^\infty \varphi(t) \cos\left(t\sqrt{\Delta_X - \tfrac{1}{4}}\right) dt\right).$$

11.2 Model wave kernel

In the following we'll use the notation $[\,\cdot\,]_+$ to denote positive part, i.e., $[x]_+ := \max\{x, 0\}$.

Proposition 11.1. *On the hyperbolic plane, the kernel of the wave operator $W_\mathbb{H}(t)$ is*

$$W_\mathbb{H}(t; z, z') = \frac{\mathrm{sgn}(t)}{4\pi}\left[\sinh^2(t/2) - \sinh^2(d(z, z')/2)\right]_+^{-1/2}.$$

Proof. We'll follow the proof by Lax–Phillips [113]. (For a different approach, see [206].) It's difficult to verify the formula directly, so the idea is to establish the easier 3-dimensional formula and then use the method of descent to reduce to $\mathbb{H}$.

Let $\mathbb{H}^3$ be the upper half-space in $\mathbb{R}^3$. In geodesic polar coordinates the Laplacian is

$$\Delta_{\mathbb{H}^3} = -\frac{1}{\sinh^2 r}\, \partial_r(\sinh^2 r\, \partial_r) - \frac{1}{\sinh^2 r}\Delta_{S^2}. \tag{11.3}$$

We claim that the 3-dimensional hyperbolic wave kernel is given by

$$W_{\mathbb{H}^3}(t; w, w') = \frac{\delta(d_3(w, w') - t)}{4\pi \sinh t},$$

where d_3 denotes distance in $\mathbb{H}^3$. To see this, we first check using (11.3) that

$$(\Delta_{\mathbb{H}^3} - 1)\frac{\delta(r - t)}{\sinh r}$$
$$= -\sinh^{-2} r\, \partial_r\left(\sinh^2 r\, \delta'(r - t) - \cosh r\, \delta(r - t)\right) - \sinh^{-1} r\, \delta(r - t)$$
$$= \frac{\delta''(r - t)}{\sinh r}.$$

This shows, for $f \in C_0^\infty(\mathbb{H}^3)$, that

$$u(t, w) := \frac{1}{4\pi \sinh t} \int_{d_3(w,w')=t} f(w') \, d\Omega(w') \tag{11.4}$$

(where $d\Omega$ is the induced area form on $\{d_3(w, w') = t\}$) is a solution of the $\mathbb{H}^3$ wave equation

$$(\partial_t^2 + \Delta_{\mathbb{H}^3} - 1)u = 0.$$

Moreover, since the surface area of a sphere of radius t in $\mathbb{H}^3$ is $4\pi \sinh^2 t$, it's also easy to confirm that

$$u(0, w) = 0, \qquad \partial_t u(0, w) = f(w).$$

If we use the upper half-space model for $\mathbb{H}^3$, with coordinates $w = (x_1, x_2, y)$, $y > 0$, then $\mathbb{H}$ is naturally embedded as the set $\{x_2 = 0\}$. In these coordinates the Laplacians are

$$\Delta_{\mathbb{H}^3} = -y^2 \partial_y^2 + y \partial_y - y^2 (\partial_{x_1}^2 + \partial_{x_2}^2)$$

and

$$\Delta_{\mathbb{H}} = -y^2 \partial_y^2 - y^2 \partial_{x_1}^2.$$

If v is a function that does not depend on x_2, then an easy calculation shows that

$$(\Delta_{\mathbb{H}^3} - 1)\left[\sqrt{y}\, v(x_1, y)\right] = \sqrt{y}\, (\Delta_{\mathbb{H}} - \tfrac{1}{4})v.$$

Thus we can use (11.4) to construct solutions of $(\partial_t^2 + \Delta_{\mathbb{H}} - \tfrac{1}{4})v = 0$ by setting $u(x_1, x_2, y) = \sqrt{y}\, v(x_1, y)$. For $f \in C_0^\infty(\mathbb{H})$, regarded as a function on $\mathbb{H}^3$ independent of x_2, we obtain the 2-dimensional wave solution

$$v(t, w) = \frac{1}{4\pi \sqrt{y} \sinh t} \int_{d_3(w,w')=t} \sqrt{y'}\, f(w') \, d\Omega(w'). \tag{11.5}$$

The remainder of the proof amounts to a calculation of $d\Omega(w')$. For $w = (x_1, 0, y)$ and $w' = (x_1', x_2', y')$, the set $\{d_3(w, w') = t\}$ is a Euclidean sphere with the equation

$$(x_1 - x_1')^2 + x_2'^2 + y^2 + y'^2 - 2yy' \cosh t = 0.$$

We can set $x_1 = 0$ for convenience. Parametrizing $\{d_3(w, w') = t\}$ by

$$(x_1', y') \mapsto \left(x_1', [2yy' \cosh t - x_1'^2 - y^2 - y'^2]^{1/2}, y'\right),$$

we compute directly from the hyperbolic metric on $\mathbb{H}^3$ that

$$d\Omega(x_1', y') = \frac{y \sinh t}{[2yy' \cosh t - x_1'^2 - y^2 - y'^2]^{1/2}} \frac{dx_1' \, dy'}{y'^2}$$

$$= \frac{y \sinh t}{[4yy'(\sinh^2(t/2) - \sinh^2(d(z, z')/2))]^{1/2}} dA(z').$$

Now we can replace the integral over a sphere in $\mathbb{H}^3$ in (11.5) with twice the integral over the projection of this sphere to $\mathbb{H}$, to obtain

$$v(t, z) = \frac{1}{4\pi} \int_{d(z, z') \le t} \frac{f(z')}{\sqrt{\sinh^2(t/2) - \sinh^2(d(z, z')/2)}} \, dA(z').$$

This formula gives the solution of the wave equation on $\mathbb{H}$ for $t > 0$. We can easily deduce the corresponding result for $t < 0$. $\square$

11.3 Wave 0-trace formula

We are now prepared to present the trace formula for the regularized wave trace, a result of Guillopé–Zworski from [88].

To obtain a trace formula valid through $t = 0$, we will need to interpret the function $\sinh^{-1}(t/2)$ as a principal value distribution. This means that the integral is regularized by removing a symmetric neighborhood of the singularity,

$$\left\langle \varphi, \mathrm{PV}\left[\sinh^{-1}(t/2)\right]\right\rangle := \lim_{\varepsilon \to 0} \int_{|t| \ge \varepsilon} \frac{\varphi(t)}{\sinh(t/2)} \, dt.$$

This definition allows us to make sense of $\coth(t/2)$ as a distributional derivative,

$$\frac{\cosh(t/2)}{\sinh^2(t/2)} = -2\frac{d}{dt} \mathrm{PV}\left[\sinh^{-1}(t/2)\right].$$

Similarly, to interpret $\coth(|t|/2)$ as a distribution, we use

$$\coth(|t|/2) = 2\frac{d}{dt}\left(\mathrm{sgn}(t) \log \sinh(|t|/2)\right).$$

Theorem 11.2 (Wave 0-trace formula). *Let X be a geometrically finite, nonelementary hyperbolic surface of infinite area. As a distribution on $\mathbb{R}$, the wave trace can be expressed in terms of the length spectrum as the distribution*

$$\Theta_X(t) = \sum_{\ell \in \Lambda} \sum_{k=1}^{\infty} \frac{\ell}{4\sinh(k\ell/2)} \delta(|t| - k\ell)$$

$$+ \frac{\chi(X)}{4} \frac{\cosh(t/2)}{\sinh^2(t/2)} + \frac{n_c}{4} \coth(|t|/2) + n_c\gamma\,\delta(t), \qquad (11.6)$$

where γ is Euler's constant.

Proof. The proof is quite similar to that of Proposition 10.10; both results are versions of the Selberg trace formula. The lift of the wave kernel $W_X(t; z, w)$ to $\mathbb{H}$ can be written as an average over Γ,

$$W_X(t; z, w) = W_{\mathbb{H}}(t; z, w) + \sum_{T \in \Pi_h} \sum_{R \in \Gamma/\langle T \rangle} \sum_{k \neq 0} W_{\mathbb{H}}(t; z, R^{-1} T^k R w)$$

$$+ \sum_{S \in \Pi_p} \sum_{R \in \Gamma/\langle S \rangle} \sum_{k \neq 0} W_{\mathbb{H}}(t; z, R^{-1} S^k R w), \qquad (11.7)$$

where Π_h and Π_p are lists of representatives of conjugacy classes of maximal hyperbolic and parabolic subgroups of Γ, respectively.

Let $\mathcal{F}$ be a fundamental domain for Γ, and $\mathcal{F}_\varepsilon$ the corresponding lift of the region $\{\rho \geq \varepsilon\}$. Then the 0-trace may be computed as

$$0\text{-tr } W_X(t) = \mathop{\mathrm{FP}}_{\varepsilon \to 0} \int_{\mathcal{F}_\varepsilon} W_X(t; z, z) \, dA(z).$$

This is interpreted in terms of its pairing with a test function $\varphi \in C_0^\infty(\mathbb{R})$,

$$\langle \varphi, 0\text{-tr } W_X \rangle := 0\text{-tr } \int_{-\infty}^\infty \varphi(t) W_X(t) \, dt. \qquad (11.8)$$

We will study the decomposition of this pairing according to (11.7).

Since $W_{\mathbb{H}}(t; z, z)$ is independent of z, the contribution from the identity term is, for $t \neq 0$,

$$\frac{\mathrm{sgn}(t)}{4\pi |\sinh(t/2)|} \, 0\text{-vol}(X) = -\frac{\chi(X)}{2 \sinh(t/2)}.$$

To make sense of this at $t = 0$, we must take the regularization into account. Since we are to integrate $\varphi(t) W_X(t)$ over t before taking the zero trace, and then restrict to the diagonal, the contribution of the $W_{\mathbb{H}}(t)$ term to the pairing (11.8) is actually

$$-\frac{\chi(X)}{2} \lim_{\varepsilon \to 0^+} \int_{-\infty}^\infty \frac{\varphi(t) \, \mathrm{sgn}(t)}{\left[\sinh^2(t/2) - \varepsilon\right]^{1/2}} \, dt = -\frac{\chi(X)}{2} \int_0^\infty \frac{\varphi(t) - \varphi(-t)}{\sinh(t/2)} \, dt$$

$$= -\frac{\chi(X)}{2} \left\langle \varphi, \mathrm{PV}\left[\sinh^{-1}(t/2)\right] \right\rangle.$$

To compute the contribution from $T \in \Pi_h$, we use the same trick as in Proposition 10.10 to write

$$\sum_{R \in \Gamma/\langle T \rangle} \sum_{k \neq 0} \int_{\mathcal{F}_\varepsilon} W_{\mathbb{H}}(t; z, R^{-1} T^k R z) \, dA(z)$$

$$= \frac{1}{4\pi} \sum_{k \neq 0} \int_{\mathcal{D}_\varepsilon} \left[\sinh^2(t/2) - \sinh^2(d(z, T^k z)/2)\right]_+^{-1/2} dA(z),$$

where $\mathcal{D}_\varepsilon$ is the $\rho \geq \varepsilon$ part of a fundamental domain $\mathcal{D}$ for the action of $\langle T \rangle$. The integrals are bounded as $\varepsilon \to 0$, with bounds that decrease rapidly in k, so there is no need to worry about the regularization for these terms. By conjugation, we may assume that $T : z \mapsto e^\ell z$ and $\mathcal{D} = \mathbb{R} \times [1, e^\ell]$, where $\ell = \ell(T)$ is the displacement length. Recall that $T \mapsto \ell(T)$ defines a correspondence between Π_h and the primitive length spectrum Λ. After inserting

$$\sinh^2(d(z, e^{k\ell}z)/2) = \sinh^2(k\ell/2)\left(1 + \frac{x^2}{y^2}\right),$$

we simply compute the integrals

$$\frac{\mathrm{sgn}(t)}{4\pi} \int_{-\infty}^{\infty} \int_{1}^{e^\ell} \left[\sinh^2(t/2) - \sinh^2(k\ell/2)\left(1 + \frac{x^2}{y^2}\right)\right]_{+}^{-1/2} \frac{dx\,dy}{y^2},$$

$$= \frac{\ell\,\mathrm{sgn}(t)}{4\sinh(|k|\ell/2)}[|t| - |k|\ell]_{+}^{0}. \tag{11.9}$$

The contribution of $S \in \Pi_p$ is similar, except that the regularization does play a role in the cusp end of $\langle S\rangle\backslash\mathbb{H}$. We start with

$$\sum_{R\in\Gamma/\langle S\rangle} \sum_{k\neq 0} \int_{\mathcal{F}_\varepsilon} W_\mathbb{H}(t; z, R^{-1}S^k Rz)\,dA(z)$$

$$= \frac{\mathrm{sgn}(t)}{4\pi} \sum_{k\neq 0} \int_{\mathcal{D}_\varepsilon} \left[\sinh^2(t/2) - \sinh^2(d(z, S^k z)/2)\right]_{+}^{-1/2} dA(z).$$

Conjugating to $S: z \mapsto z+1$ with $\mathcal{D} = [0, 1] \times \mathbb{R}_+$, we note that

$$\sinh^2(d(z, z+k)/2) = \frac{k^2}{4y^2}.$$

The x-integration is thus trivial, and we can drop the regularization at $y = 0$, leaving us with

$$\frac{\mathrm{sgn}(t)}{4\pi} \sum_{k\neq 0} \int_{0}^{1/\varepsilon} \left[\sinh^2(t/2) - \frac{k^2}{4y^2}\right]_{+}^{-1/2} \frac{dy}{y^2}$$

$$= \frac{\mathrm{sgn}(t)}{\pi} \sum_{k=1}^{\infty} \int_{k\varepsilon/2}^{\infty} \frac{1}{k}[\sinh^2(t/2) - u^2]_{+}^{-1/2}\,du.$$

We can extract the finite part by incorporating the sum into the integral,

$$= \frac{\mathrm{sgn}(t)}{\pi} \int_{0}^{\sinh(|t|/2)} [\sinh^2(t/2) - u^2]^{-1/2}\left(\sum_{1\leq k\leq 2u/\varepsilon} \frac{1}{k}\right)du,$$

and noting that

$$\sum_{1\leq k\leq 2u/\varepsilon} \frac{1}{k} = \log\frac{2u}{\varepsilon} + \gamma + O(\sqrt{\varepsilon/2u}),$$

where γ is Euler's constant. Then to extract the finite part we simply drop the $\log\varepsilon$ term and let $\varepsilon \to 0$. This leaves an easily computed integral as the final result:

$$\underset{\varepsilon \to 0}{\mathrm{FP}} \sum_{R \in \Gamma / \langle S \rangle} \sum_{k \neq 0} \int_{\mathcal{F}_\varepsilon} W_{\mathbb{H}}(t; z, R^{-1} S^k R z) \, dA(z)$$

$$= \frac{\mathrm{sgn}(t)}{\pi} \sum_{k=1}^{\infty} \int_0^{\sinh(|t|/2)} \left[\sinh^2(t/2) - u^2 \right]^{-1/2} \left(\log 2u + \gamma \right) du$$

$$= \frac{\mathrm{sgn}(t)}{2} \left(\log \sinh(|t|/2) + \gamma \right).$$

By these calculations we have obtained the trace formula

$$0\text{-tr } W_X(t) = \frac{\mathrm{sgn}(t)}{2} \sum_{\ell \in \mathcal{L}_X} \sum_{k=1}^{\infty} \frac{\ell}{\sinh(k\ell/2)} \left[|t| - k\ell \right]_+^0$$

$$- \frac{\chi(X)}{2} \mathrm{PV} \left[\sinh^{-1}(t/2) \right] + n_c \frac{\mathrm{sgn}(t)}{2} \left(\log \sinh(|t|/2) + \gamma \right).$$

The trace formula (11.6) follows by taking the distributional derivative with respect to t. □

Note that a simpler version of the computation in Theorem 11.2 applies to a hyperbolic cylinder, giving

$$\Theta_{C_\ell}(t) = \sum_{k=1}^{\infty} \frac{\ell}{4 \sinh(k\ell/2)} \delta(|t| - k\ell), \tag{11.10}$$

since $0\text{-vol}(C_\ell) = 0$. This is the expression we observed in (11.1).

For the model funnel F_ℓ, we need a slight variant. In the fundamental domain $\mathcal{D}_\ell^+ = \mathbb{R}_+ \times [1, e^\ell]$, we can use the method of images formula to write

$$W_{F_\ell}(t; z, w) = \sum_{k \in \mathbb{Z}} \left[W_{\mathbb{H}}(t; z, e^{k\ell} w) + W_{\mathbb{H}}(t; z, -e^{k\ell} \overline{w}) \right].$$

The identity term doesn't contribute, because $0\text{-vol}(F_\ell) = 0$. The 0-trace over the terms $k \neq 0$ in the first sum is computed exactly as in (11.9), except that the fundamental domain is half the size. Thus we obtain

$$\int_{\mathcal{D}_\ell^+} W_{\mathbb{H}}(t; z, e^{k\ell} z) \, dA(z) = \frac{\ell \, \mathrm{sgn}(t)}{8 \sinh(|k|\ell/2)} \left[|t| - |k|\ell \right]_+^0.$$

For the terms that include the involution, we have, for $k \neq 0$,

$$\int_{\mathcal{D}_\ell^+} W_{\mathbb{H}}(t; z, -e^{k\ell} \overline{z}) \, dA(z)$$

$$= \frac{\mathrm{sgn}(t)}{4\pi} \int_0^{\infty} \int_1^{e^\ell} \left[\sinh^2(t/2) - \sinh^2(k\ell/2) - \cosh^2(k\ell/2) \frac{x^2}{y^2} \right]_+^{-1/2} \frac{dx \, dy}{y^2},$$

$$= \frac{\ell \, \mathrm{sgn}(t)}{8 \cosh(k\ell/2)} \left[|t| - |k|\ell \right]_+^0.$$

The $k = 0$ involution term is

$$\int_{\mathcal{D}_\ell^+} W_{\mathbb{H}}(t; z, -\bar{z}) \, dA(z) = \frac{\mathrm{sgn}(t)}{4\pi} \int_0^\infty \int_1^{e^\ell} \left[\sinh^2(t/2) - \frac{x^2}{y^2} \right]_+^{-1/2} \frac{dx \, dy}{y^2}$$

$$= \frac{\ell \, \mathrm{sgn}(t)}{8}.$$

After collecting these terms, taking the distributional derivative gives

$$\Theta_{F_\ell}(t) = \sum_{k=1}^\infty \frac{\ell e^{-k\ell/2}}{2 \sinh k\ell} \delta(|t| - k\ell) - \frac{\ell}{4} \delta(t). \tag{11.11}$$

11.4 Poisson formula

We turn now to the crucial connection between the wave trace and the resonances. This version of the Poisson formula was first proven by Guillopé–Zworski in [87], under the more general assumption of a surface with hyperbolic ends but allowing the metric to be arbitrary inside some compact set. Assuming that X is strictly hyperbolic gives us a somewhat simpler proof based on the zeta function. This route was used by Perry [169] and Guillarmou–Naud [81].

Theorem 11.3 (Poisson formula). *Let X be a geometrically finite, nonelementary hyperbolic surface of infinite area. As a distribution on $\mathbb{R}_+$, the wave trace can be expressed terms of the resonance set as*

$$\Theta_X(t) = \frac{1}{2} \sum_{\zeta \in \mathcal{R}_X} e^{(\zeta - \frac{1}{2})t} + \frac{n_c}{4}. \tag{11.12}$$

Let P_{cont} be the spectral projection onto $[1/4, \infty)$, the continuous spectrum of Δ_X. The contribution to the wave trace from the continuous spectrum is given by

$$\Theta_{\mathrm{cont}}(t) := 0\text{-tr} \left[P_{\mathrm{cont}} \cos \left(t \sqrt{\Delta_X - \tfrac{1}{4}} \right) \right],$$

interpreted in the same distributional sense as $\Theta_X(t)$. The difference $\Theta_X(t) - \Theta_{\mathrm{cont}}(t)$ represents the contribution from the discrete spectrum. This is easily computed as an actual trace over the span of the L^2 eigenfunctions,

$$\Theta_X(t) - \Theta_{\mathrm{cont}}(t) = 0\text{-tr} \left[(1 - P_{\mathrm{cont}}) \cos \left(t \sqrt{\Delta_X - \tfrac{1}{4}} \right) \right]$$

$$= \sum_{\lambda \in \sigma_d(\Delta_X)} \cos \left(t \sqrt{\lambda - 1/4} \right)$$

$$= \frac{1}{2} \sum_{\substack{\zeta \in \mathcal{R}_X \\ \mathrm{Re}\, \zeta > 1/2}} \left(e^{(\zeta - \frac{1}{2})t} + e^{(\frac{1}{2} - \zeta)t} \right). \tag{11.13}$$

To establish (11.12), we must also compute $\Theta_{\mathrm{cont}}(t)$.

Lemma 11.4. *For $\varphi \in C_0^\infty(\mathbb{R})$,*

$$\int_{-\infty}^{\infty} \varphi(t) \left[P_{\text{cont}} \cos\left(t\sqrt{\Delta_X - \tfrac{1}{4}} \right) \right] dt$$
$$= \frac{1}{4\pi} \int_{-\infty}^{\infty} 2i\xi \left[R_X(\tfrac{1}{2} + i\xi) - R_X(\tfrac{1}{2} - i\xi) \right] \hat{\varphi}(\xi) \, d\xi. \qquad (11.14)$$

Proof. By the resolvent functional calculus (see Corollary A.11), we can represent the continuous part of the wave operator as

$$P_{\text{cont}} \cos\left(t\sqrt{\Delta_X - \tfrac{1}{4}} \right)$$
$$= \lim_{\varepsilon \to 0} \frac{1}{2\pi i} \int_{1/4}^{\infty} \cos(t\sqrt{z - 1/4}) \left[(\Delta_X - z - i\varepsilon)^{-1} - (\Delta_X - z + i\varepsilon)^{-1} \right] dz,$$

With the substitution $z = 1/4 + \xi^2$, we note that

$$\lim_{\varepsilon \to 0^+} (\Delta_X - z \mp i\varepsilon)^{-1} = R_X(\tfrac{1}{2} \mp i\xi).$$

Thus we have

$$P_{\text{cont}} \cos\left(t\sqrt{\Delta_X - \tfrac{1}{4}} \right) = \frac{1}{2\pi i} \int_0^{\infty} \cos(t\xi) \left[R_X(\tfrac{1}{2} - i\xi) - R_X(\tfrac{1}{2} + i\xi) \right] 2\xi \, d\xi$$
$$= \frac{1}{2\pi i} \int_{-\infty}^{\infty} e^{-it\xi} \left[R_X(\tfrac{1}{2} - i\xi) - R_X(\tfrac{1}{2} + i\xi) \right] \xi \, d\xi.$$

The result follows by rearranging the constants and integrating against φ. □

The 0-trace of the integrand on the right-hand side of (11.14) yields

$$\Upsilon_X(\tfrac{1}{2} + i\xi) = 2i\xi \; 0\text{-tr} \left[R_X(\tfrac{1}{2} + i\xi) - R_X(\tfrac{1}{2} - i\xi) \right], \qquad (11.15)$$

for $\xi \neq 0$. We will work out what happens at $\xi = 0$ in the following:

Lemma 11.5. *For $\psi \in C^\infty(\mathbb{R})$,*

$$0\text{-tr} \left(\int_{-\infty}^{\infty} 2i\xi \left[R_X(\tfrac{1}{2} + i\xi) - R_X(\tfrac{1}{2} - i\xi) \right] \psi(\xi) \, d\xi \right)$$
$$= \int_{-\infty}^{\infty} \Upsilon_X(\tfrac{1}{2} + i\xi) \psi(\xi) \, d\xi + \pi (2m_{\frac{1}{2}} - n_c) \psi(0).$$

Proof. Away from $\xi = 0$ this follows immediately from (11.15). We can deduce the $\xi = 0$ contribution from the computation of $\Upsilon_X(s) - \Upsilon_F(s)$ in the proof of Proposition 10.9. First we will show that subtracting off $\Upsilon_F(s)$ does not affect the singularity at $s = \tfrac{1}{2}$. If $\Theta_F(t)$ denotes the wave trace over F, then by (11.11) we have

$$\Theta_F(t) = \sum_{j=1}^{n_f} \left(\sum_{k=1}^{\infty} \frac{\ell_j e^{-k\ell_j/2}}{2\sinh k\ell_j} \delta(|t| - k\ell_j) - \frac{\ell_j}{4}\delta(t) \right). \tag{11.16}$$

Through the functional calculus, for $\varphi \in C_0^{\infty}(\mathbb{R})$,

$$0\text{-tr} \int_{\mathbb{R}} \varphi(t)\Theta_F(t)\, dt$$

$$= \frac{1}{4\pi} \, 0\text{-tr} \int_{-\infty}^{\infty} 2i\xi \left[R_F(\tfrac{1}{2} + i\xi) - R_F(\tfrac{1}{2} - i\xi) \right] \hat{\varphi}(\xi)\, d\xi.$$

By taking the Fourier transform of (11.16) (which is clearly a tempered distribution because of the exponential decay), we can see that

$$0\text{-tr } 2i\xi \left[R_F(\tfrac{1}{2} + i\xi) - R_F(\tfrac{1}{2} - i\xi) \right] = \Upsilon_F(\tfrac{1}{2} + i\xi), \tag{11.17}$$

in a distributional sense, even through $\xi = 0$.

Thus any singularity at $\xi = 0$ appearing in the computation of (10.30) must come from the regularized trace over R_X. The sources of the singularity are, from the funnel–funnel terms (10.37),

$$\frac{\varepsilon^{1-2s}}{2s-1} \operatorname{tr}\left[S_{jj}^{\mathrm{ff}}(1-s) - S_{F_j}(1-s) \right] - \frac{\varepsilon^{2s-1}}{2s-1} \operatorname{tr}\left[S_{jj}^{\mathrm{ff}}(s) - S_{F_j}(s) \right],$$

and from the cusp–cusp terms (10.41),

$$\frac{\varepsilon^{1-2s}}{2s-1} S_{kk}^{\mathrm{cc}}(1-s) - \frac{\varepsilon^{2s-1}}{2s-1} S_{kk}^{\mathrm{cc}}(s).$$

Since $S_{F_j}(\tfrac{1}{2}) = -I$, we have

$$\sum_{j=1}^{n_f} \operatorname{tr}\left[S_{jj}^{\mathrm{ff}}(\tfrac{1}{2} \pm i\xi) - S_{F_j}(\tfrac{1}{2} \pm i\xi) \right] = \operatorname{tr}\left[S_X^{\mathrm{ff}}(\tfrac{1}{2}) + I \right] + O(\xi).$$

Similarly, for the cusp term,

$$S_{kk}^{\mathrm{cc}}(\tfrac{1}{2} \pm i\xi) = S_{kk}^{\mathrm{cc}}(\tfrac{1}{2}) + O(\xi).$$

Note also that

$$\operatorname{tr}\left[S_X^{\mathrm{ff}}(\tfrac{1}{2}) + I \right] + \operatorname{tr} S_X^{\mathrm{cc}}(\tfrac{1}{2}) = \operatorname{tr}\left[S_X(\tfrac{1}{2}) + I \right] - n_{\mathrm{c}}$$

$$= 2m_{\frac{1}{2}} - n_{\mathrm{c}},$$

by Lemma 8.5. We can thus reduce the singular term to

$$(2m_{\frac{1}{2}} - n_{\mathrm{c}}) \lim_{\varepsilon \to 0} \frac{\varepsilon^{-2i\xi} - \varepsilon^{2i\xi}}{2i\xi}.$$

The distributional limit becomes obvious after a Fourier transform. If we let $a_\varepsilon(t)$ denote the characteristic function of the interval $[2 \log \varepsilon, -2 \log \varepsilon]$, then

$$\hat{a}_\varepsilon(\xi) = \int_{2 \log \varepsilon}^{-2 \log \varepsilon} e^{-i\xi t}\, dt = \frac{\varepsilon^{-2i\xi} - \varepsilon^{2i\xi}}{i\xi}.$$

This implies

$$\lim_{\varepsilon \to 0} \int_{-\infty}^{\infty} \frac{\varepsilon^{-2i\xi} - \varepsilon^{2i\xi}}{2i\xi}\, \psi(\xi)\, d\xi = \lim_{\varepsilon \to 0} \frac{1}{2} \int_{2 \log \varepsilon}^{-2 \log \varepsilon} \hat{\psi}(t)\, dt$$

$$= \frac{1}{2} \int_{-\infty}^{\infty} \hat{\psi}(t)\, dt$$

$$= \pi\, \psi(0). \qquad \square$$

We now want to use the relation (11.14) to connect the Fourier transform of $\Theta_{\text{cont}}(t)$ to $\Upsilon_X(\frac{1}{2} + i\xi)$. Before this we need to show that these are tempered distributions, so that the Fourier transform is well-defined. (Note that the full trace $\Theta_X(t)$ will not be tempered unless the discrete spectrum is empty.)

By (10.53) we can decompose

$$\Upsilon_X(s) = \frac{d}{ds} \log \frac{Z_X(s) G_\infty(s)^{\chi(X)}}{Z_X(1-s) G_\infty(1-s)^{\chi(X)}} + n_c \Upsilon_{C_\infty}(s). \qquad (11.18)$$

By Theorem 10.1,

$$\frac{d}{ds} \log \left[Z_X(s) G_\infty(s)^{\chi(X)} \right] = \frac{d}{ds} \log \left[e^{q(s)} \Gamma(s - \tfrac{1}{2})^{n_c} P_X(s) \right]$$

$$= q'(s) + n_c \Psi(s - \tfrac{1}{2}) + \frac{P_X'(s)}{P_X(s)}. \qquad (11.19)$$

And by Propositions 10.7 and 10.8,

$$\Upsilon_{C_\infty}(s) = -\log 4 - \Psi(s - \tfrac{1}{2}) - \Psi(\tfrac{1}{2} - s). \qquad (11.20)$$

Since $q(s)$ has degree at most 2, $q'(s) + q'(1 - s)$ is constant. So by substituting (11.19) and (11.20) in (11.18) we obtain

$$\Upsilon_X(s) = c + \frac{P_X'(s)}{P_X(s)} + \frac{P_X'(1-s)}{P_X(1-s)}. \qquad (11.21)$$

From the definition of P_X we have

$$\frac{P_X'(s)}{P_X(s)} = \frac{m_0}{s} + \sum_{\substack{\zeta \in \mathcal{R}_X \\ \zeta \neq 0}} \left[\frac{1}{\zeta} + \frac{s}{\zeta^2} + \frac{1}{s - \zeta} \right].$$

By exploiting the symmetry of $\mathcal{R}_X$ under complex conjugation, we can combine the s and $1 - s$ terms to obtain

$$\Upsilon_X(\tfrac{1}{2} + i\xi) = c + \frac{m_0}{\xi^2 + \frac{1}{4}} + 2\,\mathrm{Re} \sum_{\substack{\zeta \in \mathcal{R}_X \\ \zeta \neq 0}} \left[\frac{1}{\zeta} + \frac{s}{\zeta^2} + \frac{1}{s-\zeta} \right], \tag{11.22}$$

where $s = \frac{1}{2} + i\xi$.

Lemma 11.6. *The function $\Upsilon_X(\frac{1}{2} + i\xi)$ defines a tempered distribution on $\mathbb{R}$.*

Proof. We will follow the proof of Guillopé–Zworski [87, Lemma 4.7], which draws on a method from Melrose [134]. (We will also use this method in the proof of Theorem 12.4.) The strategy is to break the sum (11.22) into two parts that can be treated separately.

Let $\psi \in C_0^\infty(\mathbb{R})$ be a test function satisfying $0 \le \psi(t) \le 1$ and

$$\psi(t) = \begin{cases} 1, & |t| \le 2, \\ 0, & |t| \ge 3. \end{cases}$$

With $s = \frac{1}{2} + i\xi$ we define the function

$$v_1(\xi) := 2\,\mathrm{Re} \sum_{\substack{\zeta \in \mathcal{R}_X \\ \mathrm{Re}\,\zeta < 1/2}} \psi\!\left(\frac{|\zeta|}{\xi}\right) \frac{1}{s-\zeta} = \sum_{\substack{\zeta \in \mathcal{R}_X \\ \mathrm{Re}\,\zeta < 1/2}} \psi\!\left(\frac{|\zeta|}{\xi}\right) \frac{1 - 2\,\mathrm{Re}\,\zeta}{|\frac{1}{2} + i\xi - \zeta|^2}. \tag{11.23}$$

Note that this sum is locally finite, meaning that only finitely many terms are nonzero if ξ is restricted to a compact set. (The restriction to $\mathrm{Re}\,\zeta < \frac{1}{2}$ isn't important here, but it will be significant for our use of v_1 in the proof of Theorem 12.4.) A simple calculus computation,

$$\int_{-\infty}^{\infty} \frac{1 - 2\,\mathrm{Re}\,\zeta}{|\frac{1}{2} + i\xi - \zeta|^2} \, d\xi = 2\pi,$$

leads to the bound, for $\mathrm{Re}\,\zeta < \frac{1}{2}$,

$$\left| \int_0^\xi v_1(\xi)\,d\xi \right| \le \sum_{|\zeta| \le 3|\xi|} \int_0^{|\xi|} \frac{|1 - 2\,\mathrm{Re}\,\zeta|}{|\frac{1}{2} + i\xi - \zeta|^2} \, d\xi$$
$$\le 2\pi\, N_X(3|\xi|)$$
$$\le O(|\xi|^2),$$

by Theorem 9.1. In particular, $\int v_1 \, d\xi$ is a tempered distribution, so its derivative $v_1(\xi)$ is also.

We claim that the remainder,

$$\Upsilon_X(\tfrac{1}{2} + i\xi) - v_1(\xi) = c + 2\,\mathrm{Re} \sum_{\zeta \neq 0} \psi\!\left(\frac{|\zeta|}{\xi}\right) \left[\frac{1}{\zeta} + \frac{s}{\zeta^2} \right]$$

$$+ 2\,\mathrm{Re} \sum_{\zeta \neq 0} \left(1 - \psi\!\left(\frac{|\zeta|}{\xi}\right)\right) \left[\frac{1}{\zeta} + \frac{s}{\zeta^2} + \frac{1}{s-\zeta} \right]$$

$$+ \sum_{\mathrm{Re}\,\zeta > 1/2} \psi\!\left(\frac{|\zeta|}{\xi}\right) \frac{1 - 2\,\mathrm{Re}\,\zeta}{|\frac{1}{2} + i\xi - \zeta|^2}, \tag{11.24}$$

where $s = \frac{1}{2} + i\xi$, is $O(|\xi|^3)$ and hence tempered also. The third sum is finite and trivially $O(|\xi|^{-2})$. A polynomial bound for the first sum in (11.24) follows easily from Theorem 9.1,

$$\sum_{\substack{\zeta \in \mathcal{R}_X \\ \zeta \neq 0}} \psi\left(\frac{|\zeta|}{\xi}\right) \left|\frac{1}{\zeta} + \frac{s}{\zeta^2}\right| \leq C|\xi| \, N_X(3|\xi|) = O(|\xi|^3).$$

For the second sum in (11.24) we note that

$$\frac{1}{\zeta} + \frac{s}{\zeta^2} + \frac{1}{s - \zeta} = \frac{s^2}{\zeta^3(1 - s/\zeta)},$$

so this term can be bounded by

$$\sum_{|\zeta| \geq 2|\xi|} \left|\frac{s^2}{\zeta^3(1 - s/\zeta)}\right| \leq C|\xi|^2 \sum_{\zeta \neq 0} \frac{1}{|\zeta|^3}.$$

The sum of $|\zeta|^{-3}$ is finite by Theorem 9.1, and this finishes the proof. □

By (11.14) and Lemmas 11.5 and 11.6 we can write the continuous part of the wave 0-trace as

$$\Theta_{\text{cont}} = \mathcal{F}\left[\frac{1}{4\pi}\Upsilon_X(\tfrac{1}{2} + i\xi) + \frac{2m_{\frac{1}{2}} - n_c}{4}\delta(\xi)\right], \tag{11.25}$$

where $\mathcal{F}$ denotes the Fourier transform. The proof of Theorem 11.3 is now reduced to computing the distributional Fourier transform of Υ_X.

Proof of Theorem 11.3. In the formula (11.21) for $\Upsilon_X(s)$, if we combine the s and $1 - s$ terms directly (without complex conjugation), we obtain

$$\Upsilon_X(\tfrac{1}{2} + i\xi) = c + \frac{m_0}{\xi^2 + 1/4} + \sum_{\substack{\zeta \in \mathcal{R}_X \\ \zeta \neq 0}} \left[\frac{1 - 2\zeta}{\xi^2 + (\zeta - 1/2)^2} + \frac{2\zeta + 1}{\zeta^2}\right]. \tag{11.26}$$

The Fourier transform of an individual resonance term is given by a simple contour integral,

$$\int_{-\infty}^{\infty} e^{-i\xi t} \frac{1 - 2\,\text{Re}\,\zeta}{\xi^2 + (\zeta - 1/2)^2}\, d\xi = \begin{cases} -2\pi\, e^{(1/2 - \zeta)|t|}, & \text{Re}\,\zeta > \tfrac{1}{2}, \\ 2\pi\, e^{(\zeta - 1/2)|t|}, & \text{Re}\,\zeta < \tfrac{1}{2}. \end{cases} \tag{11.27}$$

Of course, the sum in (11.26) doesn't converge if we split this resonance term away from $(2\zeta + 1)/\zeta^2$. To get around this, we can simply differentiate to knock these terms out,

$$\frac{d}{d\xi}\Upsilon_X(\tfrac{1}{2} + i\xi) = -2\xi \sum_{\zeta \in \mathcal{R}_X} \frac{1 - 2\zeta}{\left[\xi^2 + (\zeta - 1/2)^2\right]^2}.$$

Taking the Fourier transform of both sides, using (11.27) and the fact that $d/d\xi$ becomes multiplication by $-it$ under the transform, we obtain

$$t\,\mathcal{F}[\Upsilon(\tfrac{1}{2}+i\xi)](t) = 2\pi t \sum_{\substack{\zeta\in\mathcal{R}_X \\ \mathrm{Re}\,\zeta<1/2}} e^{(\zeta-1/2)|t|} - 2\pi t \sum_{\substack{\zeta\in\mathcal{R}_X \\ \mathrm{Re}\,\zeta>1/2}} e^{(1/2-\zeta)|t|}.$$

Using this in (11.25) gives

$$t\,\Theta_{\mathrm{cont}}(t) = t\left[\frac{n_c}{4} + \frac{1}{2}\sum_{\substack{\zeta\in\mathcal{R}_X \\ \mathrm{Re}\,\zeta\le 1/2}} e^{(\zeta-1/2)|t|} - \frac{1}{2}\sum_{\substack{\zeta\in\mathcal{R}_X \\ \mathrm{Re}\,\zeta>1/2}} e^{(1/2-\zeta)|t|}\right].$$

In conjunction with (11.13), this proves the theorem.   □

Notes

The wave 0-trace formula was extended to higher-dimensional hyperbolic manifolds without cusps by Guillarmou–Naud [81]. For asymptotically hyperbolic manifolds there is no wave trace formula, but Joshi–Sá Barreto [103] showed in this context that the wave 0-trace has singular support within the set of lengths of closed geodesics. See Perry [170] for an overview of these results.

For a survey of methods of computing the wave invariants associated to particular closed geodesics, see Zelditch [219].

The Poisson formula of Theorem 11.3 is an analogue of earlier results in odd-dimensional Euclidean obstacle scattering, by Lax–Phillips [116] and Bardos–Guillot–Ralston [12] for large times, and for all nonzero times by Melrose [131, 133]. This was extended by Sjöstrand–Zworski [198] to the "black box" scattering framework, which covers arbitrary compact perturbations of the Laplacian on $\mathbb{R}^n$ for n odd.

Resonance Asymptotics

For an infinite-area geometrically finite hyperbolic surface X, Theorems 11.2 and 11.3 give us the following trace formula for resonances: for $t > 0$,

$$\sum_{\zeta \in \mathcal{R}_X} e^{(\zeta - \frac{1}{2})t} = \sum_{\ell \in \Lambda} \sum_{k=1}^{\infty} \frac{\ell}{2 \sinh(k\ell/2)} \delta(t - k\ell) + \Psi_{\text{top}}(t), \qquad (12.1)$$

where

$$\Psi_{\text{top}}(t) := \frac{\chi(X)}{2} \frac{\cosh(t/2)}{\sinh^2(t/2)} + \frac{n_c}{2} \coth(t/2) - \frac{n_c}{2}.$$

One application to resonance counting is immediately clear: the set $\mathcal{R}_X$ must be infinite, to account for the singularities on the right-hand side. Clearly the formula contains more information on the distribution of resonances than that, and in this chapter we will show how such information can be extracted.

12.1 Lower bound on resonances

Recall that in Theorem 9.1 we found that the resonance counting function was bounded: $N_X(r) = O(r^2)$. Our first application of (12.1) will be to complement this result with a corresponding lower bound by cr^2. This was also proven by Guillopé–Zworski [87]. The quadratic lower bound is essentially derived from the order of the singularity of $\Psi_{\text{top}}(t)$ as $t \to 0$.

Theorem 12.1 (Guillopé–Zworski). *For a geometrically finite hyperbolic surface X with $\chi(X) \neq 0$, the resonance counting function satisfies*

$$N_X(r) \asymp r^2.$$

For geometrically finite hyperbolic surfaces, $\chi(X) = 0$ only for the hyperbolic or parabolic cylinders. For the hyperbolic cylinder we noted at the beginning of Chapter 9 that $N_{C_\ell}(s) \sim (\ell/2)cr^2$. The parabolic cylinder, with its single resonance, is the only exception to the lower bound among the geometrically finite hyperbolic surfaces.

Although the wave trace formula (12.1) holds only in the exactly hyperbolic case, the small-time behavior of the wave trace is understood in much greater generality (see, e.g., Hörmander [96, Section 29]). This behavior, along with the Poisson formula, is all that the proof of the lower bound on the counting function requires. Thus Guillopé–Zworski were able to prove it in greater generality, i.e., for compactly supported perturbations of hyperbolic surfaces.

To extract the asymptotic information from (12.1), we pair this equation with a test function $\phi \in C_0^\infty(\mathbb{R}_+)$. The Fourier transform $\widehat{\phi}(\xi)$ extends to an entire function, and we can write the left-hand side of (12.1) as

$$\sum_{\zeta \in \mathcal{R}_X} \int_0^\infty e^{(\zeta-1/2)t} \phi(t)\, dt = \sum_{\zeta \in \mathcal{R}_X} \widehat{\phi}\big(i(\zeta - \tfrac{1}{2})\big).$$

The full formula thus reads

$$\sum_{\zeta \in \mathcal{R}_X} \widehat{\phi}\big(i(\zeta - \tfrac{1}{2})\big) = \sum_{\ell \in \mathcal{L}_X} \sum_{m=1}^\infty \frac{\ell}{2\sinh(m\ell/2)} \phi(m\ell) + \int_0^\infty \phi(t)\, \Psi_{\text{top}}(t)\, dt. \quad (12.2)$$

To analyze the small-t asymptotics, we will assume that $\phi(t) \geq 0$, $\phi(1) > 0$ and study the large-λ behavior of the pairing (12.2) with ϕ replaced by the rescaled function

$$\phi_\lambda(t) := \lambda \phi(\lambda t).$$

Lemma 12.2. *For ϕ as above and λ sufficiently large,*

$$\left| \sum_{\zeta \in \mathcal{R}_X} \widehat{\phi}\big(i(\zeta - \tfrac{1}{2})/\lambda\big) \right| \geq C|\chi(X)|\, \lambda^2.$$

Proof. Pairing the rescaled test function ϕ_λ with the wave trace gives

$$\sum_{\zeta \in \mathcal{R}_X} \widehat{\phi}\big(i(\zeta - \tfrac{1}{2})/\lambda\big) = \sum_{\ell \in \mathcal{L}_X} \sum_{m=1}^\infty \frac{\ell}{2\sinh(m\ell/2)} \lambda \phi(\lambda m\ell)$$

$$+ \int_{\mathbb{R}_+} \phi(t)\, \Psi_{\text{top}}(t/\lambda)\, dt.$$

The contribution from the length spectrum vanishes for λ sufficiently large, because ϕ has compact support and the length spectrum is bounded below.

It suffices then to prove a lower bound on the topological term. We can easily estimate

$$\frac{\cosh(t/2\lambda)}{\sinh^2(t/2\lambda)} \geq C\frac{\lambda^2}{t^2},$$

for $0 < t/\lambda \leq \delta$, so that

$$\int_0^\infty \phi(t) \frac{\cosh(t/2\lambda)}{\sinh^2(t/2\lambda)} \, dt \geq C\lambda^2.$$

The cusp term is of lower order,

$$\int_0^\infty \phi(t) \left(\coth(t/2\lambda) - 1\right) dt = O(\lambda).$$

Hence for λ sufficiently large the topological term has a lower bound:

$$\left| \int_{\mathbb{R}_+} \phi(t) \, \Psi_{\text{top}}(t/\lambda) \, dt \right| \geq C|\chi(X)| \, \lambda^2. \qquad \qquad \square$$

Proof of Theorem 12.1. It is only the lower bound that we are concerned with here, since the upper bound was established in Theorem 9.1. Let ϕ be a test function as in Lemma 12.2. Since ϕ is compactly supported, we can derive immediately from

$$\widehat{\phi}(\xi) = \int_0^\infty e^{-i\xi t} \phi(t) \, dt$$

an estimate for $\operatorname{Im}\xi \leq 0$ and any integer $m \geq 1$,

$$\left|\widehat{\phi}(\xi)\right| \leq C_m (1 + |\xi|)^{-m}.$$

In other words,

$$\left|\widehat{\phi}(i(\zeta - 1/2)/\lambda)\right| \leq C_m (1 + |\zeta|/\lambda)^{-m},$$

for $\operatorname{Re}\zeta \leq \frac{1}{2}$. Since $\operatorname{Re}\zeta > \frac{1}{2}$ for only finitely many $\zeta \in \mathcal{R}_X$, Lemma 12.2 implies a lower bound:

$$|\chi(X)| \, \lambda^2 \leq \sum_{\zeta \in \mathcal{R}_X} C_m (1 + |\zeta|/\lambda)^{-m}. \tag{12.3}$$

By rewriting the left-hand side of (12.3) as a Stieltjes integral with respect to the point measure defined by N_X, we can integrate by parts to obtain

$$|\chi(X)| \, \lambda^2 \leq C_m \int_0^\infty (1 + r/\lambda)^{-m} \, dN_X(r)$$

$$= C_m \frac{m}{\lambda} \int_0^\infty (1 + r/\lambda)^{-m-1} N_X(r) \, dr$$

$$= C_m \int_0^\infty (1 + r)^{-m-1} N_X(\lambda r) \, dr.$$

By Theorem 9.1, $N_X(r) = O(r^2)$. We can split the integral above at $r = a$ and use this upper bound to control the remainder term:

$$|\chi(X)| \, \lambda^2 \leq C_m \int_0^a (1 + r)^{-m-1} N(\lambda r) \, dr + C_m \lambda^2 \int_a^\infty r^2 (1 + r)^{-m-1} \, dr$$

$$\leq C_m N(\lambda a) + C_m \lambda^2 a^{-m+2}.$$

The result follows immediately by setting $m = 3$ and taking a sufficiently large. $\quad \square$

12.2 Lower bound near the critical line

From a physical point of view, the important resonances are those near the continuous spectrum—in our case near the critical line $\operatorname{Re} s = \frac{1}{2}$. Another direct application of the wave trace formula (12.1), due to Guillopé–Zworski [88], is a lower bound on the number of resonances contained in a vertical strip near the critical line. See Figure 12.1.

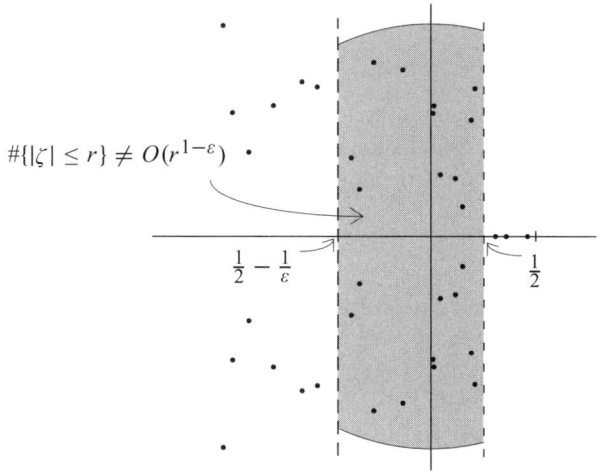

$$\#\{|\zeta| \le r\} \ne O(r^{1-\varepsilon})$$

$$\frac{1}{2} - \frac{1}{\varepsilon} \qquad \frac{1}{2}$$

Fig. 12.1. Lower bound in a vertical strip.

Theorem 12.3 (Guillopé–Zworski). *Let X be a nonelementary geometrically finite hyperbolic surface. For any $\varepsilon \in (0, \frac{1}{2})$,*

$$\#\left\{\zeta \in \mathcal{R}_X : |\zeta| \le r, \ \operatorname{Re}\zeta > \tfrac{1}{2} - \varepsilon^{-1}\right\} \ne O(r^{1-\varepsilon}). \tag{12.4}$$

Proof. Suppose that $\varphi \in C_0^\infty(\mathbb{R})$ is a positive test function supported in $(-1, 1)$, with $\varphi(0) = 1$. For $d \in \mathcal{L}_X$ and $a > 0$, define

$$\varphi_{a,d}(t) := \varphi((t - d)/a).$$

(Later we'll assume that d is large and a correspondingly small.) Using $\varphi_{a,d}$ as the test function in the Poisson formula (12.2) yields

$$\sum_{\ell \in \mathcal{L}_X} \sum_{m=1}^{\infty} \frac{\ell}{2 \sinh(m\ell/2)} \varphi_{a,d}(m\ell) = \sum_{\zeta \in \mathcal{R}_X} \widehat{\varphi}_{a,d}(i(\zeta - \tfrac{1}{2}))$$
$$- \int_{d+a}^{d-a} \varphi_{a,d}(t) \Psi_{\text{top}}(t) \, dt. \tag{12.5}$$

All terms in the sum over the length spectrum on the left-hand side of (12.5) are positive, so the sum is bounded below by the contribution from $\ell = d$ and $m = 1$:

$$\sum_{\ell \in \mathcal{L}_X} \sum_{m=1}^{\infty} \frac{\ell}{2 \sinh(m\ell/2)} \varphi_{a,d}(m\ell) \geq \frac{d}{2 \sinh(d/2)}, \tag{12.6}$$

for any $d \in \mathcal{L}_X$.

The topological term on the right-hand side of (12.5) is

$$-\frac{\chi(X)}{2} \int_{d+a}^{d-a} \varphi_{a,d}(t) \frac{\cosh(t/2)}{\sinh^2(t/2)} \, dt - \frac{n_c}{2} \int_{d+a}^{d-a} \varphi_{a,d}(t) \left(\coth(t/2) - 1\right) dt.$$

The second integral is negative, so we can ignore it. The first can be estimated by

$$-\frac{\chi(X)}{2} \int_{d+a}^{d-a} \varphi_{a,d}(t) \frac{\cosh(t/2)}{\sinh^2(t/2)} \, dt = -\frac{\chi(X)}{2} \int_{-1}^{1} \varphi(x) \frac{\cosh((ax+d)/2)}{\sinh^2((ax+d)/2)} a \, dx$$

$$\leq Cae^{-d/2}.$$

Setting $a = e^{-\beta d}$ for some $\beta > 0$, we have

$$-\int_0^\infty \varphi_{a,d}(t) \Psi_{\text{top}}(t) \, dt \leq Ce^{-(\beta+1/2)d}. \tag{12.7}$$

The spectral term on the right-hand side of (12.5) is a sum over

$$\widehat{\varphi}_{a,d}(i(\zeta - \tfrac{1}{2})) = a \, \widehat{\varphi}(ia(\zeta - \tfrac{1}{2})) \, e^{d(\zeta - \frac{1}{2})}.$$

The Fourier transform is given by

$$\widehat{\varphi}(ia(\zeta - \tfrac{1}{2})) = \int_{-1}^{1} e^{-at(\zeta - \frac{1}{2})} \varphi(t) \, dt,$$

so by integrating by parts we can estimate

$$\left| \widehat{\varphi}(ia(\zeta - \tfrac{1}{2})) \right| \leq C_m \frac{e^{a|\operatorname{Re}\zeta - \frac{1}{2}|}}{(1 + a|\zeta|)^m}$$

for any $m \in \mathbb{N}$. In particular,

$$\left| \widehat{\varphi}_{a,d}(i(\zeta - \tfrac{1}{2})) \right| \leq Ca \frac{e^{a|\operatorname{Re}\zeta - \frac{1}{2}|}}{(1 + a|\zeta|)^3} e^{d(\zeta - \frac{1}{2})}. \tag{12.8}$$

To estimate the sum over resonances, we will break the sum at $\operatorname{Re}\zeta = \frac{1}{2} - \alpha$, for some $\alpha > 0$. For $\operatorname{Re}\zeta \leq \frac{1}{2} - \alpha$ the estimate (12.8) becomes

$$\left| \widehat{\varphi}_{a,d}(i(\zeta - \tfrac{1}{2})) \right| \leq Ca \frac{e^{-(d-a)\alpha}}{(1 + a|\zeta|)^3}.$$

Using a Stieltjes integral and the upper bound $N_X(r) = O(r^2)$ from Theorem 9.1, we have

$$\left| \sum_{\text{Re}\,\zeta \leq \frac{1}{2}-\alpha} \widehat{\varphi}_{a,d}(i(\zeta - \tfrac{1}{2})) \right| \leq Ce^{-(d-a)\alpha} \int_0^\infty \frac{a\, dN_X(r)}{(1+ar)^3}$$

$$= Ce^{-(d-a)\alpha} \int_0^\infty \frac{a^2 N_X(r)}{(1+ar)^4}\, dr$$

$$\leq Ce^{-(d-a)\alpha} \int_0^\infty \frac{a^2 r^2}{(1+ar)^4}\, dr$$

$$\leq Ca^{-1} e^{-(d-a)\alpha}.$$

Inserting $a = e^{-\beta d}$ then gives the estimate

$$\left| \sum_{\text{Re}\,\zeta \leq \frac{1}{2}-\alpha} \widehat{\varphi}_{a,d}(i(\zeta - \tfrac{1}{2})) \right| \leq Ce^{(\beta-\alpha)d}. \tag{12.9}$$

Now consider $\text{Re}\,\zeta > \frac{1}{2} - \alpha$. When $\frac{1}{2} \leq \text{Re}\,\zeta < 1$, the best we can get from (12.8) is

$$\left| \widehat{\varphi}_{a,d}(i(\zeta - \tfrac{1}{2})) \right| \leq Cae^{(d+a)/2}.$$

Fortunately there are only finitely many such terms. When $\frac{1}{2} - \alpha < \text{Re}\,\zeta < \frac{1}{2}$, (12.8) gives

$$\left| \widehat{\varphi}_{a,d}(i(\zeta - \tfrac{1}{2})) \right| \leq Ca(1 + a|\zeta|)^{-3}.$$

If we define

$$P_\alpha(r) := \#\left\{ \zeta \in \mathcal{R}_X : |\zeta| \leq r,\ \text{Re}\,\zeta > \tfrac{1}{2} - \alpha \right\},$$

then with integration by parts in a Stieltjes integral we can estimate

$$\left| \sum_{\text{Re}\,\zeta \geq \frac{1}{2}-\alpha} \widehat{\varphi}_{a,d}(i(\zeta - \tfrac{1}{2})) \right| \leq C \int_0^\infty \frac{a^2 P_\alpha(r)}{(1+ar)^4}\, dr + Cae^{(d+a)/2}.$$

If we assume, for the sake of contradiction, that

$$P_\alpha(r) = O(r^{1-\varepsilon}), \tag{12.10}$$

then

$$\int_0^\infty \frac{a^2 P_\alpha(r)}{(1+ar)^4}\, dr = O(a^\varepsilon).$$

Setting $a = e^{-\beta d}$ as above, we now have

$$\left| \sum_{\text{Re}\,\zeta \geq \frac{1}{2}-\alpha} \widehat{\varphi}_{a,d}(i(\zeta - \tfrac{1}{2})) \right| \leq Ce^{-\beta\varepsilon d} + Ce^{-(\beta-1/2)d}. \tag{12.11}$$

Collecting the estimates (12.7), (12.9), and (12.11) for the terms on the right in (12.5), we see that the assumption (12.10) implies

$$\frac{d}{2\sinh(d/2)} \le C\Big[e^{(\beta-\alpha)d} + e^{-\beta\varepsilon d} + e^{-(\beta-1/2)d}\Big].$$

This leads to a contradiction for large d if we can choose the exponents on the right all strictly smaller than $-d/2$. (Note that the primitive length spectrum is infinite for X nonelementary, so arbitrarily large d is possible.) In other words, we have a contradiction to (12.10) provided that

$$\alpha - \beta > \tfrac{1}{2}, \qquad \beta\varepsilon > \tfrac{1}{2}, \qquad \beta > 1.$$

If $\varepsilon < \tfrac{1}{2}$ then the third condition is superfluous, and the first two can then be satisfied for any $\alpha > (1 + \varepsilon^{-1})/2$. $\square$

12.3 Weyl formula for the scattering phase

We have already mentioned that no analogue of the Weyl asymptotic formula is known for the resonance counting function $N_X(r)$, unless X is a finite-area hyperbolic surface or one of the elementary cases. However, asymptotics can be proven for the scattering phase, which is a continuous analogue of the counting function.

Definition. The *relative scattering phase* is the logarithm of the relative scattering determinant:

$$\sigma_X(\xi) := \frac{i}{2\pi} \log \tau_X(\tfrac{1}{2} + i\xi).$$

By (10.6) and (7.37) we have

$$\overline{\tau_X(\tfrac{1}{2} + i\xi)} = \tau_X(\tfrac{1}{2} - i\xi) = \frac{1}{\tau_X(\tfrac{1}{2} + i\xi)}.$$

Thus $\sigma_X(\xi)$ is real and $\sigma_X(-\xi) = -\sigma_X(\xi)$.

In [134] Melrose established Weyl asymptotics for the scattering phase in the case of obstacle scattering in $\mathbb{R}^n$ for $n \ge 3$ odd. Guillopé and Zworski adapted the same method to the case of surfaces with hyperbolic ends in [88].

Theorem 12.4 (Guillopé–Zworski). *For a geometrically finite hyperbolic surface X of infinite area, the relative scattering phase exhibits Weyl-type asymptotics: as $\xi \to +\infty$,*

$$\sigma_X(\xi) = -\frac{\chi(X)}{2}\xi^2 - \frac{n_c}{\pi}\xi \log \xi + O(\xi). \qquad (12.12)$$

The derivative of the scattering phase,

$$\frac{d\sigma_X}{d\xi}(\xi) = -\frac{1}{2\pi}\frac{\tau_X'}{\tau_X}(\tfrac{1}{2} + i\xi),$$

is related by Proposition 10.9 to the regularized resolvent traces:

$$\frac{d\sigma_X}{d\xi}(\xi) = \frac{1}{2\pi}\left[\Upsilon_X(\tfrac{1}{2} + i\xi) - \Upsilon_F(\tfrac{1}{2} - i\xi)\right], \tag{12.13}$$

for $\xi \neq 0$. From the formula (10.54) it's easy to see that $\Upsilon_F(\tfrac{1}{2} + i\xi)$ is bounded for real ξ, so we are really concerned with the asymptotics of Υ_X here.

In (10.53) we have a formula for $\Upsilon_X(s)$ in terms of the zeta function. If Γ has exponent of convergence $\delta < \tfrac{1}{2}$, then we could see that $Z'_X/Z_X(\tfrac{1}{2} + i\xi)$ is bounded directly from the definition of $Z_X(s)$. The asymptotic (12.12) would follow easily from Stirling's formula and well-known asymptotics of the Barnes G-function.

For $\xi \in \mathbb{R}$, let us define the smooth function

$$v(\xi) := \Upsilon(\tfrac{1}{2} + i\xi),$$

which is a tempered distribution by Lemma 11.6. For the proof of Theorem 12.4, the crucial connection is the formula

$$\frac{1}{4\pi}\widehat{v}(t) = \Theta_X(t) - \sum_{\substack{\zeta \in \mathcal{R}_X \\ \mathrm{Re}\,\zeta \geq 1/2}} \cosh\left[(\zeta - 1/2)t\right] + \frac{n_c}{4}, \tag{12.14}$$

which is a combination of (11.13) and (11.25). The Weyl asymptotics essentially follow from the behavior of the wave trace near $t = 0$, as seen in Theorem 11.2. Extracting this information is delicate, however.

Choose $\phi \in \mathcal{S}(\mathbb{R})$ with the following properties: $\phi > 0$, $\widehat{\phi}(0) = 1$, and $\widehat{\phi}$ has compact support within $(-\ell_0, \ell_0)$, where ℓ_0 is the shortest length in $\mathcal{L}_X$. We first show that convolution of v with ϕ captures the correct asymptotics.

Lemma 12.5. *As $\xi \to +\infty$,*

$$v * \phi(\xi) = -2\pi \chi(X)\, \xi - 2n_c \log \xi + O(1). \tag{12.15}$$

Proof. By (12.14), the Fourier transform of the convolution is

$$\widehat{v * \phi}(t) = \widehat{v}(t)\widehat{\phi}(t)$$
$$= \widehat{\phi}(t)\left[4\pi \Theta_X(t) - \sum_{\mathrm{Re}\,\zeta \geq 1/2} 4\pi \cosh\left[(\zeta - 1/2)t\right] + 2\pi n_c\right].$$

The final two terms are smooth, and so under the inverse Fourier transform will contribute terms that decay rapidly as $\xi \to 0$.

To analyze the contribution from $\Theta_X(t)$, we use Theorem 11.2 and the restriction on the support of $\widehat{\phi}$ to write

$$4\pi\widehat{\phi}(t)\Theta_X(t) = -2\pi \chi(X)\widehat{\phi}(t)\,\mathrm{PV}\left[\sinh^{-1}(t/2)\right]'$$
$$+ 2\pi n_c\widehat{\phi}(t)\left(\mathrm{sgn}\,t \log \sinh(|t|/2)\right)' + 4\pi n_c\gamma\,\delta(t). \tag{12.16}$$

Let us introduce distributions h_1, h_2 such that

$$\widehat{h_1}(t) = -2\pi \widehat{\phi}(t) \, \mathrm{PV} \left[\sinh^{-1}(t/2)\right]',$$

$$\widehat{h_2}(t) = 2\pi \widehat{\phi}(t) \left(\mathrm{sgn}\, t \, \log \sinh(|t|/2)\right)' + 4\pi \gamma \, \delta(t).$$

At this point we have seen that

$$v * \phi(\xi) = \chi(X) h_1(\xi) + n_c h_2(\xi) + O(\xi^{-\infty}).$$

Since $t / \sinh(t/2) \in \mathcal{S}(\mathbb{R})$, we can replace $\sinh^{-1}(t/2)$ by $2/t$ in the h_1 term to obtain

$$\widehat{h_1}(t) = -4\pi f(t) \, \mathrm{PV}\,[1/t]',$$

where $f \in \mathcal{S}(\mathbb{R})$ and $f(0) = 1$. The distribution $\mathrm{PV}\,[1/t]$ is the Fourier transform of $(i/2)\,\mathrm{sgn}(\xi)$, as shown in Lemma A.7, so the inverse Fourier transform of $\mathrm{PV}\,[1/t]'$ is $-|\xi|/2$. Thus, up to a constant factor, h_1 is the convolution of the inverse transform $\check{f}$ with the absolute-value function:

$$h_1(\xi) = -2\pi \int_{-\infty}^{\infty} |\xi - \eta| \, \check{f}(\eta) \, d\eta.$$

This is now easy to estimate, using the properties of f:

$$h_1(\xi) = -2\pi |\xi| \int_{-\infty}^{\infty} \check{f}(\eta) \, d\eta + O(1),$$

$$= -2\pi |\xi| \, f(0) + O(1).$$

$$= -2\pi |\xi| + O(1).$$

To analyze $h_2(\xi)$, we note that $\log \sinh(|t|/2) = \log |t|$ plus a smooth function. With the distributional Fourier-transform calculation from Lemma A.8

$$\mathcal{F}(\log |\xi|) = -\pi \, (\mathrm{sgn}\, t \, \log |t|)' - 2\pi \gamma \, \delta(t),$$

it's easy then to deduce that

$$h_2(\xi) = -2 \log |\xi| + O(1),$$

which completes the proof.   □

By (12.13) and the fact that $\Upsilon_F(\tfrac{1}{2} + i\xi)$ is bounded, integrating (12.15) gives

$$\sigma_X * \phi(\xi) = -\frac{\chi(X)}{2} \xi^2 - \frac{n_c}{\pi} \xi \log \xi + O(\xi). \tag{12.17}$$

Our goal is now to show that $\sigma_X - \sigma_X * \phi$ is of lower order. If σ_X were increasing, then a simple Tauberian argument would accomplish this. Melrose's method involves splitting $\sigma_X(\xi)$ into an increasing term $\sigma_1(\xi)$ plus a term $\sigma_2(\xi)$ that can be estimated more directly.

As in the proof of Lemma 11.6, we choose $\psi \in C_0^\infty(\mathbb{R})$ such that $0 \le \psi(t) \le 1$ and

$$\psi(t) = \begin{cases} 1, & |t| \le 2, \\ 0, & |t| \ge 3. \end{cases}$$

Then we set

$$v_1(\xi) := \sum_{\substack{\zeta \in \mathcal{R}_X \\ \operatorname{Re}\zeta < 1/2}} \psi\left(\frac{|\zeta|}{\xi}\right) \frac{1 - 2\operatorname{Re}\zeta}{|\frac{1}{2} + i\xi - \zeta|^2}$$

and

$$v_2(\xi) := v(\xi) - v_1(\xi).$$

By definition, $v_1(\xi) \ge 0$, so we can define the increasing term $\sigma_1(\xi)$ by integrating this function. As for the remainder, we saw in the proof of Lemma 11.6 that $v_2(\xi) = O(|\xi|^3)$, but we can now use the quadratic lower bound on resonances to improve this estimate.

Lemma 12.6. *We can bound*

$$v_2(\xi) = O(|\xi|).$$

Proof. By (11.24) and the definition of $v_2(\xi)$ we have

$$v_2(\xi) = c + 2\operatorname{Re}\sum_{\zeta \neq 0}\left[\frac{1}{\zeta} + \frac{\frac{1}{2} + i\xi}{\zeta^2} + \left(1 - \psi\left(\frac{|\zeta|}{\xi}\right)\right)\frac{1}{\frac{1}{2} + i\xi - \zeta}\right]$$
$$+ \sum_{\operatorname{Re}\zeta > 1/2} \psi\left(\frac{|\zeta|}{\xi}\right) \frac{1 - 2\operatorname{Re}\zeta}{|\frac{1}{2} + i\xi - \zeta|^2}. \tag{12.18}$$

The derivative of this expression is

$$v_2'(\xi) = -2\operatorname{Im}\sum_{\zeta \neq 0}\left(1 - \psi\left(\frac{|\zeta|}{\xi}\right)\right)\left(\frac{1}{\zeta^2} - \frac{1}{(\frac{1}{2} + i\xi - \zeta)^2}\right)$$
$$+ 2\operatorname{Re}\sum_{\zeta \neq 0}\left[\psi\left(\frac{|\zeta|}{\xi}\right)\frac{i}{\zeta^2} + \psi'\left(\frac{|\zeta|}{\xi}\right)\frac{|\zeta|}{\xi^2}\frac{1}{\frac{1}{2} + i\xi - \zeta}\right]$$
$$+ \frac{d}{d\xi}\sum_{\operatorname{Re}\zeta > 1/2} \psi\left(\frac{|\zeta|}{\xi}\right)\frac{1 - 2\operatorname{Re}\zeta}{|\frac{1}{2} + i\xi - \zeta|^2}. \tag{12.19}$$

The third sum is finite and trivially $O(|\xi|^{-3})$. The sum on the second line is restricted to $|\zeta| \le 3|\xi|$ by the definition of ψ, and hence easily bounded $O(1)$ using Theorem 9.1.

This leaves the first sum in (12.19), which we can bound by

$$\sum_{|\zeta| > 2|\xi|} \frac{a\xi|\zeta| + b\xi^2}{|\zeta|^4}, \tag{12.20}$$

for some constants a, b. If the set $\mathcal{R}_X$ is ordered so that $|\zeta_n|$ is increasing, then we have $|\zeta_n| \asymp \sqrt{n}$ by Theorem 12.1. A simple integral comparison gives the estimate

$$\sum_{|\zeta|>2\xi} \frac{a\xi|\zeta| + b\xi^2}{|\zeta|^4} \leq \sum_{n>c\xi^2} (C_1\xi n^{-3/2} + C_2\xi^2 n^{-2})$$

$$\leq \int_{c\xi^2}^{\infty} (C_1\xi n^{-3/2} + C_2\xi^2 n^{-2})\, dn$$

$$\leq C.$$

This shows that $v_2'(\xi)$ is bounded, and the result follows by integration. $\square$

We should note here that although it was convenient to make use of the lower bound for this estimate, it wasn't strictly necessary. After differentiating v_2 more than once, we can get relatively easy bounds on the derivatives, of the form $|v_2^{(k)}(\xi)| = O(|\xi|^{1-k})$. The trouble with integrating these estimates is that the bound on $v_2'(\xi)$ integrates to $O(\log|\xi|)$, not $O(1)$. Melrose's original argument [134] shows how this obstacle may be overcome using interpolation and the asymptotic result of Lemma 12.5.

Proof of Theorem 12.4. Let $\sigma_1(\xi)$ be the increasing part of $\sigma_X(\xi)$ defined by

$$\sigma_1(\xi) := \frac{1}{2\pi} \int_0^\xi v_1(\xi)\, d\xi,$$

and set $\sigma_2(\xi) := \sigma_X(\xi) - \sigma_1(\xi)$. By Lemma 12.6 and the fact that $\Upsilon_F(\frac{1}{2} + i\xi)$ is bounded,

$$\sigma_2'(\xi) = \frac{1}{2\pi}\left[v_2(\xi) + \Upsilon_F(\tfrac{1}{2} + i\xi)\right] = O(|\xi|). \tag{12.21}$$

Using the fact that $\widehat{\phi}(0) = 1$, we can write

$$\sigma_2(\xi) - \sigma_2 * \phi(\xi) = \int_{-\infty}^{\infty} [\sigma_2(\xi) - \sigma_2(\xi - \eta)]\phi(\eta)\, d\eta.$$

The term in brackets can be expressed as the integral of σ_2' from $\xi - \eta$ to ξ, and then it's easy to use (12.21) to derive

$$\sigma_2(\xi) - \sigma_2 * \phi(\xi) = O(|\xi|). \tag{12.22}$$

To estimate $\sigma_1 - \sigma_1 * \phi(\xi)$, we use Melrose's Tauberian argument from [134], which is a slightly simpler version of Hörmander's [96, Lemma 17.5.6]. By Lemma 12.5 and (12.21), we know already that

$$\sigma_1' * \phi(\xi) = O(|\xi|). \tag{12.23}$$

Since ϕ is strictly positive, for $t \in (-\frac{1}{2}, \frac{1}{2})$ we will have $\phi(t) > c_0$ for some constant $c_0 > 0$. Because $\sigma_1'(\xi)$ is also positive, this implies that

$$\sigma_1' * \phi(\xi) = \int_{-\infty}^{\infty} \sigma_1'(\xi - \eta)\phi(\eta)\, d\eta$$

$$\geq c_0 \int_{-1/2}^{1/2} \sigma_1'(\xi - \eta)\, d\eta$$

$$= c_0 \big[\sigma_1(\xi + \tfrac{1}{2}) - \sigma_1(\xi - \tfrac{1}{2})\big].$$

By (12.23), we thus have

$$\sigma_1(\xi + \tfrac{1}{2}) - \sigma_1(\xi - \tfrac{1}{2}) = O(1 + |\xi|),$$

for any $\eta \in \mathbb{R}$. This estimate can be iterated, to obtain

$$\sigma_1(\xi) - \sigma_1(\xi - \eta) \leq \sum_{k=1}^{[\eta+1]} C(1 + |\xi - k\operatorname{sgn}(\eta)|)$$

$$\leq C(1 + |\eta|)(1 + |\xi| + |\eta|).$$

Pairing with $\phi(\eta)$ and integrating both sides then gives

$$\sigma_1(\xi) - \sigma_1 * \phi(\xi) = O(|\xi|). \tag{12.24}$$

With (12.22) and (12.24), we conclude that

$$\sigma_X(\xi) - \sigma_X * \phi(\xi) = O(|\xi|).$$

In view of (12.17), this finishes the proof.   □

Notes

As noted in the text, Guillopé–Zworski's proof [87] of the lower bound in Theorem 12.1 extends to complete surfaces with compact boundary and hyperbolic ends, as long as the Euler characteristic is nonzero. The proof was based on methods from Sjöstrand–Zworski [199, 200]. The exact wave trace formula from Theorem 11.2 does not hold in this context; instead, the small-time behavior of the wave trace is derived from a result of Ivrii [100]. (Ivrii's result is also used in proving the more general version of Theorem 12.4.)

For finite-area hyperbolic surfaces with hyperbolic cusp ends, the (more precise) asymptotic results of Selberg [189], noted in (3.12), were generalized by Müller [139] and Parnovski [152]. Finite-area surfaces are considered to be one-dimensional as scattering problems, and optimal lower bounds had previously been proven in one-dimensional and radial settings by Zworski [221, 223]. But Guillopé–Zworski's result was the first optimal global lower bound on resonances in a general higher-dimensional scattering theory.

Using results of Patterson–Perry [160] and Bunke–Olbrich [35], Perry [169] proved the optimal lower bound for scattering poles of conformally compact hyperbolic manifolds in higher dimensions. See Perry [170] for a survey of bounds known in various infinite-area hyperbolic cases.

Optimal lower bounds have also been obtained in the context of Schrödinger operators by Sjöstrand [196] and Nedelec [145]. Furthermore, lower bounds in regions near the continuous spectrum, analogous to Theorem 12.3, are known in a variety of contexts. For surveys of these results see Melrose [135], Sjöstrand [195], and Zworski [224, 226, 228].

The Weyl asymptotics of the scattering phase proven by Melrose [134] followed earlier asymptotic results of Buslaev, Majda–Ralston, Jensen–Kato, and Petkov–Popov. The asymptotics were generalized to obstacle scattering in even dimensions by Robert [182]. For scattering phase asymptotics in other contexts, see, e.g., Christiansen–Zworski [44], Parnovski [153], Christiansen [42, 43], and Carron [39].

13

Inverse Spectral Geometry

The problem of determining how the spectrum depends on basic properties of a system (e.g., conditions on the metric or potential) is referred to as the "forward" spectral problem. The "inverse" problem is to deduce properties of the original system from some knowledge of the spectrum (including the resonances in the case of a scattering problem). In other words, the goal of inverse spectral theory is to determine exactly how much information about the system is contained in the spectrum. There is a very strong physical motivation for the inverse problem, since the spectrum is often the aspect of a system most accessible to experimental observation. But the question is also natural from a mathematical point of view—the resonances of a hyperbolic surface provide a set of geometric invariants, and the mathematical goal is always to understand the content of such invariants.

Definition. Two hyperbolic surfaces are *isospectral* if their resonances sets coincide with multiplicities, and *length isospectral* if their length spectra coincide with multiplicities.

The term "isopolar" is also used in the literature to describe two systems with the same set of scattering poles.

It is known that the resonance set does not determine a hyperbolic surface completely. The first examples of isospectral infinite-area surfaces were pointed out Guillopé–Zworski in [87, Remark 2.3], based on the transplantation method of Bérard [18]. In [29], Brooks–Davidovich applied the Sunada method to construct such examples (see [31] and [30] for details of this approach). Among the possibilities are:

1. Two isospectral hyperbolic surfaces of genus 0 with eight funnels.
2. Two isospectral hyperbolic surfaces of genus 2 with four funnels.
3. Two isospectral hyperbolic surfaces of genus 3 with three funnels.
4. Two isospectral hyperbolic surfaces of genus 0 with sixteen funnels.
5. Families of size $c_1 k^{c_2 \log k}$ of mutually isospectral hyperbolic surfaces of genus $c_3 k$ with $c_4 k$ funnels.

All of these examples are both isospectral and length isospectral. In fact, Sunada methods produce examples satisfying the stronger condition of having the same relative scattering phase. Such families are called "isophasal" or "isoscattering."

These examples are "negative" results showing that the resonance set does not provide sufficient information to recover the surface. In this chapter we'll present a "positive" result limiting the size of isospectral families. First we'll show that the resonance set and length spectrum determine each other and fix the topological type, up to finitely many possibilities. Then we'll use this information to prove that the number of hyperbolic surfaces with the same resonance set or length spectrum is finite.

13.1 Resonances and the length spectrum

An immediate implication of Theorem 10.1 is the analogue of Huber's theorem from the compact case, the equivalence of the resonance set and length spectrum. This was first observed explicitly in Borthwick–Judge–Perry [24]. Of course, such a connection is also evident from (12.1), which was proven earlier by Guillopé–Zworski [88].

Theorem 13.1 (Borthwick–Judge–Perry). *For geometrically finite hyperbolic surface X of infinite area, the resonance set $\mathcal{R}_X$ determines the length spectrum $\mathcal{L}_X$, the Euler characteristic $\chi(X)$, and the number of cusps n_c. The length spectrum determines $\chi(X)$ and n_c up to a finite number of possibilities. And the length spectrum, $\chi(X)$, and n_c together determine the resonance set.*

Proof. Suppose that the resonance set is fixed, which determines $P_X(s)$. We claim that $P_X(s)$ determines $\chi(X)$, n_c, and $q(s)$, in the notation of Theorem 10.1, and therefore fixes $Z_X(s)$. To see this we take the log of (10.3),

$$\log Z_X(s) = q(s) - \chi(X) \log G_\infty(s) + n_c \log \Gamma\left(s - \tfrac{1}{2}\right) + \log P_X(s), \quad (13.1)$$

and analyze the asymptotics as $\operatorname{Re} s \to \infty$. Because the sum

$$\log Z(s) = \sum_{\ell \in \Lambda} \sum_{k=0}^{\infty} \log\left(1 - e^{-(s+k)\ell}\right)$$

converges uniformly for $\operatorname{Re} s \geq 1$, it is clear that $\log Z_X(s)$ decays exponentially as $\operatorname{Re} s \to \infty$. The analogue of Stirling's formula for the Barnes G-function was given by Voros [215]: for $\operatorname{Re} z > 0$,

$$\log G(z+1) \sim z^2\left(\tfrac{1}{2}\log z - \tfrac{3}{4}\right) + \tfrac{1}{2}z \log 2\pi - \tfrac{1}{12}\log z + \zeta'(-1).$$

Together with Stirling's formula (7.41), this gives an asymptotic formula for G_∞,

$$\log G_\infty(s) \sim -\left(\tfrac{1}{2}s(s-1) - \tfrac{1}{6}\right)\log s(s-1) + \tfrac{3}{2}s(s-1) + 1 + \tfrac{1}{2}\log 2\pi - 2\zeta'(-1).$$

Stirling's formula also gives

$$\log \Gamma(s - \tfrac{1}{2}) \sim \frac{1 + \log 2\pi}{2} + s \log(s - \tfrac{1}{2}) - s.$$

These asymptotics, along with the fact that q is polynomial, imply that $\log P_X(s)$ has an asymptotic expansion as $\mathrm{Re}(s) \to \infty$. The value of $\chi(X)$ may be deduced from the coefficient of $s^2 \log s$ in this expansion. With this information we can subtract off the $\chi(X) \log G_\infty(s)$ term from (13.1). In the remainder, the coefficient of the leading $s \log s$ term determines n_c. After all terms involving n_c are subtracted off, what remains in the asymptotic expansion is precisely $q(s)$, so this is also determined by $P_X(s)$.

Thus we see that $\mathcal{R}_X$ determines the function $Z_X(s)$. Deducing the length spectrum from the zeta function is a standard argument. If ℓ_0 is the minimal length in Λ, then (13.1) implies

$$\log Z_X(s) \sim -e^{-\ell_0 \mathrm{Re}(s)},$$

so we can read off ℓ_0. Then we remove the ℓ_0 terms from the sum and read off the next length ℓ_1 as the decay exponent of the remainder, and so on.

Now assume that the length spectrum is known, giving $Z_X(s)$. The order of the zero of $Z_X(s)$ at $s = -k$, $k \in \mathbb{N}_0$, is $m_{-k} + (2k + 1)(-\chi(X))$. This gives us the bound

$$0 \le -\chi(X) \le \frac{1}{2k + 1} \operatorname*{ord}_{s=-k} Z_X(s),$$

implying that only finitely many values of $\chi(X)$ are possible, and hence finitely many choices of n_c, since $n_c \le 2 - \chi(X)$. If n_c and $\chi(X)$ were fixed, then $Z_X(s)$ would determine the divisor of $P_X(s)$ and the resonance set could be read off directly. $\square$

If X has no cusps, then the results of Bunke–Olbrich [35] and Olbrich [150, Corollary 6.9] show that when Γ is nonelementary there are no resonances at points $-\mathbb{N}$. Thus the possibility of overlap referred to in the last paragraph of the proof above is eliminated, and the length spectrum determines the resonance set completely in this case.

13.2 Hyperbolic trigonometry

One of the main tools we will draw on later in the chapter is a decomposition of the surface into simple model pieces. The building blocks for these are hyperbolic hexagons. We can extend the notion of hexagon to allow sides to have length zero, just as we did for triangles in Section 2.5. Of course, sides with length zero must be nonadjacent.

Lemma 13.2 (Right-angled hexagons). *Given any three numbers $a, b, c \ge 0$, there is a unique right-angled hexagon in $\mathbb{H}$ (up to isometry) such that a, b, c are the lengths of three nonadjacent sides.*

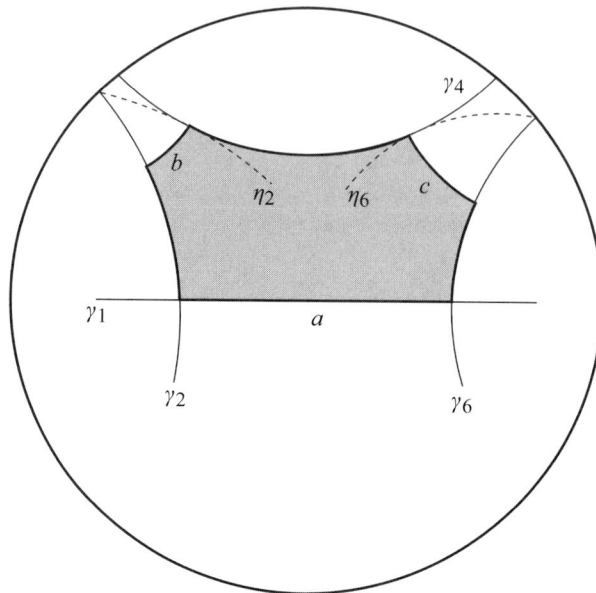

Fig. 13.1. Right-angled hexagon.

Proof. Start with an arbitrary geodesic γ_1, on which we mark off a segment of length a. From the endpoints draw geodesic arcs γ_2 and γ_6 that are perpendicular to the original segment. (If $a = 0$ we let γ_1 be a boundary point and take for γ_2 and γ_6 any two geodesic arcs meeting at γ_1.) Inside the region enclosed by these three curves, let η_2 be the locus of points whose distance to γ_2 is b, and η_6 the locus of points at distance c from γ_6, as shown in Figure 13.1. (In either the $\mathbb{H}$ or $\mathbb{B}$ model, a curve lying at a fixed distance from a geodesic is a circle meeting the boundary $\partial \mathbb{H}$ or $\partial \mathbb{B}$ at the endpoints of the geodesic.) We claim that there is a unique geodesic arc tangent to both η_2 and η_6, which we'll label γ_4. To complete the construction, we fill in γ_3 as the arc of shortest distance between γ_2 and γ_4. This meets γ_4 at its intersection point with η_2. Similarly, γ_5 is the shortest arc between γ_4 and γ_6. By the construction of γ_4, the segments γ_3 and γ_5 have lengths b and c, respectively. The hexagon obtained by this procedure is uniquely determined by the starting segment (and the choice of γ_2 and γ_6 if $a = 0$).   □

The remainder of this section is devoted to some basic formulas of hyperbolic trigonometry that will prove useful later. The simplest way to prove these is to introduce yet another model for hyperbolic space, the Minkowski or *hyperboloid model*. Minkowski 3-space is $\mathbb{R}^3$ equipped with the metric $h = dx_1^2 + dx_2^2 - dx_3^2$. This is a pseudo-Riemannian metric (not positive definite), used as a standard model in general relativity. Following the treatment in Buser [37, Section 2.1]), we restrict our attention to the hyperboloid

$$H := \left\{ x \in \mathbb{R} : \ x_1^2 + x_2^2 - x_3^2 = -1 \right\},$$

on which the restriction of h gives a positive definite metric g. The space (H, g) is isometric to $\mathbb{B}$ by stereographic projection from $(0, 0, -1)$, as shown in Figure 13.2.

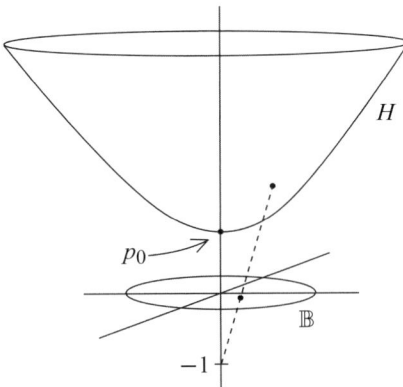

Fig. 13.2. Isometry from $\mathbb{B}$ to (H, g).

Isometries of (H, g) are generated by the linear transformations of $\mathbb{R}^3$ preserving h, so the orientation-preserving isometry group is identified with $SO(2, 1)$. In particular, we can generate all isometries using

$$L_\theta := \begin{pmatrix} \cos\theta & -\sin\theta & 0 \\ \sin\theta & \cos\theta & 0 \\ 0 & 0 & 1 \end{pmatrix}, \qquad M_r := \begin{pmatrix} \cosh r & 0 & \sinh r \\ 0 & 1 & 0 \\ \sinh r & 0 & \cosh r \end{pmatrix}, \tag{13.2}$$

for $\theta, r \in \mathbb{R}$. Fixing an origin $p_0 = (0, 0, 1)$, it is easy to check that the map $(r, \theta) \mapsto L_\theta M_r p_0$ defines a coordinate system on H in which g takes the geodesic polar form (2.10).

Lemma 13.3 (Sine rule). *For a triangle ABC with geodesic sides, let α, β, γ denote the interior angles at the vertices, and a, b, c the respective lengths of the opposite sides. Then*

$$\frac{\sin\alpha}{\sinh a} = \frac{\sin\beta}{\sinh b} = \frac{\sin\gamma}{\sinh c}.$$

Proof. Regarding the triangle as a subset of H, we may assume that vertex B is located at p_0 and that A is the point $M_{-c} p_0$. We apply first M_c to move vertex A to p_0, and then $L_{\pi-\alpha}$ to rotate so that C is located at $M_{-b} p_0$, as shown in Figure 13.3. Then we apply $L_{\pi-\gamma} M_b$ to shift C to p_0, followed by $L_{\pi-\beta} M_a$ to move B to p_0. Since this returns the triangle to its original position, we conclude that

$$L_{\pi-\beta} M_a L_{\pi-\gamma} M_b L_{\pi-\alpha} M_c = I. \tag{13.3}$$

Taking the equivalent statement

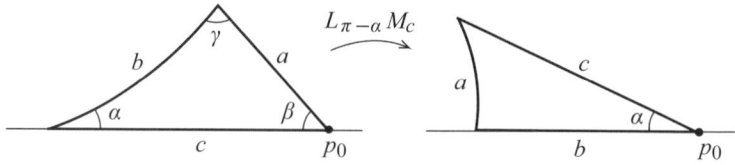

Fig. 13.3. Proving the sine rule.

$$M_a L_{\pi-\gamma} M_b = L_{\beta-\pi} M_{-c} L_{\alpha-\pi}$$

and evaluating two particular matrix elements on either side gives

$$\begin{pmatrix} * & * & * \\ * & * & \sin\gamma\,\sinh b \\ * & -\sin\gamma\,\sinh a & * \end{pmatrix} = \begin{pmatrix} * & * & * \\ * & * & \sin\beta\,\sinh c \\ * & -\sin\alpha\,\sinh a & * \end{pmatrix}.$$

This proves our identity. The matrix elements that we have omitted give other identities (cosine rules), but we won't need these.   □

We could have introduced rotations and translations in $\mathbb{B}$ or $\mathbb{H}$ and obtained an identity of the form (13.3), but it is much more difficult to read the sine and cosine rules from the matrix elements in those models. It's possible, though not easy, to prove the sine rule by more direct computation; see, e.g., [108].

Lemma 13.4 (Pentagon rule). *For a right-angled pentagon with geodesic sides, label the lengths of consecutive sides a, b, c, d, e. Then*

$$\sinh a \sinh b = \cosh d.$$

Proof. We apply the same strategy as in Lemma 13.3 to obtain the identity

$$L_{\pi/2} M_a L_{\pi/2} M_b L_{\pi/2} M_c L_{\pi/2} M_d L_{\pi/2} M_e = I.$$

The formula can then easily be read off from the matrix entries of

$$L_{\pi/2} M_a L_{\pi/2} M_b = M_{-e} L_{-\pi/2} M_{-d} L_{-\pi/2} M_{-c} L_{-\pi/2}.$$ □

13.3 Teichmüller space

In this section we will see that the resonance set or length spectrum fixes a hyperbolic metric "locally," in the sense that these sets are not preserved under small continuous deformations of the hyperbolic structure. To define continuous deformations in this context, we need to introduce the concept of Teichmüller space. Our main result will be that a set of isospectral metrics forms a discrete subset within Teichmüller space.

Teichmüller theory is an important field in its own right, and we have the space for only a relatively sketchy introduction here. Our discussion mainly follows the

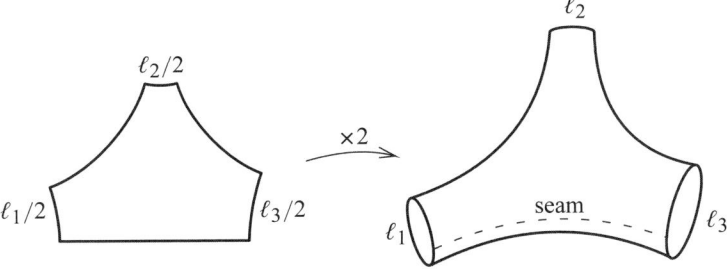

Fig. 13.4. Constructing a pair of pants.

treatment by Buser [37], and we will refer the reader to that source for some of the technical details.

Up to now, we have considered a surface X equipped with a single fixed hyperbolic metric. Now suppose we consider the space of all possible hyperbolic metrics on X, with the restriction that the type (funnel or cusp) of each end is fixed. If two such metrics are related by pullback by a diffeomorphism, then they are isometric and obviously should be considered equivalent. The *moduli space* $\mathcal{M}_X$ is the set of isometry classes of hyperbolic metrics on X with ends of the same type as those of X. The moduli space can be given a C^∞ topology, in which a sequence of isometry classes converges if and only if there exist representative metrics whose coordinate components and their derivatives converge uniformly on compact sets. Unfortunately, with this natural topology the space $\mathcal{M}_X$ is rather singular and difficult to handle directly.

The point of Teichmüller theory is to use a weaker notion of equivalence to get a nicer space. The *Teichmüller space* $\mathcal{T}_X$ is the space of complete hyperbolic metrics, modulo pullback by diffeomorphisms homotopic to the identity. There are various ways to understand this space and define its topology, for example through the holomorphic quadratic differentials on the original surface X. We will define the topology of $\mathcal{T}_X$ through a set of global coordinates called Fenchel–Nielsen coordinates.

These coordinates are based on the "pants" decomposition of X.

Definition. A *pair of pants* is a hyperbolic surface, possibly with geodesic boundary, diffeomorphic to a sphere with 3 punctures.

The Euler characteristic of such a surface is -1, so by Gauss–Bonnet (Theorem 2.15), all pairs of pants have area 2π. We can characterize each end with a boundary length ℓ, which is either the length of the closed geodesic or zero if the end is a cusp.

Lemma 13.5 (Pairs of pants). *For each triple $\ell_1, \ell_2, \ell_3 \geq 0$, there is a unique pair of pants Y with these boundary lengths.*

Proof. Start with two identical right-angled hexagons with boundary lengths $\ell_1/2$, $\ell_2/2$, $\ell_3/2$, whose existence is guaranteed by Lemma 13.2. Because of the right angles, the hexagons can be glued together along seams given by the three edges whose lengths were not specified to form a pair of pants with the appropriate boundary lengths, as shown in Figure 13.4.

To prove uniqueness we observe that taking the shortest paths between the three boundary geodesics of any pair of pants gives three unique seams that split the pair of pants into two right-angled hexagons. (For a cusp we'd use a bounding horocycle in place of the geodesic boundary.) Since three nonadjacent side lengths (the seams) of the hexagons already match, the two hexagons are identical. Hence the nonseam boundary lengths are $\ell_1/2$, $\ell_2/2$, $\ell_3/2$ and the hexagons are uniquely determined. □

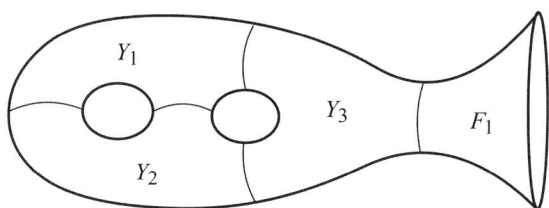

Fig. 13.5. Pants decomposition.

Theorem 13.6 (Pants decomposition). *The convex core of a geometrically finite, nonelementary hyperbolic surface X can be decomposed into a finite union of pairs of pants Y_j, $j = 1, \ldots, m$, where $m = -\chi(X)$, so that*

$$X = Y_1 \cup \cdots \cup Y_m \cup F_1 \cup \cdots \cup F_{n_f}.$$

Proof. Figure 13.5 illustrates the claimed decomposition. From Theorem 2.13 we recall that the convex core N is X with funnels removed, and the compact core K is N minus the cusps. Since area$(N) = -2\pi \chi(X)$ and each pair of pants has area 2π, it's clear from the outset that at most m pairs of pants could be used. By induction, it suffices to show that we can cut a single pair of pants from N. (We're following the argument of Buser [37, Theorem 4.4.5] here.)

The boundary of K consists of finitely many closed geodesics or horocycles. For simplicity, we'll assume that there is at least one boundary geodesic, say γ. (The argument starting from a bounding horocycle is quite similar.) The neighborhood of points within distance a of γ,

$$G_a := \{z \in K : d(z, \gamma) \leq a\},$$

is isometric for small a to a half-collar $[0, a] \times S^1$, $ds^2 = dr^2 + \ell^2 \cosh^2 r \, d\theta^2$. (If we had started with a boundary horocycle, we'd get $ds^2 = dr^2 + e^{2r} d\theta^2$ instead.) As a increases, G_a must stop being isometric to a half-collar at some point. Otherwise the limit of G_a as $a \to \infty$ would be a funnel (or the big end of a parabolic cylinder). There are only two ways for the isometry to break down; for some value of a either G_a meets itself (case 1) or G_a bumps into some other boundary curve of K (case 2).

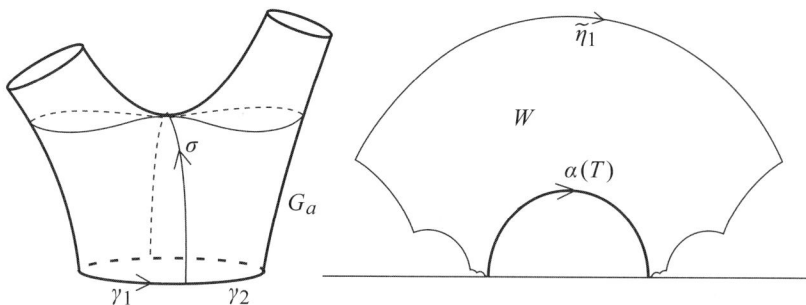

Fig. 13.6. Case 1 for pants decomposition.

Case 1: The perpendicular geodesic segments from γ to the first self-intersection point of G_a connect to form a geodesic arc σ dividing γ into two parts, γ_1, γ_2. This setup is shown on the left in Figure 13.6. From the division of γ we form two simple closed curves, $\eta_1 = \gamma_1\sigma$ and $\eta_2 = \gamma_2\sigma^{-1}$. We can focus on η_1, since the argument is the same for either. Let $\tilde{\eta}_1$ denote a lift of η_1 to $\mathbb{H}$, as shown on the right in Figure 13.6. This lift is a union of segments meeting at right angles, which project down to γ_1 and σ in alternation. Since $\tilde{\eta}_1$ covers a closed curve on X, it is preserved by some maximal cyclic subgroup of Γ. Let $T \in \Gamma$ be the generator of this subgroup.

Assume first that T is parabolic, in which case $\tilde{\eta}_1$ would be a loop meeting $\partial\mathbb{H}$ at the fixed point p of T. For $R \in \Gamma - \langle T \rangle$, $R\tilde{\eta}_1$ cannot intersect $\tilde{\eta}_1$ because η_1 is simple. Also, p could not be fixed by R. (Lemma 2.12 shows that hyperbolic and parabolic fixed points cannot coincide, and if p were a parabolic fixed point of R this would contradict $R\tilde{\eta}_1 \cap \tilde{\eta}_1 = \emptyset$.) If W denotes the region enclosed by $\tilde{\eta}_1$, then since $R\tilde{\eta}_1 \cap \tilde{\eta}_1 = \emptyset$ and R does not fix p, we have $RW \cap W = \emptyset$. This means that $\Gamma\backslash W = \langle T \rangle\backslash W$, which therefore contains a single cusp.

On the other hand, suppose T is hyperbolic, with axis $\alpha(T)$. This is the case actually shown in Figure 13.6. Reasoning as above, for $R \in \Gamma - \langle T \rangle$ we have $R\tilde{\eta}_1 \cap \tilde{\eta}_1 = \emptyset$ and R cannot fix the endpoints of $\alpha(T)$. Together these imply that if W is

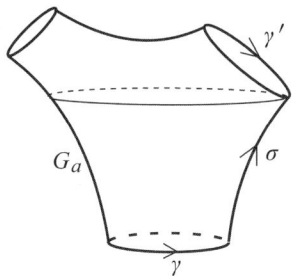

Fig. 13.7. Case 2 for pants decomposition.

the region bounded by $\tilde{\eta}_1$ and $\alpha(T)$, then $RW \cap W = \emptyset$. Hence $\Gamma \backslash W = \langle T \rangle \backslash W$ is an annulus bounded at one end by η_1 and at the other by the simple closed geodesic $\Gamma \backslash \alpha(T)$.

With these two possibilities accounted for, and a similar argument applied to η_2, we have shown that γ is one of the boundary curves of a pair of pants contained within N.

Case 2: Let γ' be the boundary curve that G_a has bumped into, as shown in Figure 13.7. By linking γ and γ' through the geodesic arc σ connecting them and then expanding slightly, we can produce a simple closed curve $\eta := \sigma^{-1}\eta'\sigma\gamma$ that encompasses both boundary curves. We can argue exactly as in case 1 that η bounds either a cusp or an annulus with a closed geodesic at the other end. Thus γ and γ' are two of the boundary curves of a pair of pants contained within N.   □

The decomposition in Theorem 13.6 gives us a family of disjoint simple closed geodesics between pairs of pants (possibly separating two legs of the same pair of pants). We will label these curves $\gamma_1, \ldots, \gamma_d$, where the number of such curves is given by

$$d = 3p - 3 + n_f + n_c.$$

For each of these curves, we have a two-parameter family of deformations. First, we can change the length ℓ_j of γ_j by changing the appropriate boundary lengths of the pairs of pants. And second, we can introduce a rotation by angle $\theta_j \in \mathbb{R}$ in the gluing map between the two legs. To make the twist angle well-defined, we fix an orientation and starting point for γ_j and parametrize by $\theta = $ (arclength)$/\ell_j$. Then the twist parameter θ_j corresponds to the gluing map $\theta \mapsto \theta + \theta_j$. This normalization means that taking $\theta_j = 1$ gives a surface isometric to the original. However, this isometry is not given by a pullback homotopic to the identity, so θ_j and $\theta_j + 1$ correspond to distinct points of Teichmüller space.

There are n_f additional geodesics where funnels attach, which we label $\gamma_{d+1}, \ldots, \gamma_{d+n_f}$. We can also change the lengths of these boundary geodesics. But since funnels are rotationally symmetric, rotating the funnel does not deform the surface. There are no twist parameters associated to funnel boundaries.

Definition. The deformation parameters $\ell_1, \ldots, \ell_{d+n_f}$ and $\theta_1, \ldots, \theta_d$ obtained through the above construction are collectively called the *Fenchel–Nielsen coordinates* (for Teichmüller space).

If we are given any hyperbolic surface diffeomorphic to X, with ends of the same type, we can apply Theorem 13.6 to decompose the new surface into pairs of pants that line up with those of X. Since the Fenchel–Nielsen coordinates describe all possible pairs of pants and all possible gluings, a direct consequence of Theorem 13.6 is the following:

Corollary 13.7. *Every hyperbolic surface diffeomorphic to X, with ends of the same type, is obtained through variation of the Fenchel–Nielsen coordinates.*

If we fix the hyperbolic surface X with a particular pants decomposition as an origin, the Fenchel–Nielsen coordinates give an alternative definition of Teichmüller space,

$$\mathcal{T}_X := \left\{ (\ell_1, \ldots, \ell_{d+n_f}; \theta_1, \ldots, \theta_d) \in \mathbb{R}_+^{d+n_f} \times \mathbb{R}^d \right\}. \tag{13.4}$$

This is a particularly simple way to understand $\mathcal{T}_X$; not only do we have a global set of coordinates, but the Euclidean topology agrees with the C^∞ topology on $\mathcal{T}_X$. To see this, one can fix a family of identification maps between right-angled hexagons, extend these to the pairs of pants, and then perturb the identifications in collar neighborhoods of the boundary geodesics to account for the twist parameters. See, e.g., Buser [37, Sections 3.2–3.3] for the details of the proof. For our purposes it suffices to accept (13.4) as the definition of $\mathcal{T}_X$ as a topological space.

We next want to show that the lengths of a certain set of closed geodesics determine the hyperbolic structure completely. For this purpose, we keep the curves $\gamma_1, \ldots, \gamma_{d+n_f}$ as above, and seek to introduce extra curves whose lengths will determine the twist angles θ_j. Assume first that the closed geodesic γ_j separates two distinct pairs of pants. The union of these pieces along γ_j is then called an *X-piece*. On the untwisted X-piece, let $\alpha_j(0)$ be the closed geodesic in the homotopy class of a curve that passes between the ends of each pair of pants and intersects γ_j once on each side (see Figure 13.8). If the X-piece is twisted by $\theta \neq 0$, we define a new geodesic $\alpha_j(\theta)$ by deforming $\alpha_j(0)$ along the twist and then taking the closed geodesic homotopic to this deformed curve. Note that at $\theta = 1$ we return to the original X-piece, but $\alpha_j(1)$ lies in a different homotopy class from $\alpha_j(0)$. We can define another family of curves $\beta_j(\theta)$ by starting from $\beta_j(0) := \alpha_j(1)$, as illustrated in Figure 13.8, and performing the same procedure.

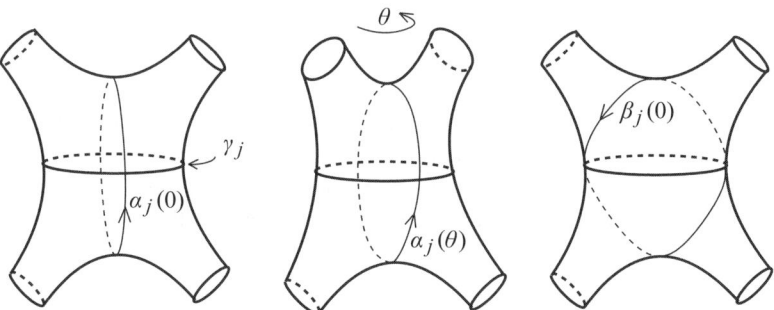

Fig. 13.8. Canonical geodesics on an X-piece.

With some hyperbolic trigonometry (see [37, Proposition 3.3.11]), one can derive explicit formulas

$$\cosh \ell(\alpha_j(\theta))/2 = a + b \cosh(\theta \ell_j),$$
$$\cosh \ell(\beta_j(\theta))/2 = a + b \cosh((\theta + 1)\ell_j),$$

where a, b are independent of θ and depend only on the boundary lengths of the two pairs of pants making up the X-piece. From the first formula we see that $|\theta|$ is determined by the length $\ell(\alpha_j(\theta))$. The family $\beta_j(\theta)$ was included just so we could use $\ell(\beta_j(\theta))$ in the second formula to fix the sign of θ. Thus θ can be written as a function solely of the lengths of certain closed geodesics.

Similar arguments apply if two boundary circles of a single pair of pants are glued together along γ_j to give a surface that is topologically a punctured torus. By taking two copies of the pair of pants, we can create an X-piece that is a double cover of the punctured torus. The curve families $\alpha_j(\theta)$, $\beta_j(\theta)$ are defined by taking the families introduced on the X-piece above and projecting them down to the punctured torus. (See, e.g., [37, Section 3.4] for the details.)

For each $j = 1, \ldots, d$, we introduce two new families of closed geodesics $\alpha_j(\theta_j)$, $\beta_j(\theta_j)$ by the above construction. The full collection $\{\gamma_j, \alpha_j(\theta_j), \beta_j(\theta_j)\}$ is called a *canonical curve system*. Because the formulas above give the twist angle at each γ_j as a function of the lengths of these curves, we have the following:

Theorem 13.8. *The lengths of closed geodesics for a canonical curve system determine the Fenchel–Nielsen coordinates. In particular, they determine a hyperbolic surface up to isometry.*

In particular, this shows that the length spectrum would determine the geometry completely if only we knew which lengths were attached to which curves.

Because the length spectrum $\mathcal{L}_X$ is discrete, we can draw the following immediate conclusion:

Corollary 13.9. *The set of points within $\mathcal{T}_X$ corresponding to hyperbolic metrics with the same length spectrum as X forms a discrete subset of $\mathcal{T}_X$.*

13.4 Finiteness of isospectral classes

A simple way to show that a set is finite is to prove that it is both discrete and compact. Having shown in Corollary 13.9 that an isospectral class of hyperbolic surfaces corresponds to a discrete subset of Teichmüller space, we will now take up the issue of compactness. This will lead to a proof of the following result of Borthwick–Judge–Perry [24]:

Theorem 13.10 (Borthwick–Judge–Perry). *Let X be a nonelementary geometrically finite hyperbolic surface of infinite area. Then either the length spectrum or the resonance set of X determines the surface up to finitely many possibilities.*

A key ingredient in the compactness argument is a beautiful estimate known as Mumford's lemma, which Mumford [140] proved originally for compact manifolds with nonpositive sectional curvature. Bers [19] adapted the proof to hyperbolic surfaces with cusps, and that is the version we give here.

Lemma 13.11 (Mumford, Bers). *Suppose X is a hyperbolic surface of finite area. Let K be the compact core of X and ℓ_0 the length of the smallest closed geodesic on X. Then there is a constant C depending only on the topological type of X such that*

$$\min(\ell_0, 1)\, \mathrm{diam}(K) \le C.$$

Proof. The length ℓ_0 is equal to the minimum displacement length for any hyperbolic element of Γ. In view of (2.7), this means that all points in $\mathbb{H}$ are translated at least a distance ℓ_0 by any hyperbolic element. If p is a parabolic fixed point, then the neighborhood O_p used to define $\widetilde{K}$ in Section 2.4 consists precisely of the points translated by a distance < 1 by the generator of the parabolic subgroup fixing p. Any points in $\widetilde{K} \subset \mathbb{H}$ are thus translated by a distance of at least 1 by the parabolic elements of Γ.

Hence $a = \min(\ell_0, 1)$ is the minimum translation distance for any point in $\widetilde{K}$ by any element of $\Gamma - \{I\}$. In particular, geodesic rays issuing from any point $z \in K$ cannot intersect within a distance $a/2$.

Find two points $p, q \in K$ separated by distance $\mathrm{diam}(K)$, and let γ be the shortest geodesic in X connecting p to q. Starting from γ, we create a new path γ' in K by cutting off any portion of γ inside a cusp and replacing it with a segment of the horocyclic cusp boundary. Figure 13.9 shows an example in which γ passed through a single cusp. Since γ was length minimizing, we can be sure that each cusp was entered at most once. This implies that γ' consists of at most $n_c + 1$ geodesic segments lying inside K, separated by at most n_c horocyclic arcs. These horocyclic arcs must have length less than 1. Therefore, if γ_1 denotes the longest of the geodesic segments of γ' lying in K, then we have

$$\mathrm{diam}(K) = \ell(\gamma) \le \ell(\gamma') \le (n_c + 1)\ell(\gamma_1) + n_c. \tag{13.5}$$

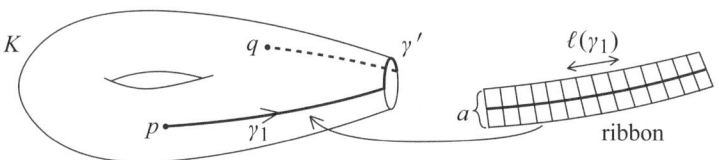

Fig. 13.9. Proof of Mumford's lemma.

From each point in γ_1 draw a perpendicular of length a centered on γ_1. We claim that the collection of these perpendiculars forms an embedded "ribbon" in X. The ends of the perpendiculars coming from the same point do not meet, except possibly at the endpoints, by the choice of a. And perpendiculars emanating from different points of γ can't meet because of the negative curvature. (Such a meeting would produce a triangle with angle sum greater than π, contradicting Gauss–Bonnet.) This shows that the ribbon is embedded in X. To compute the area of the ribbon, we introduce geodesic (Fermi) coordinates along γ_1. With t an arc-length parameter along

γ_1, and r the signed distance into the perpendiculars, the metric takes the local form $ds^2 = dr^2 + \cosh^2 r \, dt^2$. The area of the ribbon is thus given by $2\sinh(a/2)\ell(\gamma_1)$. Since the ribbon is embedded, we have

$$2\sinh(a/2)\ell(\gamma_1) < \text{area}(X) = -2\pi\chi(X).$$

The estimate (13.5) then yields

$$\text{diam}(K) \leq (n_f + 1)\frac{-\pi\chi(X)}{\sinh(a/2)} + n_f. \qquad \square$$

In the finite-area case, this implies (with some additional argument) that the set of hyperbolic metrics with $\ell_0 \geq \varepsilon$ for fixed $\varepsilon > 0$ is a compact subset of moduli space.

Unfortunately, Mumford's lemma does not extend to the infinite-area case. As a counterexample, we can take a sequence of pairs of pants with boundary lengths diverging to infinity; the diameter of the compact core diverges also, while ℓ_0 remains bounded below. To make use of Mumford's lemma in this case, our strategy is to take the convex core N and double it across its geodesic boundary components. If a sequence of hyperbolic metrics on X diverges in $\mathcal{M}_X$, then applying Mumford's lemma to the double gives us a tool for understanding the behavior of the limit.

Recall that a closed curve homotopic to the horocycle bounding a cusp was called cuspidal. Proposition 2.16 shows that if a curve $\alpha \subset X$ is noncuspidal, then for any hyperbolic metric on X (with ends of the same type) there exists a unique geodesic that is homotopic to α.

Proposition 13.12 (Geometric limit). *Let X be a geometrically finite nonelementary hyperbolic surface. Suppose that $[g_n] \in \mathcal{M}_X$ is a divergent sequence with the minimum of the length spectrum ℓ_0 uniformly bounded from below by a positive constant. Then there exist*

1. *a subsequence of metric representatives, g_n;*
2. *a geometrically finite hyperbolic surface Z, h, possibly consisting of a finite number of connected components;*
3. *a precompact neighborhood U of the convex core of (Z, h) (for a hyperbolic cylinder the convex core is interpreted as the central closed geodesic); and*
4. *a smooth embedding $f : U \to X$*

such that

(A) each metric f^*g_n on U extends to a complete hyperbolic metric h_n on Z;
(B) $h_n \to h$ in $C^\infty(U)$;
(C) for each n, the convex core of (Z, h_n) lies inside U;
(D) given $L > 0$, there exists $M > 0$ such that for any noncuspidal closed curve $\alpha \subset S$ that is not homotopic to a curve in $f(U)$, the g_n-length of α is larger than L for all $n > M$;
(E) $\chi(X) < \chi(Z)$.

The surface (Z, h) constructed in Proposition 13.12 is called a *geometric limit* of (X, g_n). Before proving this result, let's see how it implies our main theorem.

Proof of Theorem 13.10. By Theorem 13.1, fixing either the resonance set or the length spectrum leaves only finitely many possibilities for the genus and numbers of funnels and cusps. Thus it suffices to start with a fixed surface X and prove that an isospectral class within $\mathcal{M}_X$ is finite. By Corollary 13.9, such a class is already known to be discrete, so our goal is to prove that there can't exist a sequence of length isospectral surfaces that leaves every compact set in $\mathcal{M}_X$.

Suppose to the contrary that such a sequence exists. Since the shortest closed geodesic length is fixed for the sequence, Proposition 13.12 applies to give a geometric limit (Z, h) for some subsequence (X, g_n), with associated map $f : U \to Z$ and metrics h_n.

Given a noncuspidal closed geodesic β in Z, h, let β_n be the unique geodesic in the same homotopy class in (Z, h_n). We claim that we may assume, by passing to a further subsequence if necessary, that β_n converges to β. Indeed, each geodesic is a solution to an ordinary differential equation whose coefficients depend continuously on the metric. By part (C) of Proposition 13.12, the convex core of each (Z, h_n) lies in the precompact set U, and hence there exists a subsequence such that the initial conditions converge. The claim follows from part (B) of Proposition 13.12 and the continuity of solutions to ordinary differential equations with respect to coefficients and initial data.

Consider a canonical curve system as in Theorem 13.8. Since the metrics g_n all have the same length spectrum, by assumption, the lengths of closed geodesics in the canonical curve system are constant for sufficiently large n. Therefore, by dropping the first portion of the sequence if necessary, we may assume that each (Z, h_n) is isometric to (Z, h).

Let $\pi_X(x)$ denote the common length counting function for the length isospectral metrics (X, g_n), and similarly $\pi_Z(x)$ the common length counting function for the isometric family (Z, h_n). Since closed geodesics on (Z, h_n) must lie in U, they correspond to closed geodesics on (X, g_n) that are homotopic to curves lying in $f(U)$. Thus if $\kappa_n(x)$ denotes the length counting function restricted to closed geodesics on (X, g_n) that are not homotopic to curves in $f(U)$, we have

$$\pi_X(x) = \pi_Z(x) + \kappa_n(x). \tag{13.6}$$

To obtain a contradiction, it suffices to show that κ_n depends on n.

Let $\alpha \subset X$ be a noncuspidal g_1-geodesic that is not homotopic to a curve in $f(U)$. The topology change shown in (E) of Proposition 13.12 guarantees that such a curve exists. We apply part (D) of Proposition 13.12 to the curve α, choosing L to be equal to twice the g_1-length of α. Thus, there exists m such that the g_m-length of any closed g_m-geodesic that is not homotopic to a curve in $f(U)$ is greater than L.

It follows that for $x < L$, $\kappa_m(x) = 0$. But since α was a closed geodesic with g_1-length $L/2$, we also have $\kappa_1(L/2) \geq 1$. This contradicts (13.6), showing that the divergent isospectral sequence cannot exist. □

The remainder of this section is devoted to the proof of Proposition 13.12. We start with some auxiliary results. The first is the collar lemma, which tells us that a short closed geodesic has a correspondingly wide collar.

Lemma 13.13 (Collar). *Suppose that γ is a simple closed geodesic of length $\ell(\gamma)$ on a geometrically finite hyperbolic surface X. Then γ has a collar neighborhood of half-width d, such that*

$$\sinh(d) = \frac{1}{\sinh(\ell(\gamma)/2)}. \tag{13.7}$$

(The collar neighborhood is isometric to $(-d, d) \times S^1$ with the metric $ds^2 = dr^2 + \ell^2 \cosh^2 r \, dt^2$.)

As a consequence, if η is any other closed geodesic intersecting γ transversally (still assuming γ is simple), then the lengths of the two geodesics satisfy the inequality

$$\sinh(\ell(\eta)/2) > \frac{1}{\sinh(\ell(\gamma)/2)}.$$

Proof. Because γ is a simple closed geodesic, we can develop a pair of pants decomposition as in Theorem 13.6 starting from γ as the first cut. In this decomposition γ is the boundary curve either between two pairs of pants or between a pair of pants and a funnel. Since the funnel can support a half-collar of arbitrary width, the argument reduces to estimating the size of a half-collar within a pair of pants.

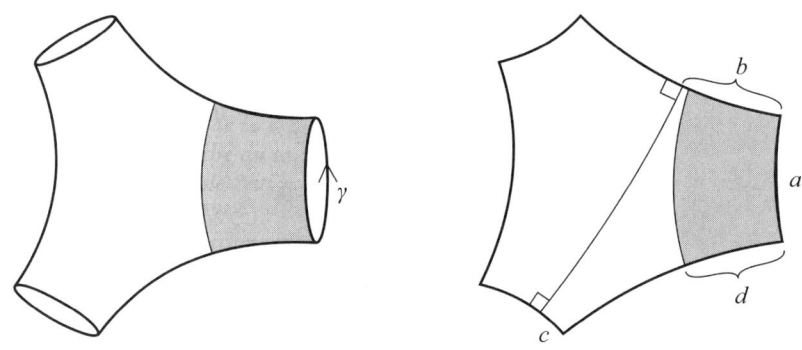

Fig. 13.10. Collar lemma.

The seams of the pants are distinguished as the shortest geodesic segments between the boundary curves, as noted in Lemma 13.5. Splitting at each seam reduces the pair of pants to two identical right-angled hexagons. The problem is to find the maximal width for the neighborhood of a side of length $a = \ell(\gamma)/2$ inside the hexagon. Draw a perpendicular geodesic segment from one of the sides adjacent to a to the opposite side, subdividing the hexagon into two right-angled pentagons. Let b be the length of the segment from a to this perpendicular, and c the corresponding length on the opposite side (see Figure 13.10). Then, by the pentagon rule (Lemma 13.4),

$$\sinh a \sinh b = \cosh c \geq 1.$$

Thus (13.7) will ensure that $d \leq b$. Applying this argument on either side of a shows that the neighborhood of a meets only a and its adjacent sides. Hence the half-collar γ is embedded in the pair of pants.

The second statement follows from the first by observing that a geodesic η intersecting γ must pass completely through the collar, so that $\ell(\eta) > 2d$. □

Our second auxiliary result is complementary to the collar lemma. It says that if one simple closed geodesic comes too close to another, then the two must intersect. We could derive this from the pants decomposition, but it is easy enough to argue directly.

Lemma 13.14 (Point of no return). *Let γ be a simple closed geodesic of length ℓ on a complete hyperbolic surface X. If α is a simple closed geodesic that does not intersect γ, then*

$$\cosh d(\gamma, \alpha) \geq \coth(\ell/2)).$$

Proof. By conjugating Γ if necessary, we can assume that the lift $\tilde{\gamma}$ of γ to $\mathbb{H}$ is the y-axis, and that $\langle T \rangle$ is the maximal subgroup of Γ preserving $\tilde{\gamma}$, generated by $T : z \mapsto e^{\ell} z$.

Let $d = d(\gamma, \alpha)$, which is nonzero since the curves are closed and do not intersect. Choose a lift $\tilde{\alpha}$ of α such that $d = d(\tilde{\gamma}, \tilde{\alpha})$. The d-neighborhood of $\tilde{\gamma}$, which we denote by

$$\Sigma := \{z \in \mathbb{H} : d(\tilde{\gamma}, z) < d\},$$

is a Euclidean sector of the form $\{\pi/2 - \theta < \arg z < \pi/2 + \theta\}$, where

$$\cos \theta = \frac{1}{\cosh d}.$$

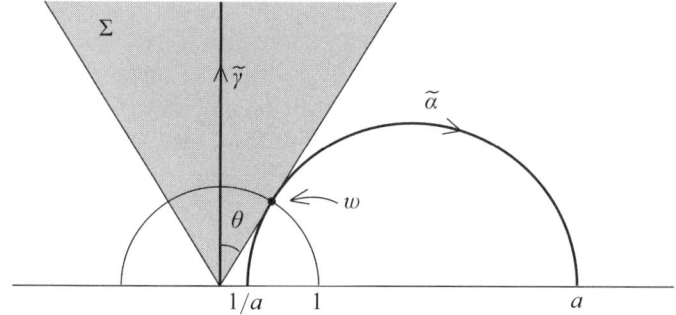

Fig. 13.11. Lifts of γ and α.

The curve $\widetilde{\alpha}$ meets $\partial\Sigma$ tangentially at one point w, as shown in Figure 13.11. The quotient $\langle T\rangle\backslash\Sigma$ is a collar neighborhood of γ of half-width d. By conjugating by a suitable dilation (allowed since dilations commute with T), we may assume that $|w| = 1$. This makes $\widetilde{\alpha}$ a semicircle orthogonal to the unit circle, so we can label its endpoints $(1/a, 0)$ and $(a, 0)$, where $a > 1$. A nice geometry exercise shows that

$$\cos\theta = \frac{a - 1/a}{a + 1/a}.$$

Since α is simple, two points of α cannot be related by T. This implies that $T(1/a) \geq a$, and hence $a \leq e^{\ell/2}$. Thus we have

$$\cosh d = \frac{a + 1/a}{a - 1/a} \geq \coth(\ell/2). \qquad \square$$

Finally, we will need Bers's theorem [20], which for a finite-area surface guarantees the existence of a pants decomposition with boundary lengths bounded by a topological constant.

Theorem 13.15 (Bers). *If X is a hyperbolic surface of finite area, then there is a constant L depending only on $\chi(X)$ such that there exists a pants decomposition of X with all boundary lengths less than L.*

Proof. The proof is by induction following the pants construction of Theorem 13.6, but we first need to find a short geodesic as a starting point. Assume first that X has no cusps. Then if γ_0 is a closed geodesic with minimal length ℓ_0, any point p on γ_0 will have a geodesic polar coordinate neighborhood of radius $\ell_0/2$. Otherwise, we could find a closed loop with length less than ℓ_0, and the homotopy class of that loop would contain a geodesic shorter than γ_0 by Proposition 2.16. (Incidentally, this argument shows that the injectivity radius of X is $\ell_0/2$.) The area of a geodesic polar neighborhood of radius r is $2\pi(\cosh r - 1)$. The fact that a neighborhood of radius $\ell_0/2$ can be embedded in X implies

$$2\pi(\cosh(\ell_0/2) - 1) < A,$$

where $A = \text{area}(X) = -2\pi\chi(X)$.

Now we cut X at γ_0 and start the construction of the pants decomposition at this cut, following the proof of Theorem 13.6. We will argue inductively, with the above bound on ℓ_0 as the first step. For the inductive step, assume that γ is a geodesic boundary curve of length ℓ. As in the proof of Theorem 13.6, we construct the half-collar G_a starting from γ, with a as large as possible.

In case 1 of that proof, the boundary of G_a meets itself, and we subdivide into two curves that are freely homotopic to two other boundary geodesics of lengths ℓ_1, ℓ_2. The sum of the lengths of these boundary geodesics must be less than the length of the boundary of G_a at the critical value of a, which is $\ell\cosh a$. On the other hand, the area of the half-collar is $\text{area}(G_a) = \ell\sinh a$, and this is bounded above by A. Hence,

$$\ell_1 + \ell_2 < \sqrt{\ell^2 + A^2}.$$

In case 2, G_a meets another geodesic boundary component γ'. Then we produce a new boundary geodesic in the homotopy class given by connecting γ and γ' by the arc between them (which has length a). If ℓ_1 denotes the length of this new boundary geodesic, then we have $\ell_1 \le \ell + \ell' + 2a$, so that

$$\ell_1 < \ell + \ell' + 2\sinh^{-1}(A/\ell).$$

Thus each time a pair of pants is removed from the surface, we can bound the new boundary lengths in terms of A and the existing set of lengths. Since the induction terminates after $-\chi(X)$ steps, this proves the result.

If X has cusps, then we choose the boundary horocycle of a cusp as our starting point. The first half-collar G_a is isometric to $[0, a] \times \mathbb{R}/\mathbb{Z}$, $ds^2 = dr^2 + e^{2r}\, dt^2$. This half-collar has boundary length e^a and area $e^a - 1$. So we start from the inequality $e^a - 1 < A$ and the induction proceeds just as above. In case 2 we could run into a horocyclic boundary component, but since these all have length 1 there is no change in the structure of the argument. $\square$

With a little more care in the proof of Bers's theorem, one could produce a bound $L \le cA$; see Buser [37, Chapter 5] for details.

Proof of Proposition 13.12. Assume that $[g_n]$ is a divergent sequence in $\mathcal{M}_X$, and let $L_0 > 0$ be the uniform lower bound on the minima of the length spectrum $\ell_0(X, g_n)$, for all n. Let us introduce a pair of pants decomposition for X, so that the isometry classes $[g_n]$ can be described with Fenchel–Nielsen coordinates. For each $[g_n]$ we use a representative metric g_n to assign coordinates $(\ell_1^{(n)}, \dots, \ell_{d+n_{\mathrm{f}}}^{(n)}; \theta_1^{(n)}, \dots, \theta_d^{(n)})$. We can assume that the twist parameters $\theta_j^{(n)}$ are contained within $[0, 1]$, since $\theta_j \to \theta_j + 1$ is an isometry. Thus, since the minimum length is bounded below, the assumption of divergence of $[g_n]$ in $\mathcal{M}_X$ means that

$$\max_j \ell_j^{(n)} \to \infty \quad \text{as } n \to \infty. \tag{13.8}$$

In the example shown in Figure 13.12, we assume that $\ell(\partial N_n) \to \infty$.

Let N_n, N_n' be two isometric copies of the convex core of (X, g_n), and let D be the surface obtained by gluing N_n and N_n' along their respective boundaries. Since the convex core has geodesic boundary, the metric g_n extends to a complete hyperbolic metric on D having finite area. We will label these extended metrics (D, g_n).

By (13.8), we have $\mathrm{diam}(D, g_n) \to \infty$. Mumford's lemma (Lemma 13.11) shows then that $\ell_0(D, g_n) \to 0$. In particular, given $\varepsilon > 0$, for all n sufficiently large (D, g_n) must have at least one simple closed geodesic of length less than ε. If $\varepsilon < L_0$, then such geodesics can't lie completely in either convex core N_n or N_n', so they must intersect ∂N_n transversally. If we take ε small enough that $\sinh \varepsilon < 1$, then the collar lemma (Lemma 13.13) guarantees that the simple geodesics of length

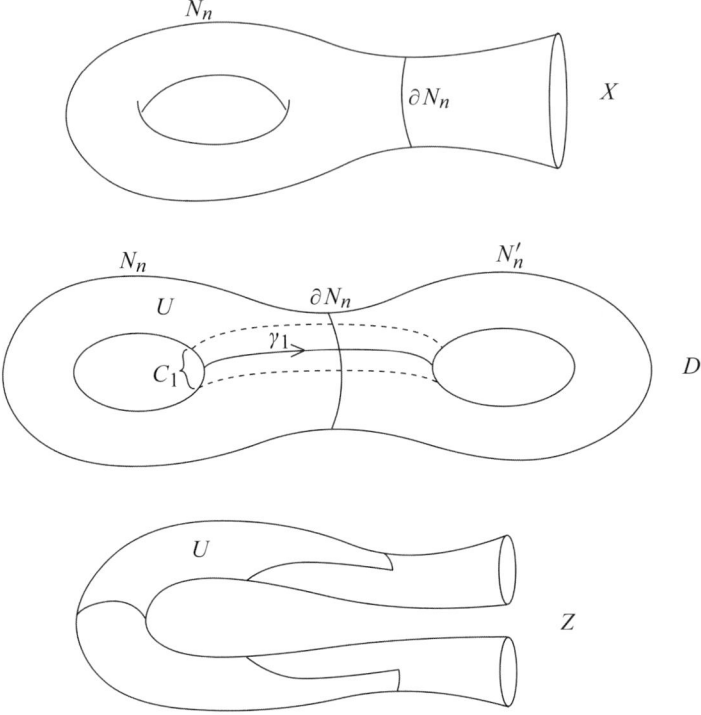

Fig. 13.12. Geometric limit example.

less than ε are disjoint. By passing to a subsequence we can assume that all (D, g_n) have simple closed geodesics shorter than ε.

Because they do not intersect, these short closed geodesics can be used as the first cuts in a pants decomposition of (D, g_n). The resulting pants decompositions may depend on n. But since there are only finitely many combinatorial types of pants decompositions of D, by passing to a subsequence again we may assume that all of these decompositions have the same combinatorial type. By Bers's theorem (Theorem 13.15), the boundary lengths for these pants decompositions lie in some fixed range $[0, L]$, with L independent of n. As above, we assume that the twist angles labeling the isometry class of (D, g_n) lie in the range $[0, 1]$.

With these assumptions, the sequence of Fenchel–Nielsen coordinates is contained in a compact set. Therefore, there exists a subsequence for which these parameters converge. Because $\ell_0(D, g_n) \to 0$, the length parameters for a certain set of boundary curves $\gamma_1, \ldots, \gamma_k$ will approach 0. (In other words, we have lifted a subsequence of the equivalence classes $[g_n] \in \mathcal{M}_X$ to a convergent sequence in the closure of $\mathcal{T}_X$.) In the example shown in Figure 13.12, there is a single shrinking boundary curve, labeled γ_1.

By identifying the pairs of pants through a regular family of diffeomorphisms [37, Lemma 3.2.6], and then perturbing in collar neighborhoods of the boundary geodesics [37, Lemma 3.3.8], we can assume that representative hyperbolic metrics (D, g_n) converge in the C^∞ topology within compact sets of $D - \{\gamma_1, \ldots, \gamma_k\}$. The limiting metric is a complete metric in which the shrinking boundary geodesics $\gamma_1, \ldots, \gamma_k$ have each given rise to a pair of cusps.

As noted above, each curve γ_j must intersect the boundary ∂N_n transversally. If $\tau : D \to D$ denotes the reflection isometry across the doubling, then $\tau \circ \gamma_j$ is a closed geodesic of the same length and intersecting γ_j. Since that length is approaching zero, these two geodesics must coincide by the collar lemma. Hence γ_j intersects ∂N_n orthogonally for each n. We can thus assume that the representatives g_n are such that $N_n = N$ independent of n, and ∂N is geodesic with respect to all of the metrics g_n.

Let $Z_0 = N - \gamma$. Then (Z_0, g_n) has piecewise geodesic boundary with interior angles equal to $\frac{\pi}{2}$. We want to define (Z, g_n) by extending (Z_0, g_n) to a complete metric. In the example shown Figure 13.12, the result of this extension is a hyperbolic cylinder. Consider a single boundary component η of Z_0, which is a simple closed curve consisting of two geodesic arcs meeting orthogonally. We can apply the arguments used in case 1 in the proof of Theorem 13.6 to show that either η bounds an annulus with a closed geodesic at the other end, or η bounds a cusp. In the first case, there is a uniquely defined extension of the annulus to a funnel. In the second, the cusp bounded by η has a unique extension to a parabolic cylinder. We thus have a uniquely defined complete hyperbolic surface (Z, h_n) (possibly a union of connected components) such that h_n is an extension of g_n and Z is homeomorphic to the interior of Z_0.

Since the metrics g_n converge uniformly on compact subsets of Z_0, it follows by analytic continuation that h_n converges to a limiting metric h uniformly on compact subsets of Z, and h is a complete hyperbolic metric. The convex core of (Z, h_n) (which we define to be the central closed geodesic for a hyperbolic cylinder component and empty for a parabolic cylinder) is clearly contained inside Z_0. By Lemma 13.14, there exists, independent of n, a collar neighborhood C_j of each shrinking geodesic γ_j such that no simple closed g_n-geodesic in $N - \gamma_j$ intersects C_j. Thus, the convex core of each (Z, h_n) lies within the compact set $U = Z_0 - \cup C_j$. Let $f : U \to X$ be the associated inclusion. Then Z, U, f, h, and h_n satisfy (A), (B), and (C).

By the collar lemma, given $L > 0$, there exists n such that if a g_n-geodesic on D intersects γ, then the g_n-length must be greater than L. Part (D) of the lemma follows.

Let k be the number of shrinking geodesics, and note that ∂Z_0 has $4k$ corners each with an interior angle equal to $\pi/2$. By Gauss–Bonnet (Theorem 2.15), we thus have

$$\chi(Z_0) = \chi(N) + k. \tag{13.9}$$

Since $\chi(Z) = \chi(Z_0)$ and $\chi(X) = \chi(N)$ and $k > 0$, this proves part (E). $\qquad\square$

Notes

For more-complete accounts of Teichmüller theory, see, e.g., Jost [104], Lehto [119], Seppälä–Sorvali [190], or Tromba [209].

In Borthwick–Judge–Perry [23], the compactness result given in Section 13.4 was proven for general surfaces with hyperbolic funnel ends, under certain restrictions. This is an analogue of the Osgood–Phillips–Sarnak [151] result for compact surfaces. The corresponding result was proven for exterior domains in $\mathbb{R}^2$ by Hassell–Zelditch [89]. Discreteness of isospectral sets remains an open question in all of these situations. For further background see the review articles by Perry [165] and Zelditch [219].

Another type of inverse scattering result involves comparison of the scattering matrices, rather than just the poles. For hyperbolic 3-manifolds without cusps, results of this type were proven by Perry [166] and Borthwick–McRae–Taylor [25]. For asymptotically hyperbolic manifolds, Joshi–Sá Barreto [102] proved that the matching of scattering matrices up to a certain order implies corresponding matching of the metrics at the boundary. And Sá Barreto [13] proved that equality of scattering matrices on the critical line implies an isometry between two asymptotically hyperbolic metrics.

14

Patterson–Sullivan Theory

The exponent of convergence δ of a Fuchsian group Γ was defined in (2.19). We also noted some basic facts about the exponent due to Beardon: $\delta = 0$ for elementary groups, $\delta = 1$ for Γ of the first kind, $0 < \delta < 1$ for Γ of the second kind, and $\delta > \frac{1}{2}$ if Γ contains parabolic elements. (We will prove these in Theorem 14.3.) In addition to these results, Beardon [14, 15] also showed that δ is bounded below by $\dim_H \Lambda(\Gamma)$, the Hausdorff dimension of the limit set. Later, Patterson [156] and Sullivan [202, 203, 204] established a remarkable set of results concerning the exponent of convergence, the limit set, the dynamical properties of the geodesic flow on X, and spectral theory. There results include the following:

Theorem 14.1 (Patterson, Sullivan). *For Γ geometrically finite, we have*

$$\delta = \dim_H \Lambda(\Gamma).$$

Furthermore, if $\delta > \frac{1}{2}$, then $\delta(1 - \delta)$ is the lowest eigenvalue of the Laplacian Δ_X acting on X.

The connection to spectral theory was later extended to the case $\delta \leq \frac{1}{2}$ by Patterson [158]. In this case the discrete spectrum of Δ_X is empty, and δ is the location of the first resonance.

In this chapter we'll develop some of the Patterson–Sullivan theory for a geometrically finite Fuchsian group of the second kind, in order to present the spectral theory applications. We'll show that the first resonance occurs at δ in Section 14.4, and in Section 14.6 we'll see that δ also determines the leading asymptotics of the length counting function $\pi_X(t)$. Further spectral applications of the Patterson–Sullivan theory will be given in Chapter 15.

14.1 A measure on the limit set

The fundamental tool developed by Patterson and Sullivan is a very special family of measures on the limit set $\Lambda(\Gamma)$. Because the measure construction takes place on the boundary of the hyperbolic plane, it is most convenient to present it using the unit disk model $\mathbb{B}$ given in (2.2), in which the boundary is represented uniformly. For the following discussion, we assume that the Fuchsian group Γ is a discrete subgroup of $PSU(1, 1)$, the orientation-preserving isometry group of $\mathbb{B}$.

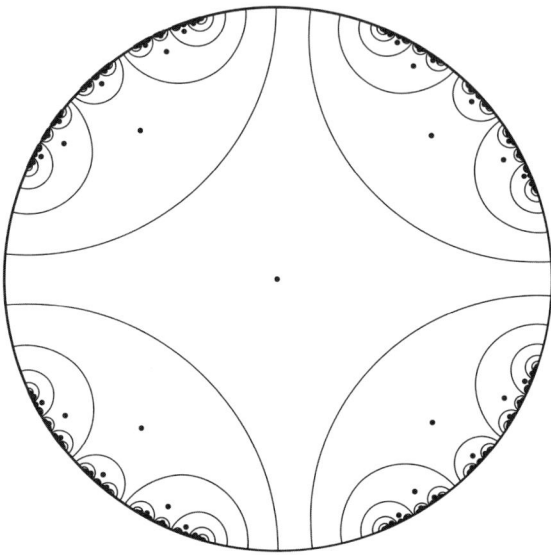

Fig. 14.1. Accumulation of an orbit.

To describe the measure construction intuitively, we consider an orbit, say $\Gamma 0$, as seen from a fixed "vantage point," say the origin. Figure 14.1 illustrates how the orbit points accumulate on $\Lambda(\Gamma)$. Suppose that the size of an orbit point $T0$ is weighted according to its distance from the vantage point, specifically by the factor $e^{-sd(0,T0)}$ for some $s > 0$. If we fix the total illumination, then for large values of s all orbit points are visible. However, at the threshold value $s = \delta$ the orbit points disappear and the limit set becomes visible. The Patterson–Sullivan measure can be pictured as the resulting pattern of illumination of $\Lambda(\Gamma)$.

To construct this measure, we start by using the weights described above to create a probability measure for $s > \delta$,

$$\mu^{(s)} := \left(\sum_{T \in \Gamma} e^{-sd(0,T0)} \right)^{-1} \sum_{T \in \Gamma} e^{-sd(0,T0)} \nu_{T0},$$

where ν_w denotes the point measure at $w \in \mathbb{B}$. We wish to extract a limit as $s \to \delta$. For this we invoke Helly's theorem, which says that the closed unit ball in the dual of a separable Banach space is weakly sequentially compact. (This is commonly stated in more general form as Alaoglu's theorem; see, e.g., [63, Theorem 5.18].) The dual space in question here is the space of Borel measures on $\overline{\mathbb{B}}$, with norm given by the total mass. Since each $\mu^{(s)}$ has total mass one, Helly's theorem implies that there exists a sequence $s_j \to \delta$ such that the $\mu^{(s_j)}$ converge weakly to some limiting measure,

$$\mu := \lim_{s_j \to \delta} \mu^{(s_j)}.$$

We will call μ the *Patterson–Sullivan measure* associated to Γ. (It's not clear at this point that the limiting measure is uniquely determined, but that is in fact the case for Γ geometrically finite.)

A crucial property in the measure construction is whether the absolute Poincaré series diverges at the critical exponent. For Γ geometrically finite and nonelementary, Patterson [156] proved the inequality

$$\sum_{T \in \Gamma} e^{-sd(z,Tw)} > \frac{C}{s - \delta} \tag{14.1}$$

for $s > \delta$. This shows in particular that the absolute Poincaré series diverges at $s = \delta$ (we may interchange limits since the terms are all positive). We won't prove this result here. Instead, in Section 14.4 we'll observe that the meromorphic continuation of the resolvent we established in Chapter 6 implies a more refined result, that the left-hand side of (14.1) has a meromorphic continuation to $s \in \mathbb{C}$ with a simple pole at $s = \delta$. (Patterson speculated about such behavior in his original paper, but meromorphic continuation of the resolvent was not proven until years later.)

Divergence at the critical exponent implies that $\mu(\mathbb{B}) = 0$. In particular, μ must be supported on $\Lambda(\Gamma)$. Note that by varying the vantage point z or the base point w of the Γ-orbit, both of which we took to be the origin, the same construction yields a family of measures $\mu_{z,w}$, all conformally related to each other. We focus on $\mu = \mu_{0,0}$ for now, just for ease of notation.

While the measure μ is not invariant under Γ, it does have nice transformation properties. If $T \in \Gamma$, then the pullback measure $T^*\mu$ is defined by

$$T^*\mu(E) := \mu(TE).$$

To analyze this we introduce the Poisson kernel for $\mathbb{B}$:

$$P(z, q) := \frac{1 - |z|^2}{|z - q|^2},$$

where $z \in \mathbb{B}$ and $q \in \partial\mathbb{B} = S^1$. The Poisson kernel is relevant here because of the following identity. In the unit disk model, the hyperbolic distance satisfies

$$4\sinh^2(d(z, w)/2) = \frac{|z - w|^2}{(1 - |z|^2)(1 - |w|^2)}. \tag{14.2}$$

From this one can see directly that for $q \in \partial\mathbb{B}$,

$$\lim_{w \to q} e^{d(z,w) - d(z',w)} = \frac{P(z', q)}{P(z, q)}. \tag{14.3}$$

Recall the identity used in the proof of Proposition 2.3: for any Möbius transformation T, and $z, w \in \mathbb{C}$,

$$|Tz - Tw|^2 = |T'(z)| \, |T'(w)| \, |z - w|^2. \tag{14.4}$$

From this formula it is a simple exercise to derive the transformation rule for the Poisson kernel,

$$P(Tz, Tq) \, |T'(q)| = P(z, q), \tag{14.5}$$

for any $z \in \mathbb{B}$, $q \in \partial\mathbb{B}$, and $T \in \mathrm{PSU}(1, 1)$.

Lemma 14.2. *For $R \in \Gamma$,*

$$R^* \mu = |R'|^\delta \mu. \tag{14.6}$$

Proof. If E is a Borel subset of $\mathbb{B}$ and $R \in \Gamma$, then by definition,

$$\mu^{(s)}(RE) = \left(\sum_{T \in \Gamma} e^{-sd(0, T0)} \right)^{-1} \sum_{T \in \Gamma: \, T0 \in RE} e^{-sd(0, T0)}.$$

The substitution $T = RS$ yields

$$\mu^{(s)}(RE) = \left(\sum_{T \in \Gamma} e^{-sd(0, T0)} \right)^{-1} \sum_{S \in \Gamma: \, S0 \in E} e^{-sd(0, RS0)}$$

$$= \left(\sum_{T \in \Gamma} e^{-sd(0, T0)} \right)^{-1} \sum_{S \in \Gamma: \, S0 \in E} e^{-sd(R^{-1}0, S0)}. \tag{14.7}$$

By (14.3), and noting that $P(0, q) = 1$, as $w \to q \in \mathbb{B}$ we have

$$e^{-sd(R^{-1}(0), w)} \sim e^{-sd(0, w)} P(R^{-1}0, q)^s.$$

And by (14.5),

$$P(R^{-1}0, q) = |R'(q)|.$$

Thus for a sequence $T_j 0 \to q \in \Lambda(\Gamma)$,

$$e^{-sd(R^{-1}0, T_j 0)} \sim |R'(q)|^s \, e^{-sd(0, T_j 0)}. \tag{14.8}$$

Using the fact that the measure concentrates on $\Lambda(\Gamma)$ in the $s \to \delta$ limit, we deduce (14.6) from (14.7) and (14.8).   □

After defining this measure, Patterson [156] introduced the related function

$$F(z) := \int_{\partial\mathbb{B}} P(z, q)^\delta \, d\mu(q). \tag{14.9}$$

By Lemma 14.2 and (14.5), F is invariant under Γ and so descends to a function on X, which is positive since $P(z, q)$ is positive. Moreover, it's not hard to check that

$$(\Delta_X - \delta(1 - \delta))F = 0,$$

since $P(z, q)^\delta$ satisfies this equation. This is the first hint of the connection to spectral theory. For $\delta > \frac{1}{2}$, F turns out to be the ground-state eigenfunction of the Laplacian.

As a first illustration of the usefulness of the Patterson–Sullivan measure, we'll prove some of Beardon's results concerning the range of δ.

Theorem 14.3 (Beardon). *For any nonelementary Fuchsian group Γ we have $\delta > 0$, and if Γ contains parabolic elements then $\delta > \frac{1}{2}$.*

Proof. Suppose that $\delta = 0$. Then Lemma 14.2 shows that μ is a Γ-invariant measure on $\Lambda(\Gamma)$. If $T \in \Gamma$ is hyperbolic, with fixed points $q_\pm \in \partial\mathbb{B}$, then we can partition $\partial\mathbb{B} - \{q_\pm\}$ into a countably infinite collection of disjoint intervals that are mapped to each other by powers of T. By invariance, and because the total mass is finite, we conclude that $\mu(\partial\mathbb{B} - \{q_\pm\}) = 0$. Then invariance of μ further implies that $\{q_\pm\}$ is a finite orbit of Γ, and hence Γ is elementary. Similar reasoning applies if Γ is assumed to contain a parabolic element.

Next assume that Γ is nonelementary and contains a parabolic element T. By conjugation we can assume that T fixes 1 and maps i to -1, so that

$$T^n = \begin{pmatrix} 1 + in/2 & -in/2 \\ in/2 & 1 - in/2 \end{pmatrix}.$$

The linear distortion of T^n is

$$|T^{n\prime}(z)| = \frac{1}{|1 + in(z - 1)|^2}. \tag{14.10}$$

Let $E = \{e^{i\theta} : \pi/2 < \theta \le \pi\}$, so that $\{T^n E\}$ forms a disjoint cover for $\partial\mathbb{B} - \{1\}$. We claim that $\mu(E) > 0$, for otherwise μ would be concentrated entirely at the point 1, which would imply that Γ was elementary. Using Lemma 14.2 and (14.10), we have

$$\mu(\partial\mathbb{B} - \{1\}) = \sum_n \mu(T^n E)$$

$$= \sum_n \int_E |T^{n\prime}(z)|^\delta \, d\mu(z)$$

$$\ge \mu(E) \sum_n (1 + 4n^2)^{-\delta}.$$

The left side is at most 1, so the series on the right must converge, implying that $\delta > \frac{1}{2}$. $\qquad\square$

We conclude this section by showing that the measure μ has no atoms, i.e., $\mu\{\text{point}\} = 0$. For this purpose we will first characterize the points in $\Lambda(\Gamma)$ by the directions from which orbit points accumulate. A point $q \in \Lambda(\Gamma)$ is called a *radial limit point* if there exists a geodesic ray η_q in $\mathbb{B}$ with endpoint q and an orbit Γw such that the set $\{z \in \Gamma w : d(z, \eta_q) < r\}$ is infinite for some $r > 0$. This is illustrated

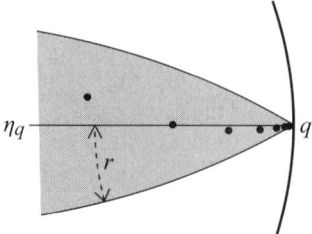

Fig. 14.2. Radial limit point.

in Figure 14.2. Radial limit points are also frequently called "conical" limit points, because one could equivalently require infinitely many orbit points lying inside a Euclidean cone.

For example, if q is the attracting hyperbolic fixed point of $T \in \Gamma$, then we can easily see that q is radial by taking η_q to be a portion of the axis of T. On the other hand, parabolic fixed points are clearly not radial. In our context these turn out to be the only exceptions.

Proposition 14.4. *For Γ a geometrically finite Fuchsian group, all points in $\Lambda(\Gamma)$ are either parabolic fixed points or radial limit points.*

Proof. Let $\mathcal{F} \subset \mathbb{B}$ be a finite-sided fundamental region for Γ, and note that $\partial\mathcal{F}$ meets $\Lambda(\Gamma)$ only at parabolic fixed points. Let $\widetilde{K} \subset \mathbb{B}$ be the truncated Nielsen region, and suppose γ is a geodesic ray originating in $\mathcal{F} \cap \widetilde{K}$ with endpoint $q \in \Lambda(\Gamma)$. Assume that γ crosses only finitely many translations of $\mathcal{F} \cap \widetilde{K}$ by an element of Γ. Then there must be a last side crossed, say in the the boundary of $T(\mathcal{F} \cap \widetilde{K})$. This implies $q \in T(\partial\mathcal{F})$, and since $q \in \Lambda(\Gamma)$ it must therefore be a parabolic fixed point.

Now assume that γ crosses infinitely many images of $\mathcal{F} \cap \widetilde{K}$. This means that there are sequences of distinct $w_n \in \mathcal{F} \cap \widetilde{K}$ and $T_n \in \Gamma$ such that $T_n w_n \in \gamma$ for $n \in \mathbb{N}$. Since $\mathcal{F} \cap \widetilde{K}$ is compact, by passing to a subsequence we can assume that $w_n \to w \in \mathcal{F} \cap \widetilde{K}$. Then $d(T_n w, \gamma)$ is bounded, which implies that q is a radial limit point. $\square$

Proposition 14.5. *For Γ nonelementary and geometrically finite, the measure μ has no atoms.*

Proof. First consider $q \in \Lambda(\Gamma)$ a radial limit point. We can assume that $q = 1$ and γ is a geodesic ray ending at 1 such that $d(T_n^{-1}0, \gamma) < C$ for some sequence of distinct $T_n \in \Gamma$. This implies that $T_n^{-1}0$ approaches 1 within a sector of the form $\{|\operatorname{Im} z| \le c \operatorname{Re}(1 - z)\}$ (see Figure 14.2). Since by (14.5),

$$|T_n'(1)| = P(T_n^{-1}0, 1) = \frac{1 - |T_n^{-1}0|^2}{|T_n^{-1}0 - 1|^2},$$

the fact that $T_n^{-1}0$ approaches 1 within a sector implies that $|T_n'(1)| \to \infty$. By Lemma 14.2,

$$\mu\{T_n 1\} = T_n^* \mu\{1\} = |T_n'(1)|^\delta \mu\{1\}.$$

Since $\delta > 0$ for a nonelementary group, and μ is a probability measure, the divergence of $|T_n'(1)|$ implies $\mu\{1\} = 0$. Hence there are no atoms at radial limit points.

To see that there can be no atoms at parabolic fixed points is slightly trickier. We follow Patterson's argument in [157] and switch back to the $\mathbb{H}$ model for the proof. Let $A : \mathbb{H} \to \mathbb{B}$ be the isometry between the two models given by

$$A : z \mapsto \frac{z - i}{z + i}.$$

The pullback by A of the function $F(z)$ defined in (14.9) is

$$F(Az) = \int_{\partial \mathbb{B}} P(Az, q)^\delta \, d\mu(q).$$

If we change variables from q to $x' = A^{-1} q \in \partial \mathbb{H}$, we can rewrite the integral in terms of the $\mathbb{H}$ version of the Poisson kernel. This is given by

$$E(z, x') := \frac{\operatorname{Im} z}{|z - x'|^2}$$

for $x' \in \mathbb{R}$, and for $x' = \infty$ by

$$E(z, \infty) := \operatorname{Im} z.$$

A direct computation shows that

$$P(Az, Ax') = E(z, x')(1 + x'^2),$$

for $x' \in \mathbb{R}$, and

$$P(Az, A\infty) = E(z, \infty).$$

Thus we have

$$F \circ A(z) = \mu(1) E(z, \infty)^\delta + \int_{\mathbb{R}} E(z, x')^\delta (1 + x'^2)^\delta \, d\mu(Ax').$$

Restricting our attention to the y-axis, $z = iy$, we have

$$F \circ A(iy) = \mu(1) y^\delta + y^\delta \int_{\mathbb{R}} \left[\frac{1 + x'^2}{y^2 + x'^2} \right]^\delta \, d\mu(Ax').$$

Since the total mass of $d\mu(Ax')$ is finite, the dominated convergence theorem shows that

$$\lim_{y \to \infty} \int_{\mathbb{R}} \left(\frac{1 + x'^2}{y^2 + x'^2} \right)^\delta \, d\mu(Ax') = 0.$$

Hence, as $y \to \infty$,

$$F \circ A(iy) = \mu(1)y^\delta + o(y^\delta). \tag{14.11}$$

Now suppose that $\Gamma \subset \mathrm{PSU}(1,1)$ has a parabolic fixed point at 1. We'll conjugate the group to $\Gamma' = A^{-1}\Gamma A \subset \mathrm{PSL}(2,\mathbb{R})$. Then Γ' has a parabolic fixed point at ∞ and we can assume by further conjugation that the associated maximal parabolic subgroup is generated by $T : z \mapsto z + 1$.

By the definition (14.9) and the weak convergence $\mu^{(s_j)} \to \mu$, we have

$$F(z) = \lim_{s_j \to \delta} \left(\sum_{S \in \Gamma} e^{-s_j d(0, S0)} \right)^{-1} \sum_{S \in \Gamma} e^{-s_j d(0, T0)} \left(\frac{1 - |z|^2}{S0 - z|^2} \right)^{s_j}.$$

From (14.3) we can then derive

$$F(z) = \lim_{s_j \to \delta} \left(\sum_{S \in \Gamma} e^{-s_j d(0, S0)} \right)^{-1} \sum_{S \in \Gamma} e^{-s_j d(z, S0)}. \tag{14.12}$$

Translating this identity to $\mathbb{H}$, with $Aw' = w$, we obtain

$$F \circ A(iy) = \lim_{s_j \to \delta} \left(\sum_{S \in \Gamma'} e^{-s_j d(i, Sw')} \right)^{-1} \sum_{S \in \Gamma'} e^{-s_j d(iy, Sw')}. \tag{14.13}$$

To estimate the sums over Γ' in (14.13), we'll break them up into sums over $\Gamma_\infty := \langle T \rangle$ and Γ'/Γ_∞. Let $\mathcal{F}_\infty := \{-\frac{1}{2} < \mathrm{Re}\, z \le \frac{1}{2}\}$ be a fundamental domain for Γ_∞. Any $S \in \Gamma'$ can be written uniquely as $T^n R$ such that $Rw' \in \mathcal{F}_\infty$ and $n \in \mathbb{Z}$. The sums can thus be decomposed as

$$\sum_{T \in \Gamma'} e^{-s_j d(iy, Tw')} = \sum_{n \in \mathbb{Z}} \sum_{\substack{R \in \Gamma': \\ Rw' \in \mathcal{F}_\infty}} e^{-s_j d(iy, Rw' + n)}.$$

By (2.6) we can estimate, for $z, z' \in \mathbb{H}$,

$$\frac{yy'}{(x - x')^2 + y^2 + y'^2} \le e^{-d(z,w)} \le \frac{2yy'}{(x - x')^2 + y^2 + y'^2}.$$

Within $\mathcal{F}_\infty$, the points Rw' can accumulate only on $\mathbb{R}$, so $\mathrm{Im}\, Rw'$ is bounded above (and of course $|\mathrm{Re}\, Rw'| \le \frac{1}{2}$). For any $Rw' \in \mathcal{F}_\infty$ we can thus bound

$$\sum_{S \in \Gamma'} e^{-s_j d(iy, Sw')} \le C \sum_{n \in \mathbb{Z}} \sum_{\substack{R \in \Gamma': \\ Rw' \in \mathcal{F}_\infty}} \left(\frac{y\, \mathrm{Im}\, Rw'}{n^2 + y^2} \right)^{s_j}.$$

In a similar way, we obtain

$$\sum_{S \in \Gamma'} e^{-s_j d(i, Sw')} \ge c \sum_{n \in \mathbb{Z}} \sum_{\substack{R \in \Gamma': \\ Rw' \in \mathcal{F}_\infty}} \left(\frac{\mathrm{Im}\, Rw'}{n^2 + 1} \right)^{s_j}.$$

Applying these estimates in (14.13), we can factor out $\sum (\operatorname{Im} Rw')^{s_j}$ and then take the $s_j \to \delta$ limit. This yields

$$F \circ A(iy) \le Cy^{\delta} \frac{\sum_n (n^2 + y^2)^{-\delta}}{\sum_n (n^2 + 1)^{-\delta}}.$$

(Interchanging the $s_j \to \delta$ limit with the sums was valid because all terms are positive.) Note that the assumption that Γ was nonelementary and contains a parabolic fixed point guarantees $\delta > \frac{1}{2}$ by Theorem 14.3, so the sums are convergent. The denominator is just a constant, while the numerator is comparable to the integral

$$\int_{-\infty}^{\infty} (t^2 + y^2)^{-\delta} \, dt = O(y^{1-2\delta}).$$

We conclude that

$$F \circ A(iy) = O(y^{1-\delta}).$$

Comparing this to (14.11), and noting $\delta > \frac{1}{2}$ again, we conclude that $\mu(1) = 0$. Hence μ has no mass at any parabolic fixed point. □

14.2 Ergodicity

A fundamental issue in the Patterson–Sullivan theory, and a crucial fact for the applications we wish to make to spectral theory, is the ergodicity of μ and other measures related to it. A group action on a measure space is *ergodic* if all measurable invariant sets have either zero or full measure. This is equivalent to the statement that the only invariant functions on the space are constant almost everywhere with respect to μ.

There are several actions for us to consider, for which the ergodicity properties are related. The most obvious candidate is the action of Γ on $\partial \mathbb{B}$, with respect to Patterson–Sullivan μ. A related measure can be defined on the space

$$(\partial \mathbb{B} \times \partial \mathbb{B})_- := \left\{ (q, q') \in \partial \mathbb{B} \times \partial \mathbb{B} : q \ne q' \right\},$$

on which Γ acts by the product action, $T : (q, q') \mapsto (Tq, Tq')$. By Lemma 14.2 and (14.4), the measure defined by

$$d\tilde{\mu}(q, q') := \frac{d\mu(q) \, d\mu(q')}{|q - q'|^{2\delta}} \tag{14.14}$$

is invariant under Γ.

We will also study the geodesic flow, on both $\mathbb{B}$ and X. The unit tangent bundle of $\mathbb{B}$ is denoted by $S\mathbb{B} \subset T\mathbb{B}$. We can identify $S\mathbb{B}$ with $(\partial \mathbb{B} \times \partial \mathbb{B})_- \times \mathbb{R}$ through a convenient set of coordinates. For $q_{\pm} \in \partial \mathbb{B}$, $q_- \ne q_+$, there is a unique oriented geodesic in $\mathbb{B}$ given by $[q_-, q_+]$. We can let s denote the signed arc length along this geodesic, using the Euclidean midpoint of the geodesic for the starting point $s = 0$. The setup is shown in Figure 14.3. The triple (q_-, q_+, s) thus specifies a unique point in $\mathbb{B}$, and we obtain a corresponding point in $\mathcal{S}\mathbb{B}$ by taking the unit tangent vector pointing along the geodesic $[q_-, q_+]$ toward q_+. In these coordinates the geodesic flow on $\mathcal{S}\mathbb{B}$ has a particularly simple form:

$$\phi_t(q_-, q_+, s) = (q_-, q_+, s + t),$$

for $t \in \mathbb{R}$. Using (14.14) we construct a Γ-invariant measure m_μ on $\mathcal{S}\mathbb{B}$,

$$dm_\mu(q_-, q_+, s) := d\widetilde{\mu}(q_-, q_+)\, ds. \tag{14.15}$$

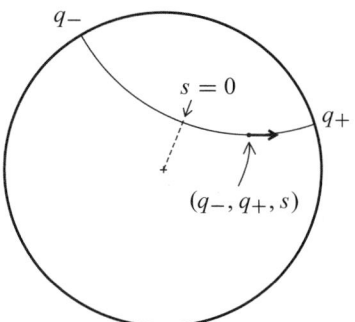

Fig. 14.3. Coordinates for $\mathcal{S}\mathbb{B}$.

For the associated hyperbolic surface $X = \Gamma\backslash\mathbb{H}$, the unit tangent bundle $\mathcal{S}X$ is naturally identified with the quotient $\Gamma\backslash\mathcal{S}\mathbb{B}$. It is easy to verify that the geodesic flow commutes with the action of Γ on $\mathcal{S}\mathbb{B}$. Hence, if π denotes the projection $\mathcal{S}\mathbb{B} \to \mathcal{S}X$, we have $\pi \circ \phi_t|_{\mathcal{S}\mathbb{B}} = \phi_t|_{\mathcal{S}X}$. Since m_μ is an invariant measure on $\mathcal{S}\mathbb{B}$, it descends to a quotient measure on $\mathcal{S}X$.

We will prove that the geodesic flow is ergodic on $\mathcal{S}\mathbb{B}$ with respect to m_μ, and from this derive the ergodicity of the other actions given above. If Γ is Fuchsian of the first kind (X has finite area), we have $\delta = 1$ and μ is just Lebesgue measure on $\partial\mathbb{B}$. In this classical case, the ergodicity of the geodesic flow on $\mathcal{S}X$ is a famous result of E. Hopf from the mid-1930s [94].

The proof of ergodicity is broken into two parts. First we show that the geodesic flow ϕ_t on $\mathcal{S}X$ is *conservative*. This means that for almost every starting point in $\mathcal{S}X$, the flow returns to a fixed compact set repeatedly, at arbitrarily large times. The second step is to apply Hopf's original argument to show that conservative implies ergodic.

Proposition 14.6. *If $X = \Gamma\backslash\mathbb{H}$ is a nonelementary geometrically finite surface, then the geodesic flow on $\mathcal{S}X$ is conservative with respect to the measure m_μ.*

Proof. Our goal is to show that almost every geodesic returns to the compact core $K = \Gamma\backslash\widetilde{K}$ infinitely often (where the truncated Nielsen region $\widetilde{K}$ is translated to the

$\mathbb{B}$ model here). Consider a geodesic in $\mathbb{B}$ that starts inside $\widetilde{K}$. By the construction of the (untruncated) Nielsen region $\widetilde{N}$ in Section 2.4, the geodesic leaves $\widetilde{N}$ only if it has an endpoint in one of the arcs U_j that make up $\partial\mathbb{H} - \Lambda(\Gamma)$. But of course $\mu(\partial\mathbb{H} - \Lambda(\Gamma)) = 0$, so almost every geodesic stays within $\widetilde{N}$.

A geodesic that remains in $\widetilde{N}$ can leave $\widetilde{K}$ only by crossing one of the horocycles σ_p associated to a parabolic fixed point. But if the geodesic remains within σ_p after that crossing, its endpoint must be p. By Lemma 14.5, the set of geodesics ending in parabolic fixed points has m_μ-measure zero. Thus almost every geodesic on X returns to K at arbitrarily large times.   □

The centerpiece of Hopf's argument is an extension of the Birkhoff ergodic theorem. Since we will be concerned with various time averages over the flow, we should first check that $f \in L^1(SX, dm_\mu)$ implies that the function $(u, t) \mapsto f(\phi_t u)$ is measurable on $SX \times \mathbb{R}$ (with Lebesgue measure on $\mathbb{R}$). This is a simple measure theory exercise using continuity of the flow; we'll omit the proof.

Lemma 14.7. *Suppose $\kappa \in L^1(SX, dm_\mu)$ is continuous and $\kappa > 0$. Then because the geodesic flow is conservative we have*

$$\lim_{T\to\infty} \int_0^T \kappa(\phi_t u)\, dt = \infty \qquad (14.16)$$

for almost every $u \in SX$.

Proof. Let F denote the compact set in SX consisting of vectors with base points inside the compact core K. For $\varepsilon > 0$, set

$$F_\varepsilon = \{u \in SX : d(u, F) \le \varepsilon\}.$$

Assume that $\phi_{t_n} u \in F$ for a sequence $t_n \to \infty$, which we know to be the case for almost every $u \in SX$. If we suppose that there is some t_0 such that $\phi_t u \in F_\varepsilon$ for all $t \ge t_0$, then $\kappa(\phi_t u)$ is uniformly bounded below for $t \ge t_0$, and the divergence (14.16) follows. On other hand, if $\phi_t u$ leaves F_ε infinitely often, then by passing to a subsequence we can assume that $\phi_t u$ leaves F_ε at least once in each interval $[t_n, t_{n+1}]$. Since it takes at least time ε to cross from F to the boundary of F_ε, we deduce that

$$\int_{t_n}^{t_{n+1}} \kappa(\phi_t u)\, dt \ge \varepsilon \inf_{u \in F_\varepsilon} \kappa(u).$$

Therefore the sum over n diverges, implying (14.16).   □

Hopf's continuous version of the Birkhoff ergodic theorem is the following result.

Theorem 14.8 (Hopf–Birkhoff). *For $f, \kappa \in L^1(SX, dm_\mu)$, with $\kappa > 0$, consider the function*

$$f_\kappa(u) := \lim_{T\to\infty} \frac{\int_0^T f(\phi_t u)\, dt}{\int_0^T \kappa(\phi_t u)\, dt}. \qquad (14.17)$$

If κ satisfies (14.16), then the limit defining f_κ exists for almost every $u \in SX$, and the result is a function invariant under the flow. Moreover, the flow is ergodic if and only if f_κ is almost everywhere constant for any choice of f.

Proof. For $a < b$, we define

$$E_{a,b} := \left\{ \liminf_{T \to \infty} \frac{\int_0^T f(\phi_t u)\, dt}{\int_0^T \kappa(\phi_t u)\, dt} < a < b < \limsup_{T \to \infty} \frac{\int_0^T f(\phi_t u)\, dt}{\int_0^T \kappa(\phi_t u)\, dt} \right\}.$$

By dropping a set of measure zero, we can assume that (14.16) holds everywhere, which implies that $E_{a,b}$ is invariant under the flow. The goal is to show that $m_\mu(E_{a,b}) = 0$, and this is essentially an application of the maximal ergodic theorem (see, e.g., [216, Theorem 1.6]).

For $k = 0, 1, 2, \ldots$, let

$$h_k(u) := \int_0^k \left[f(\phi_t u) - b\kappa(\phi_t u) \right] dt$$

and

$$H_N(u) := \max_{0 \le k \le N} h_k(u).$$

For $k \in [0, N]$, note that

$$H_N(\phi_1 u) \ge h_k(\phi_1 u) = h_{k+1}(u) - h_1(u).$$

This implies

$$H_N(\phi_1 u) + h_1(u) \ge \max_{1 \le k \le N} h_k(u),$$

for any u. If $H_N(u) > 0$, we know that some $h_k(u) > 0$, and so

$$H_N(\phi_1 u) + h_1(u) \ge H_N(u) \qquad \text{when } H_N(u) > 0.$$

Integrating the inequality over $\{u : H_N(u) > 0\}$ gives

$$\int_{\{H_N > 0\}} h_1 \, dm_\mu \ge \int_{\{H_N > 0\}} (H_N - H_N \circ \phi_1) \, dm_\mu.$$

Using $H_N \ge 0$ and the invariance of dm_μ, we then have

$$\int_{\{H_N > 0\}} h_1 \, dm_\mu \ge \int_{SX} (H_N - H_N \circ \phi_1) \, dm_\mu = 0. \tag{14.18}$$

The divergence (14.16) implies that a finite shift in the range of integration $[0, T]$ will not affect the definition of $E_{a,b}$. In particular, we can take the limit $T \to 0$ through integer values. In this case, $u \in E_{a,b}$ implies $H_N(u) > 0$ for some N. Therefore, by (14.18),

$$\int_{E_{a,b}} \int_0^1 \left[f(\phi_t u) - b\kappa(\phi_t u) \right] dt \, dm_\mu(u) \ge 0.$$

Similar reasoning on the other side shows that

$$\int_{E_{a,b}} \int_0^1 \left[a\kappa(\phi_t u) - f(\phi_t u) \right] dt \, dm_\mu(u) \geq 0.$$

Putting these two inequalities together gives

$$(b - a) \int_{E_{a,b}} \int_0^1 \kappa(\phi_t u) \, dt \, dm_\mu(u) \leq 0.$$

Since $a < b$ and $\kappa > 0$, we conclude that $m_\mu(E_{a,b}) = 0$. This implies that the limit (14.17) exists almost everywhere. And the resulting function f_κ is invariant almost everywhere because (14.16) allows us to shift the range of integration.

If the flow is ergodic, then the invariance of f_κ implies that it is constant almost everywhere. On the other hand, suppose that the flow is not ergodic. Then there exists an invariant set A such that $m_\mu(A) > 0$ and $m_\mu(A^c) > 0$. Setting f equal to the characteristic function χ_A, we see that f_κ is nonconstant; it equals zero on A^c but not on A.   □

The final ingredient in the ergodicity proof is the existence of a function κ with properties that allow us to deduce that f_κ is always constant and thereby apply Theorem 14.8. To keep the notation clean, we'll extend the hyperbolic distance function $d(\cdot, \cdot)$ to denote distance between base points on either $S\mathbb{B}$ or SX.

Lemma 14.9. *There exists a continuous integrable function $\kappa > 0$ on SX satisfying*

$$\frac{\kappa(u) - \kappa(v)}{\kappa(v)} \leq Cd(u, v), \tag{14.19}$$

whenever $d(u, v) \leq 1$.

Proof. Let $u_0 \in S\mathbb{B}$ be a reference vector whose base point is the origin. Then we define

$$\kappa(u) := \exp\{-(2\delta + 1)d(u, \Gamma u_0)\}.$$

This is clearly Γ-invariant and so defines a function on SX as well.

Consider the set $B_r := \{u \in SX : d(u, \Gamma u_0) < r\}$. We can estimate its size by lifting it to $\widetilde{B}_r$, the set of all geodesics in $\mathbb{B}$ that pass within distance r of 0. The Euclidean distance from the origin $a = |z|$ is related to r by $a = \tanh(r/2)$. An easy geometric construction (see Figure 14.4) shows that for a geodesic tangent to the circle of radius a, the Euclidean distance between the endpoints is given by

$$|q_+ - q_-| = 2 \frac{1 - a^2}{1 + a^2} = \frac{2}{\cosh r}.$$

Thus, for $(q_-, q_+, s) \in \widetilde{B}_r$ we have

$$|q_+ - q_-| \geq 4e^{-r}.$$

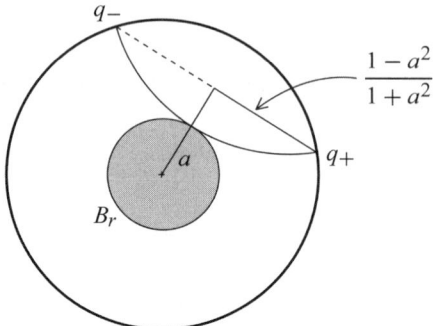

Fig. 14.4. Estimating $m_\mu(B_r)$.

The points in $\widetilde{B}_r$ also satisfy $|s| \le r$, so we can estimate directly from the definitions (14.14) and (14.15),

$$m_\mu(B_r) \le \frac{2r}{(4e^{-r})^{2\delta}} = Cre^{2\delta r}.$$

For u outside B_r, we have $\kappa(u) \le e^{-(2\delta+1)r}$. Thus

$$
\begin{aligned}
\int \kappa \, dm_\mu &= \sum_{n=0}^{\infty} \int_{B_{n+1}-B_n} \kappa \, dm_\mu \\
&\le \sum_{n=0}^{\infty} e^{-(2\delta+1)n} \mu(B_{n+1}) \\
&\le C \sum_{n=0}^{\infty} ne^{-n} < \infty.
\end{aligned}
$$

Hence κ is integrable.

As for the inequality (14.19), note that

$$\limsup_{d(u,v)\to 0} \frac{|\kappa(u) - \kappa(v)|}{d(u,v)} \le (2\delta+1)\kappa(v).$$

We can use the mean value theorem to deduce that for any $u, v \in SX$, there exists some v' with $d(v, v') \le d(u, v)$ such that

$$\frac{|\kappa(u) - \kappa(v)|}{d(u,v)} \le (2\delta+1)\kappa(v').$$

Then we have

$$
\begin{aligned}
\frac{|\kappa(u) - \kappa(v)|}{\kappa(v)d(u,v)} \le (2\delta+1)\frac{\kappa(v')}{\kappa(v)} &\le (2\delta+1)e^{(2\delta+1)d(v',v)} \\
&\le (2\delta+1)e^{(2\delta+1)d(u,v)}.
\end{aligned}
$$

$\square$

Theorem 14.10 (Ergodicity). *For* $X = \Gamma \backslash \mathbb{H}$ *a nonelementary geometrically finite hyperbolic surface, the geodesic flow is ergodic on* SX *with respect to the measure* m_μ.

Proof. Assume that κ is chosen as in Lemma 14.9. It suffices to prove that f_κ is constant for f continuous and compactly supported, because such functions are dense in $L^1(SX, dm_\mu)$. Note that since f_κ is invariant under the flow, it corresponds to the lift of a function from $(\partial \mathbb{B} \times \partial \mathbb{B})_-$.

Suppose that $u, v \in SX$ have the same endpoint $q_+ \in \mathbb{B}$. By replacing v by $\phi_{t_0} v$, if necessary, we can assume that

$$\lim_{t \to \infty} d(\phi_t u, \phi_t v) = 0. \tag{14.20}$$

Consider the difference

$$\frac{\int_0^T f(\phi_t u)\, dt}{\int_0^T \kappa(\phi_t u)\, dt} - \frac{\int_0^T f(\phi_t v)\, dt}{\int_0^T \kappa(\phi_t v)\, dt} = \frac{\int_0^T [f(\phi_t u) - f(\phi_t v)]\, dt}{\int_0^T \kappa(\phi_t u)\, dt}$$
$$- \frac{\int_0^T f(\phi_t v)\, dt}{\int_0^T \kappa(\phi_t v)\, dt} \frac{\int_0^T [\kappa(\phi_t u) - \kappa(\phi_t v)]\, dt}{\int_0^T \kappa(\phi_t u)\, dt}.$$

In the first term on the right, the numerator is bounded because of the compact support of f. Since its denominator approaches ∞, this term approaches zero as $T \to \infty$.

As for the second term, for almost every v we have

$$\lim_{T \to 0} \frac{\int_0^T f(\phi_t v)\, dt}{\int_0^T \kappa(\phi_t v)\, dt} = f_\kappa(v).$$

Furthermore, (14.19) and (14.20) imply that

$$\lim_{T \to 0} \frac{\int_0^T [\kappa(\phi_t u) - \kappa(\phi_t v)]\, dt}{\int_0^T \kappa(\phi_t u)\, dt}.$$

Thus the limit of the second term is zero almost everywhere.

We have shown that f_κ is constant almost everywhere on $\partial \mathbb{B} \times \{q_+\}$. A similar argument shows that f_κ is constant almost everywhere on $\{q_-\} \times \partial \mathbb{B}$. And then Fubini's theorem implies that f_κ is constant almost everywhere on $(\partial \mathbb{B} \times \partial \mathbb{B})_-$. By Theorem 14.8 the geodesic flow is ergodic. $\qquad \square$

Corollary 14.11. *The action of* Γ *on* $\partial \mathbb{B}$ *is ergodic with respect to* μ, *and the product action of* Γ *on* $(\partial \mathbb{B} \times \partial \mathbb{B})_-$ *is ergodic with respect to* $\tilde{\mu}$.

Proof. It's easy to see that ergodicity of the geodesic flow and ergodicity of the product action of Γ on $\tilde{\mu}$ are equivalent. Suppose A is a Γ-invariant subset of $(\partial \mathbb{B} \times \partial \mathbb{B})_-$. Then $A \times \mathbb{R} \subset S\mathbb{B}$ is invariant with respect to both Γ and the geodesic flow. Hence $\Gamma \backslash (A \times \mathbb{R})$ is an invariant subset of SX. By Theorem 14.10, either $m_\mu(A \times \mathbb{R}) = 0$ or $m_\mu(A^c \times \mathbb{R}) = 0$. Clearly these imply $\widetilde{\mu}(A) = 0$ or $\widetilde{\mu}(A^c) = 0$, respectively. (For the converse this argument is easily reversed.)

Now suppose $E \subset \partial \mathbb{B}$ is invariant with respect to Γ. Then $(E \times E)_-$ is invariant under the product action, so we have either $\widetilde{\mu}((E \times E)_-) = 0$ or $\widetilde{\mu}((E \times E)_-^c) = 0$. The first case implies that $(\mu \times \mu)((E \times E)_-) = 0$. Then, since μ has no atoms by Proposition 14.5, $\mu(E) = 0$. In the second case, note that $E^c \times E^c \subset (E \times E)^c$, and so $\mu(E^c) = 0$ by the same reasoning. $\qquad \square$

14.3 Hausdorff measure of the limit set

We will present the proof that $\delta = \dim_H \Lambda(\Gamma)$ only in the easiest case, when Γ is convex cocompact. This will simplify the presentation, and it is the only case that we will make use of later.

For $A \subset \partial \mathbb{B}$, the s-dimensional *Hausdorff measure* is

$$H^s(A) := \lim_{\varepsilon \to 0} \inf \left\{ \sum_j |I_j|^s : A \subset \cup_j I_j, \ |I_j| < \varepsilon \right\}. \qquad (14.21)$$

Here we use $| \cdot |$ to denote the Euclidean arc length in $\partial \mathbb{B}$. It's a simple exercise to show that for some threshold d,

$$H^s(A) = \begin{cases} 0, & s < d, \\ \infty, & s > d, \end{cases}$$

and then the *Hausdorff dimension* is $\dim_H A := d$.

Since for a Möbius transformation T the local distortion of Euclidean length is given by $|T'|$, we have

$$T^* H^s = |T'|^s \, H^s, \qquad (14.22)$$

for any $T \in \mathrm{PSU}(1, 1)$. Thus Lemma 14.2 shows that μ transforms under the action of Γ like the Hausdorff measure of dimension δ. This hints that μ might be related to H^δ. And indeed, for convex cocompact groups it turns out that μ is a constant multiple of $H^\delta|_{\Lambda(\Gamma)}$. Once we establish this, the fact that $\mu(\Lambda(\Gamma)) = 1$ will show immediately that $\dim_H \Lambda(\Gamma) = \delta$.

To compare the two measures, μ and H^δ, we start with a nice geometric lemma due to Sullivan [202, Section 2]. Suppose that $w \in \mathbb{B}$ and $r > 0$ are such that $0 \notin B(w; r)$. Picture a light source at the origin and consider the *shadow* cast by $B(w; r)$ on $\partial \mathbb{B}$, meaning the set

$$I(w; r) := \{q \in \partial \mathbb{B} : d([0, q], w) < r\}, \qquad (14.23)$$

where $[0, q]$ denotes the geodesic ray from 0 to q.

Consider a geodesic ray from 0 that is tangent to the boundary of $B(w; r)$. A right geodesic triangle is formed by 0, w, and this point of tangency, as shown in

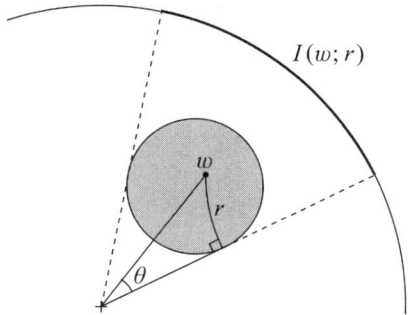

Fig. 14.5. Shadow of $B(w; r)$ on $\partial \mathbb{B}$.

Figure 14.5. If θ denotes the vertex angle of this triangle at 0, then $|I(w; r)| = 2\theta$, where $|\cdot|$ denotes the Euclidean arc length. The sine rule (Lemma 13.3) gives us the relation

$$\sin \theta = \frac{\sinh r}{\sinh d(0, w)}. \tag{14.24}$$

By the formula (14.2) for the hyperbolic distance in $\mathbb{B}$, this becomes

$$\sin \theta = \frac{1 - |w|^2}{2|w|} \sinh r. \tag{14.25}$$

Thus, with r held constant and $|w|$ bounded away from zero, we have the estimate

$$|I(w; r)| \asymp 1 - |w|. \tag{14.26}$$

Lemma 14.12 (Sullivan's shadow lemma). *For fixed r sufficiently large and all but finitely many $T \in \Gamma$, we have*

$$\mu(I(T0; r)) \asymp |I(T0; r)|^\delta$$

(with constants independent of T).

Proof. We'll follow the proof outlined by Patterson in [157]. Let $w = T0$ for some $T \in \Gamma$. Assuming that $d(w, 0) > r$, we apply T^{-1} to the shadow $I(w; r)$. As illustrated in Figure 14.6, $T^{-1}I(w; r)$ is the shadow cast by $B(0; r)$ using a light source at $T^{-1}0$. The strategy is to show that $\mu(T^{-1}I(w; r))$, which of course is bounded above by 1, can be uniformly bounded below as well. Then we use $T^*\mu = |T'|^\delta \mu$ to deduce the estimate. Note that we must assume that $T^{-1}0 \notin B(0; r)$ for $I(w; r)$ to be well-defined in the first place, which is why possibly finitely many elements of Γ must be excluded.

We claim that there exists $\varepsilon > 0$ such that for any interval $I \subset \partial \mathbb{B}$, $|I| < \varepsilon$ implies $\mu(I) < \frac{1}{2}$. Suppose this is not the case. Then there exists a sequence of closed $I_j \subset \partial \mathbb{B}$ with $|I_j| \to 0$ but $\mu(I_j) \geq \frac{1}{2}$. Since $\mu(\cup I_j) \leq 1$, infinitely many of

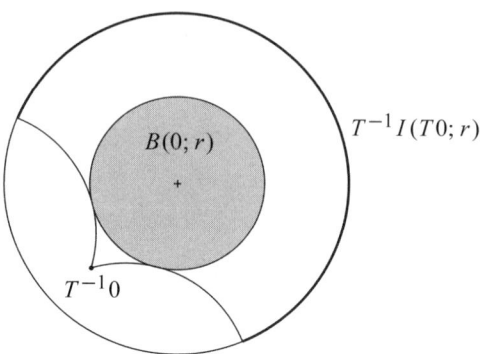

Fig. 14.6. Pullback of a shadow.

the I_j's must overlap. Passing to a subsequence, we can assume that $\cap I_j \neq \emptyset$. But because the diameters go to zero, $\cap I_j$ consists of a single point, which is therefore an atom of μ. This is forbidden by Proposition 14.5.

Fixing such an ε, we choose r large enough so that $B(0; r)$ comes within Euclidean distance $\varepsilon/2\pi$ of $\partial \mathbb{B}$. This ensures, by a simple geometric estimate, that the arc length of the complement of $T^{-1} I(w; r)$ in $\partial \mathbb{B}$ is less than ε. Hence we have bounds

$$\tfrac{1}{2} < \mu(T^{-1} I(w; r)) < 1, \tag{14.27}$$

for any $T \in \Gamma$ such that $w = T0$ lies outside of $B(0; r)$ (which excludes at most finitely many T). By the pullback formula of Lemma 14.2,

$$\mu(T^{-1} I(w; r)) = \int_{I(w;r)} |T^{-1'}|^{\delta} \, d\mu. \tag{14.28}$$

The remainder of the proof amounts to controlling the size of $|T^{-1'}|$.

By (14.5), we have

$$|T^{-1'}(q)| = P(w, q) = \frac{1 - |w|^2}{|w - q|^2}. \tag{14.29}$$

Since $|w - q| \geq 1 - |w|$, there is an obvious bound,

$$|T^{-1'}(q)| \leq \frac{2}{1 - |w|}.$$

For the bound from below, suppose that $q \in I(w; r)$, and let $\theta = |I(w; r)|/2$ be the vertex angle as in (14.25). Then, by the Euclidean triangle inequality,

$$|w - q| \leq 1 - |w| + \theta.$$

Note that by our choice of r, we consider only T for which $|w| > 1 - \varepsilon/2\pi$. In this case (14.25) implies a bound for any $q \in I(w; r)$,

$$|w - q| \leq C(1 - |w|),$$

with C independent of T. In view of (14.29) this gives the uniform bounds

$$\frac{c}{1 - |w|} \leq |T^{-1'}(q)| \leq \frac{2}{1 - |w|}. \tag{14.30}$$

From (14.27) and (14.28) we then obtain

$$\mu(I(w; r)) \asymp (1 - |w|)^{\delta}.$$

The final comparison to $|I(w; r)|$ follows immediately from (14.26). □

To use Lemma 14.12 to estimate Hausdorff measures, we need to show that we can approximate small intervals in $\partial \mathbb{B}$ by shadows of the form $I(T0; r)$. This is where the assumption that Γ is convex cocompact makes for a substantial simplification. (Note that Lemma 2.12 clearly shows that neighborhoods of a parabolic fixed point cannot be approximated by shadows of this form.)

Lemma 14.13. *Suppose Γ is a convex cocompact Fuchsian group. For $q \in \Lambda(\Gamma)$ let I_q denote an interval in $\partial \mathbb{B}$ centered at q. There exists $\varepsilon > 0$ such that for any $|I_q| < \varepsilon$,*

$$\mu(I_q) \asymp |I_q|^{\delta},$$

uniformly in q.

Proof. Convex cocompact means that the convex core $N = \Gamma \backslash \widetilde{N}$ is compact. Suppose that $\mathcal{F}$ is a finite-sided fundamental region for Γ. We can assume that $0 \in \mathcal{F}_c := \mathcal{F} \cap \widetilde{N}$. Let $d = \operatorname{diam} \mathcal{F}_c$.

Given I_q, for any $r > 0$ there is a unique point w on the segment connecting 0 to q such that $I_q = I(w; r)$. Since the Nielsen region $\widetilde{N}$ is the convex hull of $\Lambda(\Gamma)$, we know that w lies in some image $T\mathcal{F}_c$ for $T \in \Gamma$. This is illustrated in Figure 14.7. In particular, $d(T0, w) < d$, so that

$$B(T0; r - d) \subset B(w; r) \subset B(T0; r + d),$$

assuming $r > d$. Taking the shadows of these disks then gives

$$I(T0; r - d) \subset I_q \subset I(T0; r + d).$$

If we take ε sufficiently small, then $|I_q| < \varepsilon$ will imply that $|w|$ and $|T0|$ are as close to 1 as we like. In particular, through the choice of ε we can assume that T meets the requirements of Lemma 14.12 for radius $r \pm d$. This yields the estimates

$$c_1 |I(T0; r - d)|^{\delta} \leq \mu(I_q) \leq c_2 |I(T0; r + d)|^{\delta},$$

where c_1, c_2 depend on ε, but not on q or T. Because $|w|$ and $|T0|$ are bounded away from 0, we can then use (14.26) to derive that

$$c_3 |I(T0; r + d)| \leq |I_q| \leq c_4 |I(T0; r - d)|.$$

This completes the proof. □

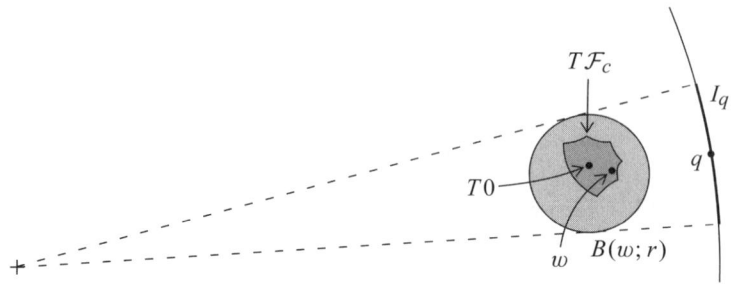

Fig. 14.7. Approximation by shadows.

Theorem 14.14 (Patterson, Sullivan). *For Γ a convex cocompact Fuchsian group, the Patterson–Sullivan measure μ is a constant multiple of the Hausdorff measure $H^\delta|_{\Lambda(\Gamma)}$. In particular,*

$$\dim_H \Lambda(\Gamma) = \delta.$$

Proof. It's a straightforward exercise to use Lemma 14.13 and the definition of Hausdorff measure to prove that for any Borel set $A \subset \Lambda(\Gamma)$,

$$\mu(A) \asymp H^\delta(A).$$

In particular, this implies that μ is absolutely continuous with respect to H^δ, so that $d\mu = f \, dH^\delta$ for some function f on $\Lambda(\Gamma)$. By Lemma 14.2 and (14.22), this function f is Γ-invariant. Then ergodicity (Corollary 14.11) implies that f is constant. $\square$

14.4 The first resonance

We have already noted Patterson's result [156] that $\delta(1 - \delta)$ is the lowest eigenvalue of Δ_X provided that $\delta > \frac{1}{2}$. The corresponding eigenvalue is the function $F(z)$ given in (14.9). Patterson later [158] generalized this result to $\delta \le \frac{1}{2}$, in terms of resonances. The full picture is given in the following:

Theorem 14.15 (Patterson). *Assume that X is a geometrically finite, nonelementary hyperbolic surface of infinite area. There is a resonance of multiplicity one at the point $s = \delta$ such that*

$$\operatorname*{res}_{s=\delta} R_X(s; z, w) = c(\Gamma) F(z) F(w).$$

No other resonances occur in the half-plane $\operatorname{Re} s \ge \delta$.

The same result holds for finite-area hyperbolic surfaces, but there is nothing to prove there since $\delta = 1$, F is constant, and 0 is clearly the lowest eigenvalue. The elementary cases, for which $\delta = 0$, are exceptional as usual. The first resonance of the hyperbolic cylinder does occur at 0, but the multiplicity is two. And the first (and only) resonance of the parabolic cylinder occurs at $\frac{1}{2}$.

Given the characterization of the divisor of the zeta function in Theorem 10.1, one could just as well phrase the result in terms of the zeta function.

Corollary 14.16. *For X a geometrically finite, nonelementary hyperbolic surface, the zeta function $Z_X(s)$ has a simple zero at $s = \delta$ and no other zeros for $\operatorname{Re} s \geq \delta$.*

Note that this parallels the behavior of the Riemann zeta function. We will exploit this parallel to prove an analogue of the prime number theorem in Section 14.5.

Naud [143] proved that there is actually a resonance-free vertical strip around δ. That is, there exists $\epsilon > 0$ such that

$$\mathcal{R}_X \cap \{\operatorname{Re} s > \delta - \epsilon\} = \{\delta\}. \tag{14.31}$$

(This holds trivially for $\delta > \frac{1}{2}$ by the discreteness of the eigenvalue spectrum, so the content of the result is the case $\delta \leq \frac{1}{2}$.) Naud uses dynamical methods, of the sort we discuss in Chapter 15, but his proof is somewhat beyond the scope of this book.

This section is devoted to the proof of Theorem 14.15, which we divide up into a series of propositions. Our first step here is to fill in the gap left in Section 14.1, when we omitted the proof of (14.1). We can do this by making explicit the connection between the resolvent kernel and the absolute Poincaré series. Let us denote the latter by

$$\Sigma(s; z, w) := \sum_{T \in \Gamma} e^{-s d(z, Tw)}, \tag{14.32}$$

which converges to an analytic function of s for $\operatorname{Re} s > \delta$.

Proposition 14.17. *Suppose $X = \Gamma \backslash \mathbb{H}$ is a nonelementary geometrically finite hyperbolic surface. For $\operatorname{Re} s > \delta$, the lift of the resolvent kernel $R_X(s; z, w)$ to $\mathbb{H} \times \mathbb{H}$ and the absolute Poincaré series are related by*

$$R_X(s; z, w) = \frac{4^{(s-1)}}{\pi} \frac{\Gamma(s)^2}{\Gamma(2s)} \Sigma(s; z, w) + H(s; z, w), \tag{14.33}$$

for $z \neq w$, where $H(s; z, w)$ is holomorphic in s for $\operatorname{Re} s > \delta - 1$. Both $R(s; z, w)$ and $\Sigma(s; z, w)$ have a pole at $s = \delta$.

Proof. In Section 4.2, we wrote $R_{\mathbb{H}}(s; z, w)$ as a function $g_s(\sigma)$, in terms of the parameter $\sigma = \cosh^2(d(z, w)/2)$. By the representation (5.3), we can write

$$R_{\mathbb{H}}(s; z, w) = \frac{4^{(s-1)}}{\pi} \frac{\Gamma(s)^2}{\Gamma(2s)} e^{-s d(z, w)} + F(s, d(z, w)),$$

where $F(s, d)$ is analytic in s for $\operatorname{Re} s > -1$ and $O(e^{-(\operatorname{Re} s + 1)d})$ for d large. For $\operatorname{Re} s > \delta$ the lifted resolvent kernel is given by the convergent series

$$R_X(s; z, w) = \sum_{T \in \Gamma} R_{\mathbb{H}}(s, z, Tw),$$

and the representation (14.33) follows immediately.

As for the existence of a pole at $s = \delta$, we can follow the proof of Landau's theorem, which says that a Dirichlet series has a pole at its abscissa of convergence (see, e.g., [6, Theorem 11.13]). Let us fix z, w and concentrate on $\Sigma(s) = \Sigma(s; z, w)$. Suppose $\Sigma(s)$ has no pole at $s = \delta$. Then, by (14.33) (and the meromorphic continuation of the resolvent), $\Sigma(s)$ is analytic in some neighborhood of $s = \delta$. This implies that the power-series expansion centered at $\delta + 1$,

$$\Sigma(s) = \sum_{k=0}^{\infty} \frac{\Sigma^{(k)}(\delta + 1)}{k!}(s - \delta - 1)^k, \tag{14.34}$$

has radius of convergence larger than 1. Since the absolute Poincaré series converges uniformly on compact sets for $\operatorname{Re} s > \delta$, we can compute $\Sigma^{(k)}$ by differentiating term by term:

$$\Sigma^{(k)}(\delta + 1) = (-1)^k \sum_{T \in \Gamma} e^{-(\delta+1)d(z,Tw)} d(z, Tw)^k.$$

The power series (14.34) can then be written as a double sum:

$$\Sigma(s) = \sum_{k=0}^{\infty} \sum_{T \in \Gamma} \frac{(\delta + 1 - s)^k}{k!} e^{-(\delta+1)d(z,Tw)} d(z, Tw)^k.$$

Because the radius of convergence of the power series is larger than 1, this formula holds in particular for some $s = \delta - \varepsilon$, with $\varepsilon > 0$:

$$\Sigma(\delta - \varepsilon) = \sum_{k=0}^{\infty} \sum_{T \in \Gamma} \frac{(1 + \varepsilon)^k}{k!} e^{-(\delta+1)d(z,Tw)} d(z, Tw)^k. \tag{14.35}$$

All terms in this series are positive, so the sums can be interchanged, yielding

$$\Sigma(\delta - \varepsilon) = \sum_{T \in \Gamma} e^{-(\delta+1)d(z,Tw)} \sum_{k=0}^{\infty} \frac{(1 + \varepsilon)^k}{k!} d(z, Tw)^k = \sum_{T \in \Gamma} e^{-(\delta-\varepsilon)d(z,Tw)}$$

The series is convergent, since (14.35) was convergent. But this contradicts the definition of δ, so we conclude that $\Sigma(s)$ must have a singularity at δ.   □

Proposition 14.17 implies in particular that the discrete spectrum $\sigma_d(\Delta_X)$ is empty if and only if $\delta \leq \frac{1}{2}$. We also clearly have that $\delta < 1$ when X is an infinite-area hyperbolic surface, because $0 \notin \sigma_d(\Delta_X)$ in this case. The inequality (14.1) now follows from the existence of a pole, and the construction of the Patterson–Sullivan measure μ in Section 14.1 is fully justified in retrospect.

Our remaining two propositions are nice spectral theory applications of the ergodicity results from Section 14.2.

Proposition 14.18. *For any geometrically finite, nonelementary hyperbolic surface, the pole of the resolvent at $s = \delta$ has multiplicity one, and near $s = \delta$ we have*

$$\operatorname*{res}_{s=\delta} R_X(s; z, w) = cF(z)F(w),$$

where c depends on Γ.

Proof. In the measure construction in Section 14.1, we used the origin as the base point for the orbit. Let us define μ_w by changing this base point to an arbitrary $w \in \partial \mathbb{B}$. To be explicit, we define

$$\mu_w^{(s)} := \left(\sum_{T \in \Gamma} e^{-sd(0,Tw)} \right)^{-1} \sum_{T \in \Gamma} e^{-sd(0,Tw)} \nu_{Tw},$$

and then μ_w is the weak limit:

$$\mu_w := \lim_{s_j \to \delta} \mu_w^{(s_j)}.$$

We claim that this measure μ_w is absolutely continuous with respect to our original $\mu = \mu_0$. This is a fairly straightforward exercise using the triangle inequality in the form (2.20). See [156, Lemma 7.1] for the details.

Note that by a direct generalization of (14.9) and (14.12) we have the formula

$$\int_{\partial \mathbb{B}} P(z,q)^\delta \, d\mu_w(q) = \lim_{s_j \to \delta} \left(\sum_{T \in \Gamma} e^{-s_j d(0,Tw)} \right)^{-1} \sum_{T \in \Gamma} e^{-s_j d(z,Tw)}.$$

In other words,

$$\int_{\partial \mathbb{B}} P(z,q)^\delta \, d\mu_w(q) = \lim_{s \to \delta} \frac{\Sigma(s;z,w)}{\Sigma(s;0,w)}. \tag{14.36}$$

We next claim that the pole of $\Sigma(s;z,w)$ at δ has order 1. By the triangle identity (2.20) it suffices to prove this for $z = 0$, $w = 0$. For $n \in \mathbb{N}$, consider the set of $T \in \Gamma$ such that $d(0,T0) \in [n-1, n]$. We can fix r so that Lemma 14.12 applies to all such T for n sufficiently large. For $q \in \Lambda(\Gamma)$, the number of shadows $I(T0;r)$ containing the point q, such that $T \in \Gamma$ with $d(0,T0) \in [n-1, n]$, is bounded independently of q and n. Since the total mass of μ is 1, this implies that

$$\sum_{\substack{T \in \Gamma \\ d(0,T0) \in [n-1,n]}} \mu(I(T0;r)) \le C. \tag{14.37}$$

By Lemma 14.12 and (14.24),

$$\mu(I(T0;r)) \asymp e^{-\delta d(0,T0)},$$

so we deduce from (14.37) that

$$\#\{ T \in \Gamma : d(0,T0) \in [n-1, n] \} = O(e^{\delta n}).$$

We can therefore bound

$$\Sigma(s;0,0) \le C \sum_{n=1}^{\infty} e^{(\delta-s)n} = O((s-\delta)^{-1}).$$

By Proposition 14.17, the pole in the resolvent at δ also has order 1, and then by Proposition 8.1 we have

$$R_X(s; z, w) = \sum_{k=1}^{q} \frac{\phi_k(z)\phi_k(w)}{s - \delta} + \text{(holo)},$$

where the ϕ_k's are linearly independent, real-valued, generalized eigenfunctions. Thus, by (14.33), the singular part of $\Sigma(s; z, w)$ at $s = \delta$ is a constant times

$$\sum_{k=1}^{q} \frac{\phi_k(z)\phi_k(w)}{s - \delta}.$$

Comparing this to (14.36), we see that

$$\int_{\partial \mathbb{B}} P(z, q)^{\delta} \, d\mu_w(q) = a(w) \sum_{k=1}^{q} \phi_k(z)\phi_k(w).$$

By the independence of the ϕ_k's, we can choose points $w_1, \ldots, w_q$ such that the matrix $[a(w_j)\phi_i(w_j)]_{i,j=1}^{q}$ is invertible. If A_{kj} denotes the inverse matrix, then

$$\phi_k(z) = \sum_{j=1}^{q} \int_{\partial \mathbb{B}} P(z, q)^{\delta} A_{kj} \, d\mu_{w_j}(q).$$

Since all of the measures are absolutely continuous with respect to μ_0, we can write

$$\sum_{j=1}^{q} A_{kj} \, d\mu_{w_j} = h_k \, d\mu_0,$$

for some set of functions h_k. Then we have

$$\phi_k(z) = \int_{\partial \mathbb{B}} P(z, q)^{\delta} h_k(q) \, d\mu_0(q). \tag{14.38}$$

For $T \in \Gamma$, we can use a substitution $q = T\eta$, with Lemma 14.2 and the formula (14.5), to see that

$$\phi_k(Tz) = \int_{\partial \mathbb{B}} P(Tz, q)^{\delta} h_k(q) \, d\mu_0(q)$$

$$= \int_{\partial \mathbb{B}} P(Tz, T\eta)^{\delta} h_k(T\eta) \, d\mu_0(T\eta)$$

$$= \int_{\partial \mathbb{B}} P(z, \eta)^{\delta} h_k(T\eta) \, d\mu_0(\eta).$$

Since ϕ_k is Γ-invariant, we have

$$\int_{\partial \mathbb{B}} P(z, q)^{\delta} h_k(q) \, d\mu_0(q) = \int_{\partial \mathbb{B}} P(z, q)^{\delta} h_k(Tq) \, d\mu_0(q),$$

for all $z \in \mathbb{B}$ and $T \in \Gamma$. We conclude that h_k is Γ-invariant almost everywhere with respect to μ_0.

Since the action of Γ on μ_0 is ergodic, by Corollary 14.11, each h_k is constant almost everywhere. By (14.38), this means that ϕ_k is a constant multiple F. Hence $q = 1$ and the residue has the form claimed.   □

Proposition 14.19. *There are no resonances on the line* $\mathrm{Re}\, s = \delta$ *with* $s \neq \delta$.

Proof. Suppose that $R_X(s)$ (and hence $\Sigma(s; z, w)$) has a pole at $\zeta \neq \delta$, with $\mathrm{Re}\, \zeta = \delta$. For $\mathrm{Re}\, s > \delta$, we have $\Sigma(s; z, w) \leq \Sigma(\mathrm{Re}\, s; z, w)$. Thus the order of the pole at ζ is limited to one by the order of the pole at δ. We can therefore use the pole to construct a complex measure σ analogous to the Patterson–Sullivan measure. For some sequence $s_j \to \zeta$, we have a weak limit

$$\sigma = \lim_{s_j \to \zeta} (s_j - \zeta) \sum_{T \in \Gamma} e^{-s_j d(0, T0)} v_{T0}. \tag{14.39}$$

Existence of the pole guarantees that σ is a finite, nonzero measure supported on $\Lambda(\Gamma)$ with the same properties as μ. In particular, for $T \in \Gamma$,

$$T^* \sigma = |T'|^\zeta \sigma.$$

This shows that the associated measure on $(\partial \mathbb{B} \times \partial \mathbb{B})_-$,

$$d\widetilde{\sigma}(q, q') := \frac{d\sigma(q)\, d\sigma(q')}{|q - q'|^{2\zeta}},$$

is invariant under the product action of Γ.

Because the weights in front of the point measures v_{T0} in (14.39) are bounded by those used to construct μ, σ will be absolutely continuous with respect to μ. For some function ψ defined on $\Lambda(\Gamma)$,

$$d\sigma(q) = \psi(q)\, d\mu(q).$$

By the pullback formulas for the two sides,

$$\psi(Tq) = |T'(q)|^{\zeta - \delta} \psi(q), \tag{14.40}$$

for any $T \in \Gamma$.

We also note that $\widetilde{\sigma}$ is absolutely continuous with respect to $\widetilde{\mu}$, and both of these measures are invariant under the action of Γ. Since the product action of Γ on $\widetilde{\mu}$ is ergodic, by Corollary 14.11, any Γ-invariant function on $(\partial \mathbb{B} \times \partial \mathbb{B})_-$ is constant almost everywhere. Hence $\widetilde{\sigma} = c\widetilde{\mu}$ for some constant c. By the definitions of these measures, we then have

$$\psi(q)\psi(q') = c\, |q - q'|^{2(\zeta - \delta)}, \tag{14.41}$$

for almost every q, q'.

We'd like to extend (14.41) to all points in $(\Lambda(\Gamma) \times \Lambda(\Gamma))_-$. Since the right-hand side is continuous (away from the diagonal), we may assume that ψ is a continuous function on $\Lambda(\Gamma)$ by altering it on a set of measure zero. Then note that the measure-zero set on which (14.41) does not hold is open. In the proof of Proposition 2.9, we saw that any point in $\Lambda(\Gamma)$ is a limit point of hyperbolic fixed points (for Γ nonelementary). Thus any nonempty open set $U \subset \Lambda(\Gamma)$ contains a hyperbolic fixed point q. If q_+ is the attracting fixed point for some $T \in \Gamma$, with q_- the repelling fixed point, then $\cup_n T^n U = \Lambda(\Gamma) - \{q_-\}$. Hence $\mu(\cup_n T^n U) = 1$, implying $\mu(U) > 0$. This argument shows that any nonempty open subset of $(\Lambda(\Gamma) \times \Lambda(\Gamma))_-$ has nonzero measure with respect to $\widetilde{\mu}$. The set on which (14.41) fails to hold must therefore be empty.

Now we can take a limit to get an easy contradiction. For any $p \in \Lambda(\Gamma)$, let $\{q_j\}$ be a sequence in $\Lambda(\Gamma)$ converging to p. Because (14.41) holds everywhere off the diagonal and ψ is continuous, taking the limit of $\psi(q_j)\psi(p)$ as $q_j \to p$ gives $\psi(p)^2 = 0$. This means that σ vanishes on $\Lambda(\Gamma)$, contradicting the existence of a resonance at ζ. □

Taken together, Propositions 14.17, 14.18, and 14.19 now establish Theorem 14.15.

14.5 Prime geodesic theorem

The prime number theorem says that

$$\#\{\text{primes} \le x\} \sim \frac{x}{\log x},$$

as $x \to \infty$. This famous result is closely related to the meromorphic continuation of the Riemann zeta function $\zeta(s)$, whose only singularity is a simple pole at $s = 1$. It is relatively straightforward to prove that the zeta function has no zeros or poles on the line $\mathrm{Re}\, s = 1$. From this one can deduce the prime number theorem through the Wiener–Ikehara Tauberian theorem.

Using Corollary 14.16, we will apply an analogous argument to the Selberg zeta function, with the primitive length spectrum $\mathcal{L}_X$ standing in for the prime numbers. (More accurately, e^ℓ plays the role of a prime number.) From (2.18) we recall the length counting function

$$\pi_X(t) := \#\{\ell \in \mathcal{L}_X : \ell \le t\},$$

and note that we showed that $\pi_X(t) = O(e^t)$ in Proposition 2.19. In this section we will work out the exact asymptotic behavior of $\pi_X(t)$. This result was established independently by Guillopé [82] and Lalley [111].

Theorem 14.20 (Prime geodesic theorem). *For X a geometrically finite nonelementary hyperbolic surface with critical exponent δ,*

$$\pi_X(t) \sim \frac{e^{\delta t}}{\delta t} \qquad as\ t \to \infty.$$

Instead of using the Wiener–Ikehara Tauberian theorem, which is somewhat more technical to prove, we will base our proof on a very simple Tauberian argument due to Newman [147] (see also [218]).

Theorem 14.21 (Newman's Tauberian theorem). *Suppose that $f(t)$ is bounded and locally integrable on $[0, \infty)$, and*

$$g(z) := \int_0^\infty f(t)e^{-zt}\, dt$$

exists for $\mathrm{Re}\, z > 0$. If g extends to a holomorphic function on a neighborhood of $\{\mathrm{Re}\, z \geq 0\}$, then f is integrable on $[0, \infty)$, and

$$\int_0^\infty f(t)\, dt = g(0).$$

Proof. If for $T > 0$ we define the entire function

$$g_T(z) := \int_0^T f(t)e^{-zt}\, dt,$$

then our goal is to show that $\lim_{T \to \infty} g_T(0) = g(0)$.

By the assumption that $g(z)$ has an analytic extension, for any $R > 0$ we can choose $\varepsilon > 0$ such that g is analytic on $\Omega = \{|z| \leq R, \mathrm{Re}\, z \geq -\varepsilon\}$. Then, by Cauchy's integral formula,

$$g(0) - g_T(0) = \frac{1}{2\pi i} \int_{\partial \Omega} [g(z) - g_T(z)]\, e^{zT} \left(1 + \frac{z^2}{R^2}\right) \frac{dz}{z}, \qquad (14.42)$$

with $\partial \Omega$ oriented counterclockwise. Let us break $\partial \Omega$ into components η_+ and η_-, according to the sign of $\mathrm{Re}\, z$, and write the corresponding pieces of the integral (14.42) as $I_\pm(T)$. Let $M = \sup |f|$.

On η_+ we use the fact that f is bounded to estimate

$$|g(z) - g_T(z)| = \left| \int_T^\infty f(t)e^{-zt}\, dt \right| \leq \frac{M}{2\pi} \frac{e^{-(\mathrm{Re}\, z)T}}{\mathrm{Re}\, z}.$$

This gives

$$|I_+(T)| \leq \frac{M}{2\pi} \int_{\eta_+} \frac{1}{\mathrm{Re}\, z} \left| 1 + \frac{z^2}{R^2} \right| \frac{|dz|}{R}.$$

For z on η_+ we have $|1 + z^2/R^2| = 2\,\mathrm{Re}\, z/R$, and of course the length of η_+ is πR. Thus,

$$|I_+(T)| \leq \frac{M}{R}.$$

To estimate I_- we separate the integrals over g and g_T. Since g_T is analytic, we can deform the contour to the arc of a circle that is the continuation of η_+,

$$\int_{\eta_-} g_T(z) e^{zT} \left(1 + \frac{z^2}{R^2} \right) \frac{dz}{z} = \int_{\eta_R} g_T(z) e^{zT} \left(1 + \frac{z^2}{R^2} \right) \frac{dz}{z},$$

where $\eta_R(\theta) = Re^{i\theta}$ for $\theta \in [\pi/2, 3\pi/2]$. Then the estimate

$$|g_T(z)| \le M \frac{e^{|\operatorname{Re} z|T} - 1}{|\operatorname{Re} z|}$$

leads directly to

$$\left| \int_{\eta_-} g_T(z) e^{zT} \left(1 + \frac{z^2}{R^2} \right) \frac{dz}{z} \right| \le \frac{M}{R}.$$

For the g contribution to I_-, because $e^{zT} \to 0$ uniformly on compact subsets of $\{\operatorname{Re} z < 0\}$, we have

$$\lim_{T \to 0} \left| \int_{\eta_-} g(z) e^{zT} \left(1 + \frac{z^2}{R^2} \right) \frac{dz}{z} \right| = 0.$$

Hence

$$\limsup_{T \to 0} |I_-(T)| \le \frac{M}{R}.$$

Combining the estimates of $I_+(T)$ and $I_-(T)$ gives

$$\limsup_{T \to 0} |g(0) - g_T(0)| \le \frac{2M}{R}.$$

Since R was arbitrary, this completes the proof.   □

If our goal were to prove the prime number theorem, we would apply Theorem 14.21 with $f(t)$ defined as a sum over $\log p$ for p prime, and with $g(z)$ given essentially by the logarithmic derivative of the Riemann zeta function. To analyze $\pi_X(t)$ we follow exactly the same course, except using the Selberg zeta function.

Proof of Theorem 14.20. Define

$$\eta(s) := \sum_{\ell \in \mathcal{L}_X} \ell e^{-s\ell}.$$

By the definition (2.22), for $\operatorname{Re} s > \delta$ we have

$$\frac{Z_X'(s)}{Z_X(s)} = \sum_{\ell \in \Gamma} \sum_{k=0}^{\infty} \frac{\ell}{e^{(s+k)\ell} - 1}.$$

We can then write

$$\frac{Z_X'(s)}{Z_X(s)} - \eta(s) = \sum_{\ell \in \Gamma} \frac{\ell}{e^{s\ell}(e^{s\ell} - 1)} + \sum_{\ell \in \Gamma} \sum_{k=1}^{\infty} \frac{\ell}{e^{(s+k)\ell} - 1},$$

and note that the first sum on the right is holomorphic for $\mathrm{Re}\, s > \delta/2$ and the second is holomorphic for $\mathrm{Re}\, s > \delta - 1$. Therefore $\eta(s)$ has a meromorphic extension to $\mathrm{Re}\, s > \delta/2$, and in this region has the same poles and residues as $Z'_X(s)/Z_X(s)$. In particular, by Corollary 14.16,

$$\eta(s) - \frac{1}{s - \delta} \quad \text{is holomorphic for } \mathrm{Re}\, s \geq \delta. \tag{14.43}$$

Now let us write $\eta(s)$ as a Stieltjes integral,

$$\eta(s) = \int_0^\infty e^{-st}\, d\theta(t),$$

where

$$\theta(t) := \sum_{\ell \in \mathcal{L}_X : \ell \leq t} \ell.$$

Since $\pi_X(t) = O(e^t)$, for $\mathrm{Re}\, s > 1$ we can justify integrating by parts to give

$$\eta(s) = s \int_0^\infty e^{-st}\theta(t)\, dt. \tag{14.44}$$

If we set

$$g(z) := \frac{\eta(z + \delta)}{z + \delta} - \frac{1}{\delta z},$$

then (14.44) implies that

$$g(z) = \int_0^\infty e^{-(z+\delta)t}\theta(t)\, dt - \frac{1}{\delta z} = \int_0^\infty e^{-zt}\left[e^{-\delta t}\theta(t) - \frac{1}{\delta} \right] dt.$$

Since (14.43) shows that $g(z)$ extends holomorphically to $\mathrm{Re}\, z \geq 0$, Theorem 14.21 applies to give

$$\int_0^\infty \left[e^{-\delta t}\theta(t) - \frac{1}{\delta} \right] dt < \infty. \tag{14.45}$$

The next step is to argue that (14.45) implies

$$\theta(t) \sim \frac{e^{\delta t}}{\delta}. \tag{14.46}$$

Suppose that for $\varepsilon > 0$, we had $\theta(t_j) \geq e^{\delta(t_j + \varepsilon)}/\delta$ for some sequence $t_j \to \infty$. Then since $\theta(t)$ is an increasing function,

$$\int_{t_j}^{t_j + \varepsilon} \left[e^{-\delta t}\theta(t) - \frac{1}{\delta} \right] dt \geq \int_{t_j}^{t_j + \varepsilon} \left[e^{-\delta t}\theta(t_j) - \frac{1}{\delta} \right] dt \geq \frac{\varepsilon}{\delta}.$$

Since $t_j \to \infty$, this contradicts (14.45). We conclude that

$$\limsup_{t \to \infty} \delta e^{-\delta t}\theta(t) \leq 1.$$

We can bound the lim inf in a similar fashion, and this proves (14.46).

For the final step, note that $\theta(t) \leq t\pi_X(t)$ by its definition, so

$$\liminf_{t\to\infty} \frac{t\pi_X(t)}{\theta(t)} \geq 1. \tag{14.47}$$

On the other hand, for any $\lambda < 1$ we have

$$\theta(t) \geq \sum_{\lambda t \leq \ell \leq t} \ell \geq \lambda t \left[\pi_X(t) - \pi_X(\lambda t)\right]. \tag{14.48}$$

By the definition of δ as the exponent of convergence, $\sum_\ell e^{-s\ell}$ converges for any $s > \delta$. Since

$$\sum_{\ell \leq t} e^{-s\ell} \geq \pi_X(t)e^{-st},$$

this implies that

$$\pi_X(t) = O(e^{(\delta+\varepsilon)t}),$$

for any $\varepsilon > 0$. Using this in (14.48) gives

$$\frac{t\pi_X(t)}{\theta(t)} \leq \lambda^{-1} + Cte^{(\delta+\varepsilon)\lambda t - \delta t}.$$

With the choice $\varepsilon < \delta(1-\lambda)/\lambda$ we see that

$$\limsup_{t\to\infty} \frac{t\pi_X(t)}{\theta(t)} \leq \lambda^{-1}, \tag{14.49}$$

for any $\lambda > 1$. Together, (14.47) and (14.49) imply

$$\theta(t) \sim t\pi_X(t),$$

and by (14.46) this completes the proof.   □

Lax–Phillips [113], Colin de Verdière [50], and Patterson [158] applied similar methods to obtain the asymptotics of the *lattice-point counting function*. This is defined for $w, w' \in \mathbb{H}$ by

$$N(r; w, w') := \#\{T \in \Gamma : d(w, Tw') \leq r\}.$$

In this case the argument is based on the absolute Poincaré series $\Sigma(s; w, w')$ as defined in (14.32). By Theorem 14.15 and Proposition 14.17, for some constant c (depending only on δ),

$$\Sigma(s; w, w') - \frac{cF(w)F(w')}{s-\delta}$$

is holomorphic for $\mathrm{Re}\, s \geq \delta$. We can write

$$\Sigma(s; w, w') = s \int_0^\infty e^{-sr} N(r; w, w') \, dr,$$

and then apply Theorem 14.21 to the function

$$
\begin{aligned}
g(z) &:= \frac{\Sigma(z + \delta; w, w')}{z + \delta} - \frac{cF(w)F(w')}{\delta z} \\
&= \int_0^\infty e^{-zr} \left[e^{-\delta r} N(r; w, w') - \frac{cF(w)F(w')}{\delta} \right] dr.
\end{aligned}
$$

Repeating the argument above gives the asymptotics of $N(r; w, w')$:

Theorem 14.22 (Patterson). *For X a geometrically finite nonelementary hyperbolic surface, there exists $c > 0$ such that*

$$N(r; w, w') \sim cF(w)F(w')e^{\delta r},$$

where $F(w)$ is the generalized eigenfunction appearing in the residue of the resolvent at $s = \delta$.

14.6 Refined asymptotics of the length spectrum

For $\delta > \frac{1}{2}$, Theorem 14.20 shows that the first eigenvalue of the Laplacian, $\lambda_0 = \delta(1 - \delta)$, gives the leading asymptotic behavior of the length counting function. It turns out that this beautiful connection can be extended even further, to include the contribution from each eigenvalue in the discrete spectrum. In the infinite-area case this was proven by Naud [144], whose argument we will follow here. The main ingredient is the Poisson formula for the wave trace, Theorem 11.3.

As in the actual prime number theorem, one obtains sharper asymptotics by replacing $x / \log x$ by the *log integral* function,

$$\mathrm{li}(x) := \int_2^x \frac{dt}{\log t} \sim \frac{x}{\log x} \left[1 + \frac{1}{\log x} + \cdots \right].$$

Theorem 14.23 (Naud). *Let X be a nonelementary geometrically finite hyperbolic surface of infinite area such that $\delta > \frac{1}{2}$. Label the resonances corresponding to discrete eigenvalues by $\{\zeta_j\}$, where*

$$\delta = \zeta_1 > \zeta_2 \geq \cdots \geq \zeta_n > \frac{1}{2},$$

repeated according to multiplicity. Then

$$\pi_X(t) = \sum_{j=1}^n \mathrm{li}(e^{\zeta_j t}) + O(e^{(\delta/2 + 1/4)t}).$$

Naud [143] also proved a version of Theorem 14.23 for $\delta \leq \frac{1}{2}$. In this case, (14.31) can be used to show that

$$\pi_X(t) = \mathrm{li}(e^{\delta t}) + O(e^{(\delta+\beta)t/2}),$$

where $\beta \in [0, \delta)$ is either the real part of the next resonance beyond δ or zero, whichever is greater.

In order to prove Theorem 14.23, we introduce a function related to π_X,

$$\psi(r) := \sum_{\substack{\ell \in \mathcal{L}_X, m \in \mathbb{N} \\ m\ell \leq \log r}} \ell.$$

Proposition 14.24. *If X is a geometrically finite hyperbolic surface with $\delta > \frac{1}{2}$, then*

$$\psi(r) = \sum_{j=1}^{n} \frac{r^{\zeta_j}}{\zeta_j} + O(r^{\delta/2+1/4}).$$

The proof relies on the Poisson formula for resonances. Before we present it, let us see how it implies the theorem.

Proof of Theorem 14.23. By the definition of $\psi(r)$, we can write the Stieltjes integral

$$\int_{2}^{e^t} \frac{d\psi(r)}{\log r} = \sum_{\substack{\ell \in \mathcal{L}_X, m \in \mathbb{N} \\ 2 \leq m\ell \leq t}} \frac{1}{m}.$$

On the other hand, we can evaluate this sum in terms of the length counting function,

$$\sum_{\substack{\ell \in \mathcal{L}_X, m \in \mathbb{N} \\ 2 \leq m\ell \leq t}} \frac{1}{m} = \sum_{m=1}^{\infty} \frac{\pi_X(t/m)}{m} + O(1).$$

The sum over m on the right has only finitely many nonzero terms, because $\pi_X(t) = 0$ for $t < \ell_0$. Therefore, we can use the asymptotic result from Theorem 14.20 to bound

$$\sum_{m=2}^{\infty} \frac{\pi_X(t/m)}{m} = O(e^{\delta t/2}),$$

and conclude that

$$\int_{2}^{e^t} \frac{d\psi(r)}{\log r} = \pi_X(t) + O(e^{\delta t/2}). \tag{14.50}$$

If we set

$$f(r) := \psi(r) - \sum_{j=1}^{n} \frac{r^{\zeta_j}}{\zeta_j},$$

then (14.50) implies

$$\pi_X(t) = \sum_{j=1}^{n} \int_2^{e^t} \frac{d(r^{\zeta_j})}{\zeta_j \log r} + \int_2^{e^t} \frac{df(r)}{\log r} + O(e^{\delta t/2})$$

$$= \sum_{j=1}^{n} \mathrm{li}(e^{\zeta_j t}) + \int_2^{e^t} \frac{df(r)}{\log r} + O(e^{\delta t/2}).$$

Proposition 14.24 gives the estimate $f(r) = O(r^{\delta/2+1/4})$, so the error term is easily controlled with an integration by parts,

$$\int_2^{e^t} \frac{df(r)}{\log r} = \frac{f(r)}{\log r}\Big|_2^{e^t} + \int_2^{e^t} \frac{f(r)}{r(\log r)^2}\, dr = O(e^{(\delta/2+1/4)t}). \qquad \square$$

Proof of Proposition 14.24. Theorems 11.2 and 11.3 together gave us the trace formula (12.2), which we can write in the form

$$\sum_{\zeta \in \mathcal{R}_X} \int_0^\infty e^{(\zeta - \frac{1}{2})t} \phi(t) = \sum_{\ell \in \mathcal{L}_X} \sum_{m=1}^\infty \frac{\ell}{2\sinh(m\ell/2)} \phi(m\ell)$$

$$+ \int_0^\infty \phi(t)\Psi_{\mathrm{top}}(t)\, dt, \qquad (14.51)$$

for $\phi \in C_0^\infty(\mathbb{R}_+)$. The strategy of the proof is quite familiar from Chapter 12; we seek a family of choices for ϕ that will allow us to extract asymptotics and bound the error terms.

Let ℓ_0 be the minimal length of a closed geodesic on X. For r large and $\sigma > 0$, let $h_{r,\sigma} \in C_0^\infty(\mathbb{R}_+)$ be a function satisfying

$$h_{r,\sigma}(t) = \begin{cases} 0, & t \le \ell_0/2, \\ 1, & \ell_0 \le t \le \log r, \\ 0, & \log t \ge \log(r + r^{1-\sigma}), \end{cases}$$

and then set

$$\phi_{r,\sigma}(t) := e^{t/2} h_{r,\sigma}(t).$$

In the range $[\log r, \ \log(r + r^{1-\sigma})]$, we can assume that

$$h_{r,\sigma}(t) = f\left(\frac{t - \log r}{\log(r + r^{1-\sigma}) - \log r}\right),$$

for some function f that is independent of r and σ. If we observe that $1/(\log(r + r^{1-\sigma}) - \log r) \sim r^\sigma$, this implies estimates on derivatives

$$\sup_{t \in [\log r, \ \log(r + r^{1-\sigma})]} |h_{r,\sigma}^{(k)}(t)| \le C_k r^{k\sigma}, \qquad (14.52)$$

where C_k is independent of r.

On the spectral (left) side of (14.51), consider the contribution from a single ζ_j,

$$\int_0^\infty e^{(\zeta_j - \frac{1}{2})t} \phi_{r,\sigma}(t) = \int_0^\infty e^{\zeta_j t} h_{r,\sigma}(t)\,dt.$$

Splitting the integral according to the assumptions on $h_{r,\sigma}(t)$, we obtain

$$\int_0^\infty e^{(\zeta_j - \frac{1}{2})t} \phi_{r,\sigma}(t)\,dt = \int_0^{\ell_0} e^{\zeta_j t} h_{r,\sigma}(t)\,dt + \int_{\ell_0}^{\log r} e^{\zeta_j t}\,dt$$

$$+ \int_{\log r}^{\log(r + r^{1-\sigma})} e^{\zeta_j t} h_{r,\sigma}(t)\,dt$$

$$= O(1) + \frac{1}{\zeta_j}\left(r_j^\zeta - e^{\zeta_j \ell_0}\right) + O(r^{\zeta_j - \sigma})$$

$$= \frac{r_j^\zeta}{\zeta_j} + O(r^{\delta - \sigma}). \tag{14.53}$$

If $\mathrm{Re}\,\zeta \le \frac{1}{2}$ and $\zeta \neq 0$, then we use repeated integration by parts to obtain, for $k = 1, 2, \ldots,$

$$\int_0^\infty e^{(\zeta - \frac{1}{2})t} \phi_{r,\sigma}(t)\,dt = \frac{(-1)^k}{\zeta^k} \int_0^\infty e^{\zeta t} h_{r,\sigma}^{(k)}(t)\,dt.$$

Hence, using the structure of h as well as (14.52), we can obtain the estimate

$$\int_0^\infty e^{(\zeta - \frac{1}{2})t} \phi_{r,\sigma}(t)\,dt \le \frac{(-1)^k}{\zeta^k} \int_0^{\ell_0} e^{\zeta t} h_{r,\sigma}^{(k)}(t)\,dt$$

$$+ \frac{(-1)^k}{\zeta^k} \int_{\log r}^{\log(r + r^{1-\sigma})} e^{\zeta t} h_{r,\sigma}^{(k)}(t)\,dt$$

$$= O\left(|\zeta|^{-k}\right) + O\left(|\zeta|^{-k} r^{\zeta + (k-1)\sigma}\right)$$

$$= O\left(|\zeta|^{-k} r^{1/2 + (k-1)\sigma}\right). \tag{14.54}$$

For $\zeta = 0$, then, the contribution would be $\int_0^\infty h_{r,\sigma}(t)\,dt$, which is easily seen to be $O(r^{1/2})$.

We need to use the estimates (14.54) to control the sum over $\mathrm{Re}\,\zeta \le \frac{1}{2}$ on the left-hand side of (14.51). To accomplish this, we break the sum at $|\zeta| = r^\sigma$ and apply (14.54) with $k = 1$ for $0 < |\zeta| < r^\sigma$ and (14.54) with $k = 3$ for $|\zeta| > r^\sigma$. This gives

$$\sum_{\mathrm{Re}\,\zeta \le 1/2} \left| \int_0^\infty e^{(\zeta - \frac{1}{2})t} \phi_{r,\sigma}(t)\,dt \right| \le C_0 r^{1/2} + C_1 r^{1/2} \sum_{0 < |\zeta| < r^\sigma} |\zeta|^{-1}$$

$$+ C_2 r^{1/2 + 2\sigma} \sum_{|\zeta| > r^\sigma} |\zeta|^{-3}. \tag{14.55}$$

With a Stieltjes integral, the first sum on the right can be estimated as

$$\sum_{0<|\zeta|<r^\sigma} |\zeta|^{-1} = \int_0^{r^\sigma} \frac{dN_X(t)}{t}$$

$$= r^{-\sigma} N_X(r^\sigma) + \int_0^{r^\sigma} \frac{N_X(t)}{t^2} \, dt$$

$$= O(r^\sigma),$$

using $N_X(t) = O(t^2)$ from Theorem 9.1. Similarly,

$$\sum_{|\zeta|>r^\sigma} |\zeta|^{-3} = \int_{r^\sigma}^\infty \frac{dN_X(t)}{t^3} = O(r^{-\sigma}).$$

With these estimates, (14.55) becomes

$$\sum_{\mathrm{Re}\,\zeta<1/2} \left| \int_0^\infty e^{(\zeta-\frac{1}{2})t} \phi_{r,\sigma}(t) \, dt \right| = O(r^{1/2+\sigma}). \tag{14.56}$$

Next we turn to the right-hand side of (14.51). The topological term is easily dealt with:

$$\int_0^\infty \phi_{r,\sigma}(t) \Psi_{\mathrm{top}}(t) \, dt$$

$$= \int_0^{\log(r+r^{1-\sigma})} \left[\frac{\chi}{2} \frac{\cosh(t/2)}{\sinh^2(t/2)} + \frac{n_c}{2}(\coth(t/2) - 1) \right] e^{t/2} h_{r,\sigma}(t) \, dt$$

$$= O(r^{1/2} \log r). \tag{14.57}$$

Finally, we examine the sum over the length spectrum in (14.51):

$$\sum_{\ell \in \mathcal{L}_X} \sum_{m=1}^\infty \frac{\ell}{2 \sinh(m\ell/2)} \phi_{r,\sigma}(m\ell) = \psi(r) + \sum_{m\ell \leq \log r} \frac{\ell e^{-m\ell}}{1 - e^{-m\ell}}$$

$$+ \sum_{\log r < m\ell \leq \log(r+r^{1-\sigma})} \frac{\ell}{1 - e^{-m\ell}} h_{r,\sigma}(m\ell).$$

Because $\delta < 1$, we know that $\sum_{\ell \in \mathcal{L}_X} e^{-\ell} < \infty$, so it's easy to bound the second term on the right by $O(\log r)$. As for the third term, the asymptotic $\pi_X(t) \sim e^{\delta t}/(\delta t)$ from Theorem 14.20 implies that

$$\#\{\log r < \ell \leq \log(r+r^{1-\sigma})\} \leq C \left[\frac{(r+r^{1-\sigma})^\delta}{\log(r+r^{1-\sigma})} - \frac{r^\delta}{\log r} \right] = O\left(\frac{r^{\delta-\sigma}}{\log r} \right).$$

This allows us to bound the third term by $O(r^{\delta-\sigma})$, so that the full estimate of the sum over lengths is

$$\sum_{\ell \in \mathcal{L}_X} \sum_{m=1}^{\infty} \frac{\ell}{2 \sinh(m\ell/2)} \phi_{r,\sigma}(m\ell) = \psi(r) + O(r^{\delta-\sigma}). \tag{14.58}$$

Applying the estimates (14.53), (14.56), (14.57), and (14.58) to the trace formula (14.51), we find that

$$\psi(r) = \sum_{j=1}^{n} \frac{r^{\zeta_j}}{\zeta_j} + O(r^{\delta-\sigma}) + O(r^{1/2+\sigma}).$$

As a final step, we set $\sigma = (\delta - 1/2)/2$ to optimize the error term.   □

Notes

We have described Patterson–Sullivan theory only in the simplest case; the theory applies in much greater generality. The construction of the measure supported on the limit set applies to discrete groups of isometries of $\mathbb{H}^n$ for any $n \geq 2$, even for geometrically infinite groups. The critical exponent is always equal to the Hausdorff dimension of the set of radial limit points. For geometrically finite groups one can show that the nonradial part of the limit set has measure zero, so that $\delta = \dim_H \Lambda(\Gamma)$. Ergodicity holds as long as $m_\mu(\mathcal{S}X) < \infty$, which is always the case for geometrically finite groups.

There is also much more to the ergodic theory part of the story than we have been able to cover here. In addition to the original references of Patterson [156, 158] and Sullivan [202, 203, 204], we refer the reader to the expository accounts of Patterson [157] and Nicholls [149]. See also the compilation [17].

The results of Sections 14.4–14.6 extend to higher-dimensional hyperbolic manifolds without cusps (i.e., convex cocompact groups). In particular, Patterson [158] proved the analogue of Theorem 14.15 for such manifolds. Guillarmou [76] proved the absence of resonances near the critical line for conformally compact manifolds under certain conditions.

The problem of the asymptotic behavior of the lattice-point counting function is a classical one, in both the Euclidean and non-Euclidean contexts. For compact hyperbolic surfaces the asymptotics are a result of Huber [97], and these were extended to finite-area surfaces by Patterson [155]. Lax–Phillips [113] established the asymptotics for geometrically finite quotients of $\mathbb{H}^{n+1}$ with $\delta > \frac{n}{2}$. For geometrically finite hyperbolic surfaces, Colin de Verdière [50] proved the meromorphic continuation of the absolute Poincaré series and derived the asymptotics of the lattice-point counting function, for any $\delta \geq 0$. Patterson [158] extended the asymptotics to higher dimensions for convex cocompact groups, using ideas from ergodic theory as suggested by earlier work of Nicholls [148].

The prime geodesic theorem was first proven for compact hyperbolic surfaces by Huber [97], and refined (in the sense of Theorem 14.23) by Hejhal [91] and Randol [178]. The result was extended to finite-area hyperbolic surfaces by Sarnak [184]. For higher-dimensional hyperbolic manifolds without cusps, the prime geodesic theorem was proven by Perry [167] and refined by Guillarmou–Naud [81].

The Selberg trace formula can also be used to study the asymptotics of lengths of closed geodesics within homology classes, and in particular to prove that closed geodesics are asymptotically evenly distributed among homology classes. There are results of Katsuda–Sunada [109] and Phillips–Sarnak [173] for compact manifolds of negative curvature, Epstein [56] and Sharp [191] for finite-volume hyperbolic manifolds, McGowan–Perry [129] and Babillot–Peigné [8, 9, 10] for infinite-volume hyperbolic manifolds.

Dynamical Approach to the Zeta Function

The definition (2.22) of Selberg's zeta function as a product over the length spectrum was in some sense very convenient, because of the link between conjugacy classes of Γ and closed geodesics furnished by Proposition 2.17. On the other hand, existence of a meromorphic continuation was far from obvious and bounding the growth of the zeta function was quite difficult (see the proof of Theorem 10.1). There is an alternative framework for zeta functions that is much more general than Selberg's original definition, and these basic properties are much easier to prove from this point of view. The essential idea is associate the zeta function to the dynamics of the geodesic flow on the surface (rather than to the geometry of the surface). This viewpoint leads to a definition that can be applied to any dynamical system under certain conditions on the flow. The main drawback to this approach, in the context of hyperbolic surfaces, is that these methods do not apply when the surface has cusps because of the noncompactness of the convex core. In this section we restrict our attention to geometrically finite hyperbolic surfaces without cusps (convex cocompact Fuchsian groups).

The "dynamic formalism" for zeta functions was introduced by Ruelle [183], who defined them in terms of transfer matrices for families of expanding maps. Ruelle's theory, with subsequent development by Fried [65], applies in particular to compact hyperbolic manifolds and allows the Selberg zeta function to be realized in terms of determinants of transfer matrices constructed from the geodesic flow. The extension of this approach from the compact case to convex cocompact hyperbolic groups was clarified by Patterson–Perry [160, Section 2]. In particular, they showed that in dimension $n \geq 2$ the zeta function admits a meromorphic extension as a ratio of entire functions of order n and finite type.

The general Ruelle–Fried theory encompasses too much material for us to review here. However, everything simplifies beautifully in the case of hyperbolic surfaces without cusps. This is because convex cocompact Fuchsian groups are all of a particularly nice type called Schottky groups. (We'll explain these groups in Section 15.1.) The definition of Ruelle transfer matrices and subsequent application to the Selberg zeta function is quite direct for Schottky groups. Developing this theory will lead us to an alternative proof of analytic continuation of the zeta function, as well as an application to resonance counting from Guillopé–Lin–Zworski [84].

15.1 Schottky groups

Schottky groups are distinguished by the existence of a set of generators of a particular type. Working in the $\mathbb{H}$ model, suppose $D_1, \dots D_{2r}$ are open Euclidean disks in $\mathbb{C}$ with mutually disjoint closures and centers on the real axis, as shown in Figure 15.1. The disks can be arranged in any order on the real axis. For each pair, D_j, D_{j+r}, we let $S_j \in \mathrm{PSL}(2, \mathbb{R})$ denote a transformation that sends ∂D_j to ∂D_{j+r} and maps the exterior of D_j to the interior of D_{j+r}. For convenience, assume that the indices are defined cyclically, so that for any $j \in \mathbb{Z}$,

$$S_{j+2r} := S_j \quad \text{and} \quad S_{j+r} = S_j^{-1}.$$

Definition. A Fuchsian group Γ is a *Schottky group* if there exists a set of disks $\{D_j\}$ as above such that Γ is generated by the corresponding S_j's.

(More specifically, this is a classical Fuchsian Schottky group; see Button [38] for clarification.)

It obvious that each transformation S_j is hyperbolic, with a repelling fixed point inside D_j and an attracting fixed point inside D_{j+r}. The region $\mathcal{F} := \mathbb{H} - \cup D_j$ is a natural fundamental domain for the action of Γ, and the assumption that the closures of the disks are mutually disjoint means that $X = \Gamma \backslash \mathbb{H}$ has infinite area with no cusps, so Γ is convex cocompact. The fundamental region $\mathcal{F}$ has Euler characteristic $\chi = 1$, so after gluing r pairs of edges together to form X, we see that

$$\chi(X) = 1 - r.$$

The case $r = 1$ is elementary (a hyperbolic cylinder), so we will be interested in $r \geq 2$. The example shown in Figure 15.1 has genus 1 with two funnels, which is the only possibility when $r = 3$.

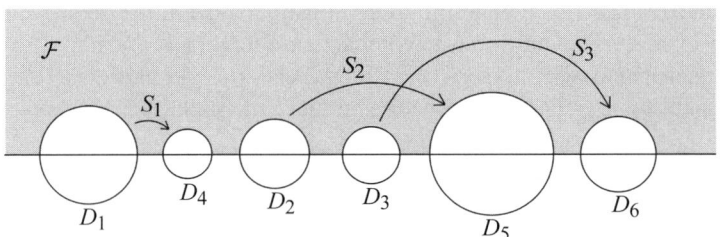

Fig. 15.1. Schottky disks and generators.

Lemma 15.1. *There are no relations among the S_j's. (Schottky groups are free.)*

Proof. This simple proof is adapted from [179]. For $j = 1, \ldots, r$, let $U_j = D_j \cup D_{j+r}$. We can write any element $T \in \Gamma$ as $T_n \cdots T_1$, with each $T_i \in \langle S_{k_i} \rangle - \{I\}$ for some $k_i \in \{1, \ldots, 2r\}$ such that $k_i \neq k_{i+1}$. Starting with a point z_0 in the interior of $\mathcal{F}$, we will prove by induction that

$$T_n \cdots T_1 z_0 \in U_{k_n}. \tag{15.1}$$

Note that $T_1 z_0 \in U_{k_1}$ is obvious. Assuming that $T_i \cdots T_1 z_0 \in U_{k_i}$ for some $i < n$, the induction step is proven by observing that

$$T_{i+1}(T_i \cdots T_1 z_0) \in T_{i+1} U_{k_i} \subset U_{k_{i+1}}.$$

Since (15.1) implies in particular that $T_n \cdots T_1 \neq I$, we conclude that there are no relations among the different S_j's. $\square$

In [38], Button showed that convex cocompact Fuchsian groups are all classical Schottky groups. We will state this result in terms of hyperbolic surfaces.

Theorem 15.2 (Button). *If X is a conformally compact hyperbolic surface (i.e., geometrically finite, infinite-area, and without cusps), then $X \cong \Gamma \backslash \mathbb{H}$ for some Schottky group Γ.*

Proof. Suppose that $X = \Gamma \backslash \mathbb{H}$ has genus g with n_f funnels. We first produce a reference Schottky group Γ_{g,n_f} that will yield a quotient of the same topological type as X. First consider the case $n_f = 1$, so that $r = 2g$. We simply line up $4g$ Schottky disks and label them $D_1, \cdot, D_{4g}$ from left to right, and then let $\Gamma_{g,1}$ be the Schottky group generated by these disks. With the assumption that S_j maps $\partial D_j \to \partial D_{j+2g}$, this ordering guarantees that there is only one cycle of boundary intervals. That is, the portions of the fundamental domain meeting $\partial \mathbb{H}$ are glued together into a single funnel. For $n_f > 1$ we simply adjust the ordering to produce n_f boundary cycles. Let $r = 4g + 2n_f - 2$ and take $2r$ Schottky disks. Label the first $4g$ disks consecutively $D_1, \ldots, D_{2g}, D_{1+r}, \ldots, D_{2g+r}$. The next $n_f - 1$ are labeled in order $D_{2g+1}, D_{2g+2}, \ldots, D_r$, but the last set is labeled in reverse order $D_{2r}, D_{2r-1}, \ldots, D_{2g+r+1}$. It is straightforward to see that this ordering results in n_f boundary cycles and so produces a Schottky group Γ_{g,n_f} such that $X_{g,n_f} := \Gamma_{g,n_f} \backslash \mathbb{H}$ is homeomorphic to X.

Define the family of curves $\eta_1, \ldots, \eta_r$ on X by projecting the boundary geodesics $\partial D_i \cap \mathbb{H}$ from the construction of Γ_{g,n_f} down to X_{g,n_f} and then pulling them back to X by some fixed homeomorphism between the surfaces. Note that the η_i's are continuous arcs that begin and end in the funnels and do not intersect each other. If we cut X along these curves, the resulting surface lifts to a domain Ω in $\mathbb{H}$ bounded by curves $\widetilde{\eta}_1, \ldots, \widetilde{\eta}_{2r}$. These are labeled so that $\widetilde{\eta}_j$ and $\widetilde{\eta}_{j+r}$ are both lifts of the same curve η_j on X. Then we can define S_j as the element of Γ mapping $\widetilde{\eta}_j$ to $\widetilde{\eta}_{j+r}$. (See Figure 15.2.) For each $j = 1, \ldots, 2r$, draw a Schottky disk D_j meeting $\partial \mathbb{H}$ at the endpoints of $\widetilde{\eta}_j$. It's easy to deduce that S_j maps the exterior of D_j to the interior of

D_{j+r} from the fact that it maps $\widetilde{\eta}_j$ to $\widetilde{\eta}_{j+r}$. Since the Schottky group generated by the S_j's glues the edges of Ω together to form a surface isometric to X, this group must coincide with Γ. $\square$

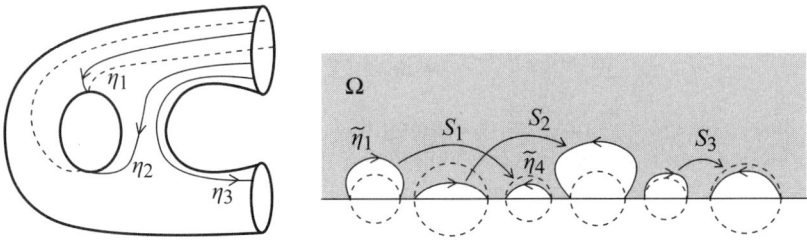

Fig. 15.2. Finding Schottky generators.

15.2 Symbolic dynamics

It turns out that there's a beautiful way to "encode" the structure of closed geodesics on X through a discrete dynamical system. The coding method we will use comes from Bowen–Series [28], but the idea traces all the way back to work of Hadamard, Morse, and Koebe.

Assume that X is a conformally compact hyperbolic surface, with Γ the corresponding Schottky group. Setting $I_j = D_j \cap \partial \mathbb{H}$, we define the *Bowen–Series map*

$$B : \cup_{j=1}^{2r} I_j \to \cup_{j=1}^{2r} I_j,$$

by

$$Bq := S_j q \text{ for } q \in I_j.$$

The definition might appear innocuous at first glance, but the following result gives some hint at its usefulness.

Lemma 15.3 (Orbit equivalence). *Given $p, q \in \cup_{j=1}^{2r} I_j$, we have $p = Tq$ for some $T \in \Gamma$ if and only if $B^k p = B^l q$ for some $k, l \geq 0$.*

Proof. This was proven in greater generality by Bowen–Series [28]; for Schottky groups the proof is simplified considerably. First note that since B is built from the action of generators of Γ, it is obvious that $B^k p = B^j q$ implies $p = Tq$ for some $T \in \Gamma$.

For the converse, it suffices to prove the case that T is one of the generators of the group. Hence we suppose that $p = S_j q$ for some j. Then, by the definition of the S_j's, either $p \in I_{j+r}$ or $q \in I_j$. In the first case, $Bp = q$, while in the second case we have $p = Bq$. $\square$

A crucial feature of the Bowen–Series map is that B is "eventually expanding" near the limit set. For I_1, I_2 intervals in $\mathbb{R}$, a differentiable map $f : I_1 \to I_2$ is said to be *expanding* at $x \in I_1$ if $|f'(x)| > 1$, and *contracting* if $|f'(x)| < 1$. Suppose that $T \in \mathrm{PSL}(2, \mathbb{R})$ is hyperbolic and doesn't fix ∞, which means that $c \neq 0$ if T is written in the standard matrix form

$$T = \begin{pmatrix} a & b \\ c & d \end{pmatrix}. \tag{15.2}$$

Then $|T'(z)| = |cz + d|^{-2}$, so $T|_{\mathbb{R}}$ is expanding on an interval of width $2/|c|$, centered on $-d/c = T^{-1}(\infty)$ and containing the repelling fixed point of T. The circle in $\mathbb{C}$ spanned by this expanding interval, namely $\{z \in \mathbb{C} : |cz + d| = 1\}$, is called the *isometric circle* of T.

If $q_- \in \mathbb{R}$ is the repelling fixed point of T, then we claim that

$$|T'(q_-)| = e^{\ell(T)}. \tag{15.3}$$

This is easily demonstrated by conjugating T to move q_- to the origin. After conjugation T takes the form $z \mapsto e^{\ell(T)}z/(cz + 1)$, and we simply evaluate $T'(0) = e^{\ell(T)}$ directly.

To analyze B further, we introduce families of intervals that will be images of the original intervals I_j under the action of Γ. It will be convenient to index them according to the action of B, as follows. For $T \in \Gamma$, we define J_T to be the interval on which the action of some some iterate of B is given by T. For instance, the original intervals correspond to the generators

$$J_{S_i} := I_i.$$

A general $T \in \Gamma$ may be decomposed as a product over the generators,

$$T = S_{i_1} S_{i_2} \cdots S_{i_n}, \tag{15.4}$$

with indices $1 \leq i_1, \ldots, i_n \leq 2r$. Since a Schottky group is free, by Lemma 15.1, this decomposition is unique provided we don't allow any generator to be adjacent to its own inverse. The length of the sequence of generators is called the *word length* of T and denoted by $|T| := n$. With T written in the form (15.4), we define

$$J_T := S_{i_n}^{-1} \cdots S_{i_2}^{-1} I_{i_1}. \tag{15.5}$$

This labeling scheme is illustrated in Figure 15.3. Since $S_{i_{k+1}}^{-1}$ is not allowed to be equal to S_{i_k}, we have $S_{i_{k+1}}^{-1} I_{i_k} \subset I_{i_{k+1}}$ for each $k = 1, \ldots, n-1$. This shows that

$$B^n|_{J_T} = T, \tag{15.6}$$

as desired. It also shows that the J_T's are nested,

$$J_T \subset J_{S_{i_2} \cdots S_{i_n}} \subset \cdots \subset J_{S_{i_n}} = I_{i_n}. \tag{15.7}$$

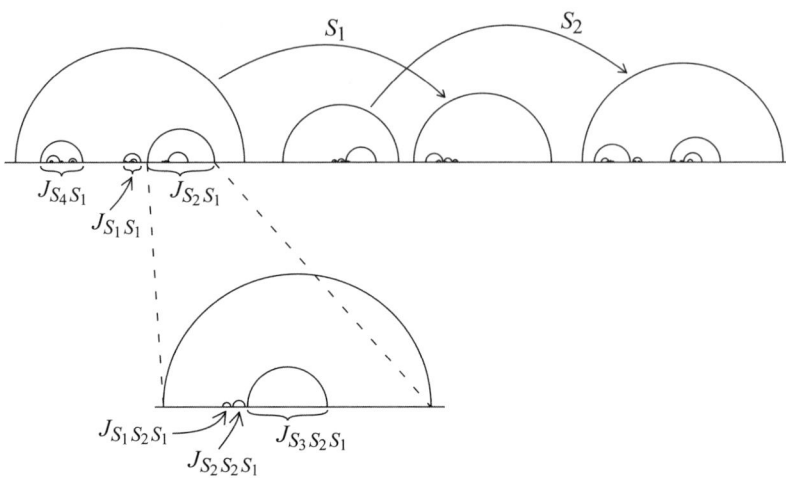

Fig. 15.3. Intervals for the Bowen–Series map.

Proposition 15.4 (Eventually expanding). *There exists a finite collection of intervals* $\{J_i\} \subset \mathbb{C}$, *covering* $\Lambda(\Gamma)$, *such that for some fixed* $n \in \mathbb{N}$ *and* $\beta > 1$,

$$|(B^n)'(p)| > \beta \qquad \text{for all } p \in \cup J_i.$$

Proof. Following the argument from [84], we will build the intervals out of the J_T's defined above. Assume that T has the form (15.4). Since $T J_T = S_{i_1} I_{i_1}$, which contains ∞ in particular, we have

$$T^{-1}(\infty) \subset J_T \subset I_{i_n}. \tag{15.8}$$

Recall that $T^{-1}(\infty)$ is the Euclidean center of the isometric circle of T. By (15.8) this is always contained in one of the original intervals I_j. We claim that the radii of the isometric circles of elements of Γ accumulate only at zero. This follows by discreteness of Γ, since the radius of the isometric circle of T is $1/|c|$ and $ab - cd = 1$. Because the points $T^{-1}(\infty)$ cannot accumulate away from $\Lambda(\Gamma)$, we deduce the existence of n_0 such that the expanding interval of any T with word length $|T| \geq n_0$ will be contained in one of the I_j.

Now consider any $T \in \Gamma$ with $|T| = n_0$, written in the form (15.4). Since $|T^{-1}| = n_0$ also, the choice of n_0 ensures that the expanding interval of T^{-1} will be strictly contained in I_{i_1+r}. Thus T^{-1} is contracting on the complement of I_{i_1+r}, with linear distortion $|(T^{-1})'(p)|$ bounded above by $1/\beta < 1$ for some $\beta > 1$ depending only n_0. Now $T J_T = S_{i_1} I_{i_1} \subset \mathbb{H} - I_{i_1+r}$, so we conclude that $|T'(p)| \geq \beta$ for $p \in J_T$. By (15.6) this establishes the claim. $\qquad\square$

Lemma 15.3 allows us to make the connection between closed geodesics and periodic orbits of B. A periodic orbit of B is called primitive if it is not the iteration of an orbit with shorter length.

Proposition 15.5. *There is a one-to-one correspondence between primitive periodic orbits* $\{q, Bq, \ldots, B^n q = q\}$ *and primitive closed geodesics on* X *of displacement length*

$$\ell(T) = \log |(B^n)'(q)|. \tag{15.9}$$

Proof. Proposition 2.17 already gives the connection between closed geodesics and conjugacy classes of hyperbolic elements in Γ, so our goal here is to link the conjugacy classes to periodic orbits of B. We will see that to each conjugacy class $[T]$ there corresponds a periodic orbit of length n, where n is the minimum word length for elements of $[T]$.

Suppose first that $\{q, Bq, \ldots, B^n q = q\}$ is a primitive periodic orbit. By Lemma 15.3, $B^n q$ corresponds to Tq for some $T \in \Gamma$, and this implies $q \in J_T$ by (15.6). Proposition 15.4 shows that q must be the repelling fixed point of T and then $e^{\ell(T)} = |(B^n)'(q)|$ by (15.3). Note that we could have taken any of the points in the orbit as a base point instead of q. For example, if we had used Bq as the starting point, then because $Bq = S_{i_n} q$, we would have obtained $S_{i_n} T S_{i_n}^{-1}$ for the associated transformation instead of T. Iterating this argument shows that the map from periodic orbits into conjugacy classes is injective.

Given a conjugacy class $[T] \subset \Gamma$, we can assume that the chosen representative T has minimal word length. If $T = S_{i_1} \cdots S_{i_n} \in \Gamma$, this means simply that $S_{i_n} \neq S_{i_1}^{-1}$. Let q be the repelling fixed point of T. Arguing as in the proof of Lemma 15.3, the fact that

$$S_{i_1}(S_{i_2} \cdots S_{i_n} q) = q$$

implies either $q \in I_{i_1+r}$ or $S_{i_2} \cdots S_{i_n} q \in I_{i_1}$. We claim that the former case cannot occur, while in the latter case we are done. This is because $S_{i_2} \cdots S_{i_n} q \in I_{i_1}$ means precisely that $q \in J_T$, and so $B^n q = Tq = q$ by (15.6). Since p was the repelling fixed point, (15.9) would then follow from (15.3).

What remains is to rule out the case $q \in I_{i_1+r}$. Under this assumption we have

$$Bq = S_{i_1}^{-1} q = S_{i_2} \cdots S_{i_n} q.$$

Applying the same reasoning as above, we see that either $Bq \in I_{i_2+r}$ or $S_{i_3} \cdots S_{i_n} q \in I_{i_2}$. The latter would imply in particular that $q \in I_{i_n}$, by (15.7). But this contradicts $q \in I_{i_1+r}$ because of the assumption that T has minimal word length. Therefore, $q \in I_{i_1+r}$ implies $Bq \in I_{i_2+r}$. Proceeding inductively from this assumption, we find that

$$B^{k-1} q = S_{i_{k-1}}^{-1} \cdots S_{i_1}^{-1} q \in I_{i_k+r} \text{ for } k = 1, \ldots, n.$$

But this would mean that $q \in J_{T^{-1}}$, which contradicts the fact that B is eventually expanding, because q is the attracting fixed point of T^{-1}. We conclude that the case $q \in I_{i_1+r}$ could not occur. $\square$

15.3 Dynamical zeta function

In Proposition 15.5, we saw that the Bowen–Series map B encodes the structure of closed geodesics on X. In this section we'll associate to B a dynamical zeta function that turns out to be equal to the Selberg zeta function $Z_X(s)$. Our treatment is based on Guillopé–Lin–Zworski [84], but will also rely on the simplified presentation given by Zworski in [227].

The first step is to associate to B a transfer operator acting on a certain L^2 space of holomorphic functions. Taking the union of the original Schottky disks,

$$U := \bigcup_{j=1}^{2r} D_j,$$

we define

$$\mathcal{H}(U) := \{u \in L^2(U) : u \text{ is analytic on } U\},$$

with respect to Lebesgue measure on $\mathbb{C}$. We can naturally extend B to a map $U \to \mathbb{C} \cup \{\infty\}$ by setting $B|_{D_j} = S_j$. The *Ruelle transfer operator* is the map $L(s) : \mathcal{H}(U) \to \mathcal{H}(U)$ defined for $s \in \mathbb{C}$ by

$$L(s)u(z) := \sum_{w \in U: \, Bw=z} B'(w)^{-s} u(w). \tag{15.10}$$

Note that $B'(w)$ is positive for $w \in \mathbb{R}$, so the power $B'(w)^{-s}$ is well-defined for $w \in U$ by analytic continuation.

If $z \in D_j$, then z has exactly one preimage under B in each of the sets $S_i^{-1}D_j \subset D_i$, for $i \neq j+r$. Since B acts by S_i on D_i, the definition (15.10) could be rephrased as

$$L(s)u(z)\Big|_{z \in D_j} = \sum_{i \neq j+r} \left[(S_i^{-1})'(z)\right]^s u(S_i^{-1}z). \tag{15.11}$$

Lemma 15.6. *The singular values of $L(s)$ can be estimated by*

$$\mu_k(L(s)) \leq e^{C|s|-ck}, \tag{15.12}$$

for some constants $C, c > 0$. In particular, $L(s)$ is trace class for any $s \in \mathbb{C}$.

Proof. Under the decomposition

$$\mathcal{H}(U) = \bigoplus_{j=1}^{2r} \mathcal{H}(D_j), \tag{15.13}$$

we can use (15.11) to separate $L(s)$ into its components,

$$L_{ji}(s) : \mathcal{H}(D_i) \to \mathcal{H}(D_j), \qquad i \neq j+r,$$

where for $u \in \mathcal{H}(D_i), z \in D_j$,

$$L_{ji}(s)u(z) := \left[(S_i^{-1})'(z)\right]^s u(S_i^{-1}z).$$

By applying the additive Fan inequality (A.20) to the sum $L = \oplus L_{ji}$, we deduce that

$$\mu_k(L(s)) \le \max_{1 \le j, i \le 2r} 2\mu_{[k/N]}(L_{ji}(s)),$$

for some N sufficiently large. Hence it suffices to prove the claimed estimate for $\mu_k(L_{ji}(s))$.

Suppose that D_i is a disk of radius a centered at $c \in \mathbb{R}$. Transplanting the standard basis of normalized monomials from the unit disk, we obtain a basis for $\mathcal{H}(D_i)$,

$$\phi_n(z) := \sqrt{\frac{n+1}{\pi a^2}} \left(\frac{z-c}{a}\right)^n. \tag{15.14}$$

For $i \ne j + r$, $S_i^{-1}(D_j)$ is contained within some Euclidean disk $B_{\mathbb{C}}(c; a - \varepsilon) \subset D_i$, for $\varepsilon > 0$. Thus for $z \in D_j$,

$$\|\phi_n \circ S_i^{-1}\|_{\mathcal{H}(D_j)} \le (n+1) \left(1 - \frac{\varepsilon}{a}\right)^n. \tag{15.15}$$

We can thus produce a bound

$$\|L_{ji}(s)\phi_n\|_{\mathcal{H}(D_j)} \le \sup_{z \in D_j} |(S_i^{-1})'(z)^s| \, e^{-cn} \le e^{C|s|-cn}. \tag{15.16}$$

From the min-max estimate (Theorem A.16) we can immediately derive that

$$\mu_k(L_{ji}(s)) \le \sum_{n=k}^{\infty} \|L_{ji}\phi_n\|_{\mathcal{H}(D_j)}.$$

This allows us to deduce (15.12) directly from (15.16).   □

Definition. The *dynamical zeta function* associated to the transfer operator $L(s)$ is

$$d_X(s) := \det(I - L(s)).$$

Note that the Fredholm determinant of $I - L(s)$ is well-defined since $L(s)$ is trace class. Moreover, $L(s)$ is holomorphic so $d_X(s)$ is an entire function. We will see below that the dynamical zeta function is equal to Selberg's zeta function.

To make the connection, we will use the following result, a classical case of the holomorphic Lefschetz fixed point formula, to compute traces of powers of $L(s)$. (See [52, 205] for more general versions and references.)

Lemma 15.7 (Fixed point formula). *For a disk $D \subset \mathbb{C}$, let $h : D \to \mathbb{C}$ be a bounded analytic function and $f : D \to D$ an analytic map such that $\overline{f(D)} \subset D$. Assume that f has a single fixed point q inside D. Then for the operator A on $\mathcal{H}(D)$ defined by*

$$A\phi(z) := h(z)\phi(f(z)),$$

we have

$$\operatorname{tr} A = \frac{h(q)}{1 - f'(q)}.$$

Proof. By a simple linear change of coordinates, it suffices to consider the case that D is the unit disk. Note that the assumptions that h is bounded and $\overline{f(D)} \subset D$ guarantee that A is trace class. This is easily checked using the orthonormal basis $\phi_k(z) := \sqrt{(k+1)/\pi}\, z^n$, as in the proof of Lemma 15.6.

The classical *Bergman kernel* $K(z, w)$ is the integral kernel of the orthogonal projection $L^2(D, dm) \to \mathcal{H}(D)$,

$$K(z, w) = \sum_{k=1}^{\infty} \phi_k(z)\overline{\phi_k(w)} = \frac{1}{\pi(1 - z\bar{w})^2}.$$

This gives a nice explicit formula for the kernel of A,

$$A(z, w) = \frac{h(z)}{\pi(1 - f(z)\bar{w})^2}.$$

Since A is trace class, the trace is given by

$$\operatorname{tr} A = \frac{1}{\pi} \int_D \frac{h(z)}{(1 - f(z)\bar{z})^2}\, |dz|^2,$$

where $|dz|^2$ denotes Lebesgue measure. We can simplify by applying the complex form of Stokes's theorem, which reads

$$2i \int_D \frac{\partial u}{\partial \bar{z}}\, |dz|^2 = \oint_{\partial D} u\, dz,$$

for any $u \in C^1(\overline{D})$. The trace becomes

$$\operatorname{tr} A = \frac{1}{2\pi i} \oint_{\partial D} \frac{\bar{z} h(z)}{1 - f(z)\bar{z}}\, dz.$$

Since the integration variable is now restricted to $|z| = 1$, this can be further reduced to

$$\operatorname{tr} A = \frac{1}{2\pi i} \oint_{\partial D} \frac{h(z)}{z - f(z)}\, dz,$$

and the result then follows immediately from the Cauchy integral formula.   □

Theorem 15.8. *The Selberg and dynamical zeta functions are equal:*

$$Z_X(s) = d_X(s).$$

(In particular, this gives an independent proof that $Z_X(s)$ extends to an entire function for a geometrically finite hyperbolic surface without cusps.)

Proof. For $z \in \mathbb{C}$, define

$$d(z, s) = \det(1 - zL(s)),$$

so that $d_X(s) = d(1, s)$. If $|z|$ is sufficiently small then we can expand

$$d(z,s) = \exp\left(-\sum_{n=1}^{\infty} \frac{z^n}{n} \operatorname{tr} L(s)^n\right). \tag{15.17}$$

Now consider the trace,

$$\operatorname{tr} L(s)^n = \sum_{i_1,\dots,i_n} \operatorname{tr}\left[L_{i_1 i_2}(s) L_{i_2 i_3}(s) \cdots L_{i_n i_1}(s)\right], \tag{15.18}$$

with the restriction $i_j \neq i_{j+1} + r$ for $j = 1,\dots,n-1$ and $i_1 \neq i_{n+r}$. The first restriction implies that the terms in this sum correspond to elements of Γ of word length exactly n, and the second implies that these elements have minimal word length in their respective conjugacy classes. If we write $T = S_{i_1} \cdots S_{i_n}$, then repeated application of the chain rule shows that

$$\left[L_{i_1 i_2}(s) L_{i_2 i_3}(s) \cdots L_{i_n i_1}(s)\right] u(z) = \left[(T^{-1})'(z)\right]^s u(T^{-1}z), \tag{15.19}$$

where $u = \mathcal{H}(D_{i_1})$, $z \in D_{i_1}$.

Let $q \in D_{i_1}$ denote the attracting fixed point of T^{-1}. Applying Lemma 15.7 to D_{i_1}, we deduce

$$\operatorname{tr}\left[L_{i_1 i_2}(s) L_{i_2 i_3}(s) \cdots L_{i_n i_1}(s)\right] = \frac{\left[(T^{-1})'(q)\right]^s}{1 - (T^{-1})'(q)} = \sum_{k=0}^{\infty}\left[(T^{-1})'(q)\right]^{s+k}.$$

Since B^n acts by T on $T^{-1} D_{i_1}$, and q is also a fixed point of B^n, we can substitute this calculation into (15.18). With this, (15.17) becomes

$$\operatorname{tr} L(s)^n = \sum_{q=B^n q} \sum_{k=0}^{\infty}\left[(B^n)'(q)\right]^{-(s+k)}.$$

With this computation, (15.17) can be written

$$d(z,s) = \exp\left(-\sum_{n=1}^{\infty} \frac{z^n}{n} \sum_{q=B^n q} \sum_{k=0}^{\infty}\left[(B^n)'(q)\right]^{-(s+k)}\right).$$

Let us regroup the sum in terms of primitive orbits. For each primitive orbit of length m, there are m distinct fixed points of B^{mj} for each iteration j. Thus,

$$d(z,s) = \exp\left(-\sum_{m,j=1}^{\infty} \frac{z^{mj}}{j} \sum_{\substack{\{q,\dots,B^{m-1}q\} \\ \text{primitive}}} \sum_{k=0}^{\infty}\left[(B^m)'(q)\right]^{-j(s+k)}\right).$$

By Proposition 15.5 this can be written as a sum over the primitive length spectrum,

$$d(z,s) = \exp\left(-\sum_{\ell \in \mathcal{L}_X} \sum_{j=1}^{\infty} \frac{z^{jm(\ell)}}{j} \sum_{k=0}^{\infty} e^{-j(s+k)\ell}\right),$$

where $m(\ell)$ denotes the minimal word length of the conjugacy class associated to the geodesic of length ℓ (a slight abuse of notation, since this is really a function of the conjugacy class). The sum over j can now be evaluated:

$$
d(z, s) = \exp\left(\sum_{\ell \in \mathcal{L}_X} \sum_{k=0}^{\infty} \log\left[1 - z^{m(\ell)} e^{-(s+k)\ell} \right] \right),
$$

$$
= \prod_{\ell \in \mathcal{L}_X} \prod_{k=0}^{\infty} \left(1 - z^{m(\ell)} e^{-(s+k)\ell} \right).
$$

For $\mathrm{Re}\, s > \delta$, the sum converges at $z = 1$ and we obtain $d(1, s) = Z_X(s)$. $\square$

15.4 Growth estimates

Writing $Z_X(s)$ as a dynamical zeta function gives additional benefits beyond the simple proof of analytic continuation in Theorem 15.8. Since the determinant $\det(1 - L(s))$ is well-defined for all $s \in \mathbb{C}$, it is much easier to estimate directly.

Proposition 15.9. *For a geometrically finite hyperbolic surface without cusps,*

$$
|Z_X(s)| \le e^{C|s|^2}. \tag{15.20}
$$

In other words, $Z_X(s)$ is an entire function of order 2 and finite type.

Proof. Using Lemma 15.6 and Weyl's inequality for the determinant (Theorem A.19), we can estimate

$$
|Z_X(s)| = |\det(1 - L(s))| \le \prod_{k=1}^{\infty} \left(1 + e^{c_1|s| - c_2 k} \right).
$$

We can bound the product over $k \ge c_1|s|/c_2$ by a constant C_1 independent of $|s|$. This gives

$$
|Z_X(s)| \le C_1 \prod_{k=1}^{\lceil c|s| \rceil} \left(1 + e^{c_1|s|} \right) \le e^{C|s|^2}. \qquad \square
$$

A comparison of this proof of the order bound to our more general version, in Section 10.5, amply demonstrates the advantages of the dynamical approach. Of course, there is also a price paid—the connection between zeros of the zeta function and the resonance set is obscured from this viewpoint.

In [84], Guillopé–Lin–Zworski gave a dramatic sharpening of the bound (15.20) when s is restricted to a vertical strip. The strategy is essentially to refine the proof of Lemma 15.6. The factor $\exp(C|s|)$ appearing in (15.16) comes from the bound on $|(S_i^{-1})'(z)^s|$ for $z \in D_j$. If we assume a fixed bound on $|\mathrm{Re}\, s|$, then we can control the size of $\left| \left[(S_i^{-1})'(z) \right]^s \right|$, for large $|\mathrm{Im}\, s|$, by keeping the imaginary part of

$(S_i^{-1})'(z)$ small. Since $(S_i^{-1})'(z)$ is real on the real axis, this can be accomplished by redefining $L(s)$ on a set of very small disks in place of the original Schottky disks. The appropriate radius will turn out to be on the order of $1/|\operatorname{Im} s|$. Of course, for the identification of Theorem 15.8 to hold, the smaller disks must still cover $\Lambda(\Gamma)$. Shrinking the disks means that more will be needed to cover $\Lambda(\Gamma)$, and the Patterson–Sullivan theory gives us a precise understanding of this relationship. This explains the appearance of $\delta = \dim_H \Lambda(\Gamma)$ in the following:

Theorem 15.10 (Guillopé–Lin–Zworski). *For X a geometrically finite nonelementary hyperbolic surface without cusps, we have for any $M > 0$,*

$$|Z_X(s)| \leq e^{C|s|^{\delta}} \quad for \ |\operatorname{Re} s| \leq M.$$

Before presenting the proof, we note the immediate application to resonance counting, which is illustrated in Figure 15.4.

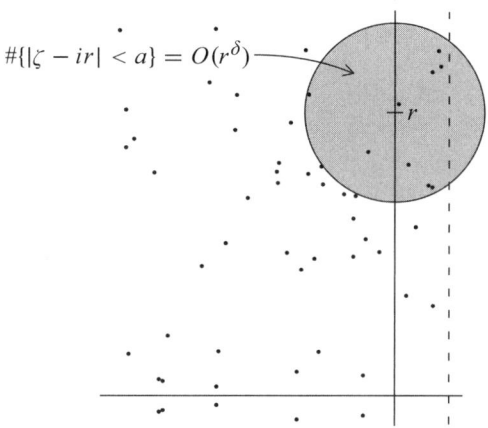

$$\#\{|\zeta - ir| < a\} = O(r^{\delta})$$

Fig. 15.4. Resonances in a vertical strip.

Corollary 15.11. *Suppose X is a geometrically finite nonelementary hyperbolic surface without cusps. Then for any fixed $a > 0$,*

$$\#\{\zeta \in \mathcal{R}_X : |\zeta - ir| < a\} = O(r^{\delta}). \tag{15.21}$$

Proof. We apply Jensen's formula (Theorem A.1) to a circle of radius $R > 2(a+1)$, centered at $1 + ir$, to obtain

$$\sum_{|\zeta - (1+ir)| < R} \log \frac{R}{|\zeta - (1+ir)|}$$

$$= -\log |Z_X(1 + ir)| + \frac{1}{2\pi} \int_0^{2\pi} \log \left| Z_X(1 + ir + Re^{i\theta}) \right| d\theta.$$

Let $N_a(r)$ denote the left-hand side of (15.21). By restricting the sum to $|\zeta - ir| < a$, Jensen's formula gives the estimate

$$N_a(r) \log 2 \leq -\log |Z_X(1 + ir)| + \sup_{|s-(1+ir)|=R} \log |Z_X(s)|.$$

The second term on the right is $O(r^\delta)$ by Theorem 15.10. From the defining product (2.22) for $Z_X(s)$, which converges absolutely for $\mathrm{Re}\, s = 1$, it's easy to see that $|Z_X(1 + ir)| \geq |Z_X(1)|$. Hence $-\log |Z_X(1 + ir)|$ is bounded above by the constant $-\log |Z_X(1)|$, and the estimate on $N_a(r)$ follows. □

Integrating the estimate (15.21) in vertical strips gives a slightly weaker averaged result, for any $M > 0$,

$$\#\{\zeta \in \mathcal{R}_X : \mathrm{Re}\, \zeta > -M, |\mathrm{Im}\, \zeta| < r\} = O(r^{\delta+1}), \tag{15.22}$$

which we can compare to the global result $N_X(r) \asymp r^2$. By the work of Sjöstrand [194] on upper bounds on resonances in chaotic quantum scattering, one expects the density of resonances in strips parallel to the continuous spectrum to be approximated by a power law with power given by half the dimension of the trapped set. For a hyperbolic surface, the trapped set consists of the geodesics whose lifts to $\mathbb{H}$ begin and end at points in the limit set $\Lambda(\Gamma)$. A point in the tangent space $T\mathbb{H}$ is specified by giving the endpoints of the corresponding geodesic, along with the position along the geodesic and the speed. Since the dimension of $\Lambda(\Gamma)$ is δ, the total dimension of the trapped set is $2\delta + 2$. Thus (15.22) agrees with the expected behavior for chaotic scattering. Guillopé–Lin–Zworski [84] give numerical evidence showing that this upper bound is sharp, and suggesting that the counting function in a vertical strip may actually be asymptotic to $r^{\delta+1}$.

To implement the strategy described above for the proof of Theorem 15.10, we first prove a lemma, concerning the set

$$\Lambda(\Gamma) + [-h, h] = \{q \in \mathbb{R} : d(q, \Lambda(\Gamma)) \leq h\},$$

which will assist in the covering of $\Lambda(\Gamma)$ by disks of a controlled size.

Lemma 15.12. *Suppose Γ is convex cocompact, with $\infty \notin \Lambda(\Gamma)$. There exists C such that for all $h > 0$ sufficiently small, the connected components of $\Lambda(\Gamma) + [-h, h]$ have length at most Ch.*

Proof. Suppose $q_0 \in \Lambda(\Gamma)$. We first claim that for any $\varepsilon > 0$ sufficiently small, we can find $T \in \Gamma$ such that

$$|Tp - Tq| \asymp \varepsilon^{-1}|p - q| \quad \text{for } p, q \in [q_0 - \varepsilon, q_0 + \varepsilon], \tag{15.23}$$

with constants independent of ε and q_0. It suffices to prove the corresponding result in the $\mathbb{B}$ model, since the distortion of the map $\partial \mathbb{H} \to S^1$ may be bounded above and below in any interval of finite Euclidean length.

Switching to $\mathbb{B}$ allows us to use shadows as in Section 14.3. Suppose that $q_0 \in S^1$. Let R_0 be the diameter of the truncated Nielsen region $\widetilde{K} \subset \mathbb{B}$, and assume $0 \in \widetilde{K}$. Given ε, find the (unique) point $z \in [0, q]$ such that the neighborhood $B_{S^1}(q_0, \varepsilon)$ corresponds with the shadow $I(z, 1)$. By the assumptions that Γ is convex cocompact and $q \in \Lambda(\Gamma)$, there exists $T \in \Gamma$ such that $Tz \in \widetilde{K}$. Setting $z_0 = T^{-1}0$, we have

$$d(z_0, z) \leq R_0.$$

This shows that $B_{S^1}(q, \varepsilon)$ is contained in the shadow $I(z_0, R_0 + 2)$. By the sine rule (14.24),

$$\sin \varepsilon = \frac{\sinh 1}{\sinh d(0, z)}$$

and

$$\sin |I(z_0, R_0 + 2)|/2 = \frac{\sinh(R_0 + 2)}{\sinh d(0, z_0)}.$$

This shows that

$$1 - |z_0| \asymp \varepsilon, \tag{15.24}$$

with constants depending only on R_0. By keeping ε sufficiently small, we may thus bound $|z_0|$ away from zero. Then we can repeat the argument from Lemma 14.12 leading up to (14.30) to obtain

$$|T'(q)| \asymp \frac{1}{1 - |z_0|}, \quad \text{for } q \in I(z_0, R_0 + 2).$$

The result (15.23) now follows by (15.24) and the distortion formula (14.4).

Returning now to the $\mathbb{H}$ model with $\Lambda(\Gamma) \subset \mathbb{R}$, let c_1, c_2 be the upper and lower constants for the comparison (15.23). Since $\Lambda(\Gamma)$ is nowhere dense, the maximal length of a connected component of $\Lambda(\Gamma) + [-h, h]$ goes to zero as $h \to 0$. So we can choose h_0 such that the maximal connected component of $\Lambda(\Gamma) + [-h_0, h_0]$ has length less than c_1. Assume that h is small enough that $\varepsilon = c_2 h/h_0$ meets the requirements for (15.23). Then for any $q_0 \in \Lambda(\Gamma)$, we can find $T \in \Gamma$ such that

$$[Tq_0 - c_1, Tq_0 + c_1] \subset T([q_0 - \varepsilon, q_0 + \varepsilon])$$

and

$$T(\Lambda(\Gamma) + [-h, h]) \subset \Lambda(\Gamma) + [-h_0, h_0].$$

Thus if $[q_0 - \varepsilon, q_0 + \varepsilon] \subset \Lambda(\Gamma) + [-h, h]$, we would have

$$[Tq_0 - c_1, Tq_0 + c_1] \subset \Lambda(\Gamma) + [-h_0, h_0],$$

contradicting the choice of h_0. This shows that the maximal length of a component of $\Lambda(\Gamma) + [-h, h]$ is less than $2\varepsilon + h = Ch$. $\qquad \square$

Proof of Theorem 15.10. For $i = 1, \ldots, n(h)$, let $\{I_i(h)\}$ be the connected components of $\Lambda(\Gamma) + [-h, h]$. By translating Lemma 14.13 from S^1 to $\mathbb{R}$, we find that

$$\mu(I_i(h)) > ch^\delta$$

for some constant c, where μ is Patterson–Sullivan measure. Since $\mu(\Lambda(\Gamma)) = 1$ and the $I_i(h)$'s cover $\Lambda(\Gamma)$, this shows that the number of connected components is bounded by

$$n(h) = O(h^{-\delta}). \tag{15.25}$$

Let $D_i(h)$ be the disk centered on $\mathbb{R}$ and spanned by $I_i(h)$, for $i = 1, \ldots, n(h)$. For h sufficiently small, the set of disks $U(h) := \cup D_i(h)$ will be contained within our original region U, and we can define the transfer operator $L^{(h)}(s) : \mathcal{H}(U(h)) \to \mathcal{H}(U(h))$,

$$L^{(h)}(s)u(z) := \sum_{w \in U(h):\, Bw=z} B_h'(w)^{-s} u(w).$$

Since $\Lambda(\Gamma) \subset U(h)$, we can apply the same fixed-point analysis as in Theorem 15.8 to conclude that

$$Z_X(s) = \det(I - L^{(h)}(s)). \tag{15.26}$$

Assume that $|\operatorname{Re} s| \leq M$ and set $h = 1/|s|$. In essence we want to repeat the proof of (15.20) using an improved bound in place of Lemma 15.6. There's one significant issue remaining. In the estimate (15.15) we can see that the constants depended on ε/a, where a was the radius of one of the disks D_i and ε the minimal spacing between the disk ∂D_i and the image $S_i^{-1}(D_j)$ inside it. Now that we've switched to $D_j(h)$, we have $a = a(h) \asymp h$, so we need uniform control as $h \to 0$ over the spacing $\varepsilon(h)$ between the boundary of $D_j(h)$ and any embedded disk of the form $S_i^{-1}(D_m(h))$. If B were uniformly expanding, then this would be a simple matter.

Fortunately, we can exploit the fact that B is eventually expanding for the same purpose. The strategy is to measure distances with respect to a metric for which B is uniformly expanding, instead of the Euclidean metric. The new metric, introduced by Mather [124], is defined as follows. Choose N according to Proposition 15.4, so that $|(B^N)'| > \beta > 1$ in some collection of intervals $J \supset \Lambda(\Gamma)$. If h is sufficiently small, we can assume that $\cup I_i(h) \subset J$. On J we introduce the function

$$\kappa(x) := \sum_{k=0}^{N-1} |(B^k)'(x)|,$$

and define the metric

$$ds_\kappa := \kappa(x)\, dx.$$

Note that $(B^{k+1})'(x) = (B^k)'(Bx)\, B'(x)$, by the chain rule, so that

$$\kappa(Bx)\, |B'(x)| = \kappa(x) + |(B^N)'(x)| - 1 > \kappa(x) + \beta.$$

Measuring the distortion with respect to this new metric thus gives

$$\|dB_q\|_\kappa = \frac{\kappa(Bx)}{\kappa(x)}|B'(x)| > 1 + \frac{\beta}{\kappa(x)} > \beta_0,$$

for some $\beta_0 > 1$. In other words, B is uniformly expanding with respect to ds_κ. Let d_κ denote the distance function defined by ds_κ, which is comparable to the Euclidean distance d since $\overline{J}$ is compact.

As in the proof of Lemma 15.6, we break $L^{(h)}(s)$ down into a direct sum of operators

$$L_{mj}^{(h)}(s) : \mathcal{H}(D_j(h)) \to \mathcal{H}(D_m(h)), \qquad (15.27)$$

for pairs i, j such that $\Lambda(\Gamma) \cap B^{-1}D_m(h) \cap D_j(h) \neq \emptyset$. For each pair of original Schottky disks D_k, D_l, $l \neq k+r$, so that $(B|D_k)^{-1}$ will pair each $D_m(h) \subset D_l$ with some $D_j(h) \subset D_k$. Thus the number of operators $L_{mj}^{(h)}(s)$ is bounded by $r^2 n(h) = O(h^\delta)$.

Now focus on some particular $L_{mj}^{(h)}(s)$. Let $q_0 \in \Lambda(\Gamma)$ be a point inside $D_j(h) \cup B^{-1}D_m(h)$ that is as close as possible to $\partial B^{-1}D_m(h)$. This implies that

$$d(Bq_0, \partial D_m(h)) = h,$$

since $I_m(h)$ was a component of $\Lambda(\Gamma) + [-h, h]$. By the expanding property of B with respect to d_κ,

$$d_\kappa(Bq_0, \partial D_m(h)) \geq \beta' d_\kappa(q_0, \partial B^{-1}D_m(h)).$$

Since d and d_κ are comparable, with constants independent of h, we obtain a uniform bound

$$d(q_0, \partial B^{-1}D_m(h)) \leq Ch.$$

Since $d(q_0, D_j(h)) \geq h$, this implies

$$d(B^{-1}D_m(h), \partial D_j(h)) \geq ch,$$

which is the control we need on the separation $\varepsilon(h)$. Arguing as in (15.15), for the standard basis $\{\phi_n\}$ for $\mathcal{H}(D_m(h))$, we have

$$\left\|\phi_n \circ S_j^{-1}\right\|_{\mathcal{H}(D_m(h))} \leq e^{-cn}.$$

This implies

$$\left\|L_{mj}^{(h)}(s)\phi_n\right\|_{\mathcal{H}(D_j)} \leq \sup_{z \in D_m(h)}\left|(S_j^{-1})'(z)^s\right|e^{-cn}.$$

Since the radius of $D_j(h)$ is $O(h)$ and $(S_j^{-1})'(z)^s$ is real on the real axis, and by the assumption that $h = 1/|s|$ with $\mathrm{Re}\,s$ bounded, we have

$$\sup_{z \in D_m(h)}\left|(S_j^{-1})'(z)^s\right| = O(1).$$

The min-max estimate gives

$$\mu_l(L_{ji}^{(h)}(s)) \leq e^{-cl}.$$

And since there are $O(h^{-\delta})$ operators $L_{ji}(s)$ making up $L^{(h)}(s)$, combining these estimates gives

$$\log\left|\det(1 - L^{(h)}(s))\right| = O(|s|^\delta). \qquad \square$$

Notes

In higher dimensions, not all convex cocompact groups of isometries of $\mathbb{H}^n$ are Schottky. For $n = 3$, Maskit [121] showed that a convex cocompact Kleinian group is Schottky if and only if the group is free.

The compilation [17] contains a set of expository articles providing additional background on symbolic dynamics and the geodesic flow on hyperbolic spaces. For the connection between the Selberg and dynamical zeta functions in the more complicated convex cocompact setting, see Pollicott [175] or Pollicott–Rocha [176].

The bound of Theorem 15.10 was proven by Guillopé–Lin–Zworski for convex cocompact Schottky groups in any dimension, and the method would extend to any group with an expanding Markov partition (see, e.g., Anderson–Rocha [5]).

Bounds on the number of resonances near the continuous spectrum are known in various settings; see, e.g., Sjöstrand–Zworski [198], Zerzeri [220], Stefanov [201].

A

Appendix

In the main body of the text, we have tried to include proofs of specialized results as much as possible, whereas background results of a more general nature were frequently cited without proof. In this appendix we will sketch in some of this background material and point the reader to sources where the full details can be found. We include proofs or at least sketches of proofs where possible. However, the goal here is not a complete exposition, but rather an outline that could serve as a guide to further study.

The topics covered here are all relatively standard, and the references cited generally represent a small fraction of those available. We give specific references for the benefit of inexperienced readers who may not know where to look, but these should not be taken as exclusive.

A.1 Entire functions

As a simple consequence of estimates on derivatives obtained from the Cauchy integral formula, one can argue that an entire function satisfying a polynomial bound $O(|z|^n)$ must in fact be a polynomial. Thus, the simplest growth restrictions that one considers for entire functions are exponential.

Definition. An entire function f has *order* λ if for any $\varepsilon > 0$,

$$|f(z)| = O(e^{|z|^{\lambda+\varepsilon}}),$$

for large $|z|$. If the bound still holds for $\varepsilon = 0$, then f is said in addition to be of *finite type*.

For functions of bounded order, the connection between growth rate and the distribution of zeros is furnished by Jensen's formula, which is derived as follows. Suppose f is analytic on a neighborhood of $\{|z| \leq r\}$, with no zeros on $|z| = r$. Define the function

$$F(z) := f(z) \prod_\zeta \frac{r^2 - \bar\zeta z}{r(z - \zeta)},$$

where $\{\zeta\}$ are the zeros of f inside $\{|z| < r\}$, repeated according to multiplicity. Because F has no zeros for $|z| \le r$, the function $\log |F(z)|$ is well-defined and harmonic. By the mean value property,

$$\log |F(0)| = \frac{1}{2\pi} \int_0^{2\pi} \log |F(re^{i\theta})| \, d\theta.$$

Substituting the definition of $F(z)$ into this formula then gives the following:

Theorem A.1 (Jensen's formula). *If f is analytic on a neighborhood of $\{|z| \le r\}$, with no zeros on $|z| = r$, and $f(0) \ne 0$, then*

$$\log |f(0)| + \sum_\zeta \log \frac{r}{|\zeta|} = \frac{1}{2\pi} \int_0^{2\pi} \log |f(re^{i\theta})| \, d\theta,$$

where $\{\zeta\}$ are the zeros of f inside $\{|z| < r\}$, repeated according to multiplicity.

Suppose that f is entire and $|f(z)| = O(\exp(|z|^\beta))$. Let

$$N(r) := \#\{\zeta : f(\zeta) = 0, |\zeta| \le r\}, \tag{A.1}$$

counted with multiplicities as always. By moving the origin if necessary, we can assume $f(0) \ne 0$, and then applying Jensen's formula at $|z| = 2r$ gives

$$\sum_{\substack{|\zeta| < 2r \\ f(\zeta) = 0}} \log \frac{2r}{|\zeta|} \le Cr^\beta,$$

for r sufficiently large. Restricting the sum over zeros to $|\zeta| \le r$, we deduce that

$$N(r) \log 2 \le Cr^\beta.$$

This proves the following:

Corollary A.2. *If f is an entire function of order λ, then for any $\varepsilon > 0$,*

$$N(r) = O(r^{\lambda + \varepsilon}).$$

If f is finite type, then we can take $\varepsilon = 0$.

The Weierstrass factorization theorem expresses an analytic function as a product over its zero set. The key to ensuring the convergence of this product is the notion of an *elementary factor*, defined for $p \in \mathbb{N}_0$ by

$$E_p(z) := (1 - z) \exp\left(z + \frac{z^2}{2} + \cdots + \frac{z^p}{p}\right).$$

The exponential term is added so that $1 - E_p(z)$ vanishes to higher order as $z \to 0$,

$$|1 - E_p(z)| \le |z|^{p+1}. \tag{A.2}$$

Suppose we are given a set of zeros $\{a_n\}$ such that $a_n \ne 0$, $|a_n| \to \infty$. Then if $p_n \in \mathbb{N}$ are chosen such that

$$\sum_{n=1}^{\infty} \left| \frac{r}{a_n} \right|^{p_n+1} < \infty, \tag{A.3}$$

for any $r > 0$ (for example, $p_n = n$ always works), the product

$$\prod_{n=1}^{\infty} E_{p_n}\left(\frac{z}{a_n}\right)$$

will converge absolutely and define an entire function with $\{a_n\}$ as zero set. Thus the Weierstrass factorization theorem: any entire function f can be written in the form

$$f(z) = z^m e^{g(z)} \prod_{n=1}^{\infty} E_{p_n}\left(\frac{z}{a_n}\right),$$

where $m \in \mathbb{N}_0$, $g(z)$ is entire, and a_n, p_n satisfy (A.3).

For functions of bounded order we can be more specific about the choice of p_n.

Theorem A.3 (Hadamard factorization). *If f is an entire function of order λ, then for $p = [\lambda]$, we can write f in the form*

$$f(z) = z^m e^{q(z)} \prod_{n=1}^{\infty} E_p\left(\frac{z}{a_n}\right), \tag{A.4}$$

where $q(z)$ is a polynomial of degree at most p. Conversely, if f is given by (A.4), assuming that $\sum_n |a_n|^{-p-1} < \infty$, then the order of f is at most $p + 1$.

Proof. Suppose f is entire of order λ, with $m \ge 0$ the multiplicity of the zero at $z = 0$ and $\{a_n\}$ the set of zeros $a_n \ne 0$, repeated according to multiplicity. Corollary A.2 implies that we can set all p_n equal to $p = [\lambda]$ in (A.3). Dividing (A.4) by the product $z^m \prod E_p(z/a_n)$ gives a nonzero entire function, which we can write as $e^{q(z)}$ for some q entire. To show that q is polynomial, we use the Poisson formula in place of the mean value property in the derivation of Jensen's formula, to obtain the Poisson–Jensen formula

$$\log |f(z)| + \sum_{|a_n|<r} \log \left| \frac{r^2 - \bar{a}_n z}{r(z - a_n)} \right| = \frac{1}{2\pi} \int_0^{2\pi} \mathrm{Re}\left(\frac{re^{i\theta} + z}{re^{i\theta} - z} \right) \log |f(re^{i\theta})| \, d\theta.$$

Differentiating $p + 1$ times and then using $N(r) = O(r^\lambda)$ to take the limit $r \to \infty$ gives

$$\frac{d^p}{dz^p}\frac{f'}{f}(z) = -p!\sum_{n=1}^{\infty}\frac{1}{(a_n - z)^{p+1}}.$$

If we apply this to (A.4), we see immediately that $q^{(p+1)} \equiv 0$.

Now for the converse. From the definition of $E_p(z)$, we see that $\log|E_p(z)| = O(|z|^p)$ for large z. On the other hand, (A.2) implies that $\log|E_p(z)| = O(|z|^{p+1})$ as $z \to 0$, so by continuity we can bound

$$\log|E_p(z)| \le C|z|^{p+1}, \tag{A.5}$$

for all $z \in \mathbb{C}$. This implies that

$$\sum_{n=1}^{\infty}\log\left|E_p\left(\frac{z}{a_n}\right)\right| \le C|z|^{p+1}\sum_{n=1}^{\infty}\frac{1}{|a_n|^{p+1}}, \tag{A.6}$$

which shows that

$$\prod_{n=1}^{\infty}E_p\left(\frac{z}{a_n}\right) \le e^{C|z|^{p+1}}.$$

Since $q(z)$ is assumed to have degree p, the right-hand side of (A.4) has order at most $p + 1$.   □

Most graduate-level complex analysis texts include the Hadamard factorization theorem; our development follows Conway [46, Section XI.3]. With a little more care, we could see that the exact order of the product $\prod E_p(z/a_n)$, and hence of f given by (A.4), is

$$\lambda = \limsup_{r\to\infty}\frac{\log N(r)}{\log r}.$$

See, e.g., Boas [21, Theorem 2.6.5].

For the upper bound argument in Section 9.3, we needed to know that a particular Weierstrass product with $p = 2$ had order 2 and finite type. The key result along these lines is Lindelöf's theorem [21, Theorem 2.10.1], which says that for $p \in \mathbb{N}$, an entire function f of order p is of finite type if and only if $N(r) = O(r^p)$ and we can bound

$$\left|\sum_{|a_n|\le R}\frac{1}{a_n^p}\right| \le C, \tag{A.7}$$

with C independent of R. For the sake of completeness, we'll present a proof of the portion of the theorem relevant to our application.

Theorem A.4 (Lindelöf). *Assume that the bound (A.7) holds for $\{a_k\}$ with $p \in \mathbb{N}$, and the corresponding zero-counting function satisfies $N(r) = O(r^p)$. Then*

$$\left|\prod_{n=1}^{\infty}E_p\left(\frac{z}{a_n}\right)\right| \le e^{C|z|^p}.$$

Proof. Let $P(z)$ denote the product in question, and assume that the zeros are ordered so that $|a_n|$ is increasing. Noting that $E_p(z) = E_{p-1}(z)e^{z^p/p}$, we can expand

$$\log|P(z)| = \sum_{k=1}^{n}\left[\log\left|E_{p-1}\left(\frac{z}{a_k}\right)\right| + \frac{1}{p}\mathrm{Re}\left(\frac{z}{a_k}\right)^p\right]$$

$$+ \sum_{k=n+1}^{\infty} \log\left|E_p\left(\frac{z}{a_k}\right)\right|, \tag{A.8}$$

where n will be chosen later. By the assumption (A.7), we have

$$\frac{1}{p}\mathrm{Re}\sum_{k=1}^{n}\left(\frac{z}{a_k}\right)^p \le C|z|^p,$$

with C independent of n.

Since the first sum in (A.8) involves only finitely many a_k's, we can apply the simple estimate $\log|E_{p-1}(z)| \le C|z|^{p-1}$, for z bounded away from 0, to deduce

$$\sum_{k=1}^{n}\log\left|E_{p-1}\left(\frac{z}{a_k}\right)\right| \le C\sum_{k=1}^{n}\left|\frac{z}{a_n}\right|^{p-1}.$$

For the final sum in (A.8), we apply the general estimate (A.5) to deduce

$$\sum_{k=n+1}^{\infty}\log\left|E_p\left(\frac{z}{a_k}\right)\right| \le C\sum_{k=n+1}^{\infty}\left|\frac{z}{a_n}\right|^{p+1}.$$

By the assumption $N(r) = O(r^p)$, $k < c|a_k|^p$ for some constant c. With this fact, the estimates above give

$$\log|P(z)| \le C\left[|z|^p + |z|^{p-1}\sum_{k=1}^{n}k^{-(p-1)/p} + |z|^{p+1}\sum_{k=n+1}^{\infty}k^{-(p+1)/p}\right].$$

If we now set $n = \left[|z|^p\right]$, then a simple integral estimate of the sums finishes the proof.   □

As a final result we present the Phragmén–Lindelöf theorem, in the version we have used in the text to extend polynomial estimates on an entire function into a vertical strip. (The theorem exists in various other forms, but all have essentially the same proof.)

Theorem A.5 (Phragmén–Lindelöf). *Suppose $f(z)$ is analytic on $a \le \mathrm{Re}\, z \le b$ and satisfies an exponential bound*

$$|f(z)| \le e^{C|z|^m}.$$

If f satisfies a polynomial bound

$$|f(z)| \le M\langle z\rangle^n$$

on the edges $\mathrm{Re}\, z = a$ and $\mathrm{Re}\, z = b$, then the polynomial bound holds on all of $a \le \mathrm{Re}\, z \le b$.

Proof. By rescaling if necessary, we can assume that $[a, b] \subset [-\pi/2, \pi/2]$. This means that for any $\varepsilon > 0$,

$$|f(z)e^{-\varepsilon \cos z}| \le M\langle z \rangle^n \tag{A.9}$$

for $\operatorname{Re} z = a$ or $\operatorname{Re} z = b$. For $|\operatorname{Im} z| = R$, we have

$$|f(z)e^{-\varepsilon \cos z}| \le \exp[CR^m - \varepsilon \cos(\operatorname{Re} z) \cosh R]. \tag{A.10}$$

For $\operatorname{Re} z \in [a, b]$, $\cos(\operatorname{Re} z)$ is bounded below by a positive constant. Thus the right-hand side of (A.10) approaches 0 as $R \to \infty$. For R sufficiently large, the bound (A.9) holds on the boundary of the rectangle $[a, b] \times [-R, R]$, and it extends to the interior by the maximum modulus principle. Since R was arbitrary, we have

$$|f(z)| \le M\langle z \rangle^n |e^{\varepsilon \cos z}|.$$

Because M is independent of ε, we can simply take $\varepsilon \to 0$ for each fixed value of z to obtain the result. □

A.2 Distributions and Fourier transforms

A *Schwartz function* on $\mathbb{R}$ is a smooth function that decreases rapidly along with all its derivatives. More precisely, φ is Schwartz if the norms

$$\rho_k(\varphi) := \sum_{0 \le l, m \le k} \sup_{x \in \mathbb{R}} |x^l \partial_x^m \varphi(x)|$$

are finite for any k. The space of Schwartz functions on $\mathbb{R}$ is denoted by $\mathcal{S}$ (or $\mathcal{S}(\mathbb{R})$ if we need to be more specific). The family of norms $\{\rho_k\}$ endows $\mathcal{S}$ with a Fréchet topology.

Definition. The space $\mathcal{S}'$ of *tempered distributions* is the topological dual of $\mathcal{S}$, i.e., the set of continuous linear functionals $\mathcal{S} \to \mathbb{C}$.

The pairing of $u \in \mathcal{S}'$ with $\varphi \in \mathcal{S}$ is denoted by $\langle u, \varphi \rangle$. The condition that the linear functional $\langle u, \cdot \rangle$ be continuous with respect to the Fréchet topology is quite simple: there exists k such that

$$|\langle u, \varphi \rangle| \le C\rho_k(\varphi), \quad \text{for all } \varphi \in \mathcal{S}.$$

For example, an integrable function u satisfying a polynomial bound $|u(x)| \le C\langle x \rangle^n$, where $\langle x \rangle := \sqrt{1 + x^2}$, defines a distribution by

$$\langle u, \varphi \rangle := \int_{-\infty}^{\infty} u(x)\varphi(x) \, dx. \tag{A.11}$$

The word "tempered" refers to the need for some control on the growth of u at infinity. (The full space of distributions, usually denoted by $\mathcal{D}'$, is the dual of $C_0^\infty(\mathbb{R})$.) See, e.g., [70] or [66] for introductory accounts of the theory.

The prototypical distribution is the Dirac delta function $\delta(x)$, defined by

$$\int_{-\infty}^{\infty} \delta(x)\varphi(x)\,dx := \varphi(0).$$

A crucial feature of the theory of distributions is that any continuous operation on $\mathcal{S}$ extends immediately to $\mathcal{S}'$. For example, to define the derivative of a distribution, we simply take a formal integration by parts as our definition:

$$\langle u', \varphi \rangle := -\langle u, \varphi' \rangle.$$

The convolution of a tempered distribution with a Schwartz function is defined similarly: for $u \in \mathcal{S}'$ and $\varphi, \psi \in \mathcal{S}$,

$$\langle u * \psi, \varphi \rangle := \langle u, \psi^- * \varphi \rangle,$$

where $\psi^-(x) := \psi(-x)$.

The main reason for specializing to distributions that are tempered is that the Fourier transform $\mathcal{F} : \varphi \mapsto \widehat{\varphi}$ is a continuous bijection $\mathcal{S} \to \mathcal{S}$. Our convention for the Fourier transform is

$$\widehat{\varphi}(\xi) := \int_{-\infty}^{\infty} e^{-ix\xi}\varphi(x)\,dx,$$

so that $\mathcal{F}^{-1}$ is given by

$$\varphi(x) = \frac{1}{2\pi} \int_{-\infty}^{\infty} e^{ix\xi}\widehat{\varphi}(\xi)\,d\xi.$$

The Fourier transform of a distribution is defined by

$$\langle \widehat{u}, \varphi \rangle := \langle u, \widehat{\varphi} \rangle.$$

For example,

$$\langle \widehat{\delta}, \varphi \rangle = \widehat{\varphi}(0) = \int_{-\infty}^{\infty} \varphi(x)\,dx,$$

so $\widehat{\delta}(\xi) = 1$.

For $u \in L^2(\mathbb{R})$, which defines a distribution by (A.11), the distributional Fourier transform agrees with the usual definition. The extensions to distributions also preserves the important properties of the Fourier transform with respect to differentiation,

$$\widehat{(u')}(\xi) = -i\xi\,\widehat{u}(\xi),$$

and convolution,

$$\widehat{u * \psi} = \widehat{u}\,\widehat{\psi}.$$

Theorem A.6 (Poisson summation). *As a relation between tempered distributions on* $\mathbb{R}$,

$$\sum_{k \in \mathbb{Z}} \delta(x - k) = \sum_{k \in \mathbb{Z}} e^{2\pi k x}.$$

This is equivalent to the statement that for $\varphi \in \mathcal{S}$,

$$\sum_{k \in \mathbb{Z}} \varphi(k) = \sum_{k \in \mathbb{Z}} \widehat{\varphi}(2\pi k).$$

Proof. Define the periodic function

$$f(x) = \sum_{k \in \mathbb{Z}} \varphi(x + k).$$

The discrete Fourier series for $f(x)$ is

$$f(x) = \sum_{m \in \mathbb{Z}} a_m e^{2\pi i m x}, \tag{A.12}$$

where

$$a_m = \int_0^1 e^{-2\pi i m x} f(x) \, dx.$$

Substituting the definition of $f(x)$ gives

$$\begin{aligned}
a_m &= \int_0^1 \sum_{k \in \mathbb{Z}} \varphi(x + k) e^{-2\pi i m x} \, dx \\
&= \int_{-\infty}^{\infty} \varphi(x) e^{-2\pi i m x} \, dx \\
&= \widehat{\varphi}(2\pi m).
\end{aligned}$$

The result then follows from (A.12) with $x = 0$. $\square$

We finish this section with two examples of Fourier-transform calculations that are needed in Section 12.3. The first concerns the principal value distribution $\mathrm{PV}\,[1/x]$, defined by

$$\langle \mathrm{PV}\,[1/x], \varphi \rangle := \lim_{\varepsilon \to 0} \int_{|x| \geq \varepsilon} \frac{\varphi(x)}{x} \, dx. \tag{A.13}$$

Lemma A.7. *For the distribution defined by integration against* $\mathrm{sgn}(x)$ *as in (A.11), we have*

$$\widehat{\mathrm{sgn}}(\xi) = -2i\,\mathrm{PV}\,[1/\xi].$$

Proof. For $a > 0$, let $u_a(x) = e^{-a|x|}\,\mathrm{sgn}(x)$. It is a simple exercise to check that $u_a \to \mathrm{sgn}$ as $a \to 0$, in the topology of $\mathcal{S}'$. The Fourier transform of u_a exists in the ordinary sense,

$$\widehat{u_a}(\xi) = \int_0^\infty e^{-ix\xi - ax} \, dx - \int_{-\infty}^0 e^{-ix\xi + ax} \, dx = \frac{-2i\xi}{\xi^2 + a^2}.$$

Therefore,

$$\langle \widehat{u}, \varphi \rangle = \lim_{a \to 0} \int_{-\infty}^\infty \frac{-2i\xi}{\xi^2 + a^2} \varphi(\xi) \, d\xi.$$

By subtracting

$$\lim_{a \to 0} \int_{|\xi| \leq 1} \frac{-2i\xi}{\xi^2 + a^2} \varphi(0) \, d\xi = 0,$$

we can rewrite this as

$$\langle \widehat{u}, \varphi \rangle = -2i \left[\int_{|\xi| \leq 1} \frac{\varphi(\xi) - \varphi(0)}{\xi} \, d\xi + \int_{|\xi| \geq 1} \frac{\varphi(\xi)}{\xi} \, d\xi \right].$$

It's then easy to see that this agrees with the definition of PV $[1/\xi]$ in (A.13).   □

Our second example is a relation between $1/|x|$ and $\log |x|$. Since $\log |x|$ is integrable near the singularity at $x = 0$ and slowly growing at infinity, it defines a tempered distribution. Motivated by the principal value, we can define a regularization FP $[1/|x|]$ by taking the finite part,

$$\langle \text{FP}\,[1/|x|], \varphi \rangle := \text{FP}_{\varepsilon \to 0} \int_{|x| \geq \varepsilon} \frac{\varphi(x)}{x} \, dx$$

$$= \int_{-1}^1 \frac{\varphi(x) - \varphi(0)}{|x|} \, dx + \int_{|x| \geq 1} \frac{\varphi(x)}{|x|} \, dx.$$

The function $\text{sgn}\,x \log |x|$ is also a tempered distribution, and it's easy to check that

$$(\text{sgn}\,x \log |x|)' = \text{FP}\,[1/|x|],$$

in a distributional sense.

Lemma A.8. *We have*

$$\mathcal{F}(\log |x|) = -\pi\, \text{FP}\,[1/|\xi|] - 2\pi\, \gamma\, \delta(\xi),$$

where γ is Euler's constant.

Proof. Taking the inverse Fourier transform of both sides, it suffices to prove that

$$\mathcal{F}\,(\text{FP}\,[1/|\xi|]) = -2 \log |x| - 2\gamma.$$

This calculation is drawn from Kanwal [106]. By definition,

$$\langle \mathcal{F}\,(\text{FP}\,[1/|\xi|]), \varphi \rangle = \int_{-1}^1 \frac{\widehat{\varphi}(\xi) - \widehat{\varphi}(0)}{|\xi|} \, d\xi + \int_{|\xi| \geq 1} \frac{\widehat{\varphi}(\xi)}{\xi} \, d\xi$$

$$= \int_{-1}^1 \int_{-\infty}^\infty \frac{\varphi(x)(e^{-ix\xi} - 1)}{|\xi|} \, dx \, d\xi$$

$$+ \int_{|\xi| \geq 1} \int_{-\infty}^\infty \frac{\varphi(x)e^{-ix\xi}}{\xi} \, dx \, d\xi.$$

By switching the order of integration, and substituting $t = x\xi$ in place of ξ, we can see that

$$\mathcal{F}\left(\text{FP}\left[1/|\xi|\right]\right)(x) = 2 \int_0^{|x|} \frac{\cos t - 1}{|t|}\, dt + 2 \int_{|x|}^\infty \frac{\cos t}{|t|}\, dt.$$

The computation reduces to a standard identity for the cosine-integral function: for $x > 0$, namely

$$\text{Ci}(x) := -\int_x^\infty \frac{\cos x}{x}\, dx = \gamma + \log x + \int_0^x \frac{\cos t - 1}{t}\, dt. \qquad \square$$

A.3 Spectral theory

The spectral theorem for self-adjoint compact operators on a Hilbert space is a close analogue of the finite-dimensional diagonalization theorem. A particularly nice method of proof (adapted from Reed–Simon [180, Section VI.5]) is provided by the analytic Fredholm theorem.

Theorem A.9 (Compact self-adjoint spectral theorem). *If A is compact and self-adjoint on a Hilbert space $\mathcal{H}$, then there exists an orthonormal basis $\{\phi_j\}$ for $\mathcal{H}$ such that $A\phi_j = \lambda_j \phi_j$ where the eigenvalues $\lambda_j \in \mathbb{R}$ accumulate only at 0.*

Proof. On the domain $\mathbb{C} - \{0\}$ we can apply the analytic Fredholm theorem (Theorem 6.1) to the operator $I - z^{-1}A$. Since A is bounded, we can guarantee that the inverse exists for $|z| > 1/\|A\|$; hence $(I - z^{-1}A)^{-1}$ is a finitely meromorphic family on $z \neq 0$. From the proof of that theorem, we can see that poles occur precisely where $(I - z^{-1}A)u = 0$ has a nontrivial solution, i.e., at nonzero eigenvalues of A. Because the family is finitely meromorphic, the eigenvalues are discrete away from 0, and the corresponding eigenspaces are finite-dimensional. We can choose an orthonormal basis for each eigenspace, and since self-adjointness implies that eigenspaces for different eigenvalues are orthogonal, these eigenvectors form an orthonormal set $\{\phi_n\}$. If $U \subset \mathcal{H}$ denotes the span of the ϕ_n, then $A|_{U^\perp}$ is still compact. But since any nonzero eigenvalues of $A|_{U^\perp}$ would have been eigenvalues of A, we must have $A|_{U^\perp} = 0$. Supplementing $\{\phi_n\}$ with an orthonormal basis of $U^\perp$ completes the proof. $\qquad \square$

To understand how this relates to the general spectral theorem, we can interpret the diagonalization in the following way. Suppose that the eigenvalues of a compact self-adjoint operator A are all nonzero and have multiplicity one. Then we can define a measure supported on the spectrum,

$$\mu = \sum_{j=1}^\infty \nu_{\lambda_j},$$

where ν_λ denotes a unit point measure at λ. One can then check that the map $W : L^2(\mathbb{R}, d\mu) \to \mathcal{H}$ given by

$$W(f) := \sum_{j=1}^{\infty} f(\lambda_j)\phi_j$$

is unitary. Moreover, W conjugates A to a multiplication operator:

$$W^{-1}AWf(x) = xf(x).$$

We could accommodate multiplicities in the spectrum by taking measures on multiple copies of $\mathbb{R}$. This is the form that the general spectral theorem takes.

An unbounded operator A defined on some dense domain $\mathcal{D}(A) \subset \mathcal{H}$ is *self-adjoint* if $A = A^*$, with the same domain. Giving the precise domain can be difficult in practice, but in some cases we can fully specify the operator using a smaller, more convenient domain. The operator A is *essentially self-adjoint* on $\mathcal{D}(A)$ if $\mathcal{D}(A) \subset \mathcal{D}(A^*)$, $A = A^*$ on $\mathcal{D}(A)$, and the closure of A is self-adjoint. This implies that the extension of A from the domain $\mathcal{D}(A)$ is uniquely determined. If X, g is a complete Riemannian manifold, then Δ is essentially self-adjoint on $C_0^{\infty}(X) \subset L^2(X, dg)$. If X, g is a region with smooth boundary inside a complete manifold, then a self-adjoint extension of Δ requires boundary conditions. For example, for Dirichlet boundary conditions we take the domain $\{f \in C_0^{\infty}(X) : f|_{\partial X} = 0\}$, on which Δ is essentially self-adjoint. See, e.g., [208, Section 8.2] for details.

In these cases the Friedrichs method can be used to actually produce the self-adjoint extension, for which the domain is a Sobolev space (see, e.g., [207, Section A.8]). But for our purposes it is enough to know that the choice of self-adjoint extension is unambiguous. (Note that if X had conical singularities, i.e., if we had allowed quotients $\Gamma \backslash \mathbb{H}$ by a group with elliptic elements, then the choice of self-adjoint extension would be a serious issue.)

The self-adjointness of A implies that the resolvent $(A - z)^{-1}$ exists as a bounded operator on $\mathcal{H}$ for $z \notin \mathbb{R}$.

Definition. The *spectrum* of A is the set

$$\sigma(A) := \{\lambda \in \mathbb{C} : A - \lambda \text{ has no bounded inverse } (A - \lambda)^{-1}\}.$$

Theorem A.10 (Spectral theorem). *For A a self-adjoint operator on a separable Hilbert space $\mathcal{H}$, there exists a measure space Ω, μ, where Ω is a union of copies of $\mathbb{R}$, a unitary map $W : L^2(\Omega, d\mu) \to \mathcal{H}$, and a real-valued measurable function a on Ω such that*

$$W^{-1}AWf(x) = a(x)f(x),$$

for $f \in L^2(X, dg)$ such that $Wf \in \mathcal{D}(A)$ (which is equivalent to the condition $af \in L^2(\Omega, d\mu)$).

Proof. We will summarize the very clear treatment by Taylor [208, Section 8.1] First, assume that A is a bounded self-adjoint operator. Then we can define a group of unitary operators $U(t) = e^{itA}$, for $t \in \mathbb{R}$, through the power series

$$U(t) := \sum_{k=0}^{\infty} \frac{(itA)^k}{k!}.$$

Because $\mathcal{H}$ is separable, we can decompose $\mathcal{H}$ into a direct sum of cyclic subspaces of the form $H_v = \{U(t)v : t \in \mathbb{R}\}$ for $v \in \mathcal{H}$. To see this, start with a basis $\{w_j\}$, let $H_1 = H_{w_1}$, find the first $w_j \notin H_1$ (if any), let v_2 be the projection of this w_j into $H_1^\perp$, set $H_2 = H_{v_2}$, and iterate.

Now focus on a single cyclic subspace H_v. The function

$$\zeta(t) = \langle v, e^{itA}v \rangle$$

is bounded and hence defines a tempered distribution. Let $\mu = \widehat{\zeta} \in S'(\mathbb{R})$. Define the map $W : S(\mathbb{R}) \to H_v$ by

$$Wf := \frac{1}{2\pi} \int_{-\infty}^{\infty} \widehat{f}(t)e^{itA}v \, dt$$

for $f \in S(\mathbb{R})$. Then for $f, g \in S(\mathbb{R})$ we compute that

$$\langle Wf, Wg \rangle = \frac{1}{(2\pi)^2} \left\langle \int \overline{\widehat{f}(t)}e^{-itA}v \, dt, \int \widehat{g}(t')e^{it'A}v \, dt' \right\rangle$$

$$= \frac{1}{(2\pi)^2} \iint \overline{\widehat{f}(t)} \, \widehat{g}(t') \, \langle v, e^{i(t'-t)A}v \rangle \, dt \, dt'$$

$$= \frac{1}{2\pi} \iint \overline{\widehat{f}(t)} \, \widehat{g}(t') \, \widehat{\mu}(t) \, dt \, dt'$$

$$= \frac{1}{2\pi} \left(\overline{\widehat{f}(t)}, \widehat{g} * \widehat{\mu} \right)$$

$$= (\overline{f}, g\mu)$$

$$= (\overline{f}g, \mu).$$

From this it's easy to deduce that μ is positive, and hence defines a measure on $\mathbb{R}$. Moreover, because $\langle Wf, Wg \rangle$ agrees with the inner product on $L^2(\mathbb{R}, d\mu)$, W extends to an isometry $L^2(\mathbb{R}, d\mu) \to H_v$. The range of W is dense in H_v because v is a cyclic vector; hence W is unitary. To see that A is conjugated to a multiplication operator, we note that for $f \in S(\mathbb{R})$,

$$U(t)Wf = \frac{1}{2\pi} \int_{-\infty}^{\infty} \widehat{f}(t')e^{i(t+t')A}v \, dt'$$

$$= \frac{1}{2\pi} \int_{-\infty}^{\infty} \widehat{f}(t' - t)e^{it'A}v \, dt'$$

$$= W(e^{itx}f(x)).$$

Hence, by differentiating,

$$W^{-1}AWf = xf(x).$$

To complete the proof in the bounded case, we apply this argument to each cyclic subspace and define Ω as a disjoint union of the measure spaces $(\mathbb{R}, \mu)$.

The unbounded case can be reduced to the bounded case by a simple trick due to von Neumann. The key observation is that self-adjointness of A implies (indeed, is equivalent to) the unitarity of the operator

$$B := (A - i)(A + i)^{-1}$$

on $\mathcal{H}$. We can then split $B = B_1 + i B_2$, where B_1, B_2 are commuting bounded self-adjoint operators (setting $B_1 := \frac{1}{2}(B + B^*)$ and $B_2 = \frac{1}{2i}(B - B^*)$). Then we form the two-parameter unitary group

$$U(t_1, t_2) := e^{it_1 B_1 + it_2 B_2},$$

and proceed just as above except that for each cyclic subspace we construct μ as a measure on $\mathbb{R}^2$ using the two-dimensional Fourier transform. $\qquad\square$

The spectral theorem gives us a *functional calculus* for operators: given a Borel measurable function $h : \mathbb{R} \to \mathbb{C}$, we can define

$$h(A) := W h(a(x)) W^{-1}. \qquad (A.14)$$

It turns out we can realize this functional calculus explicitly in terms of the resolvent.

Corollary A.11 (Resolvent functional calculus). *If A is a self-adjoint operator on $\mathcal{H}$, then for $h : \mathbb{R} \to \mathbb{C}$ bounded and continuous,*

$$h(A) = \lim_{\varepsilon \to 0} \frac{1}{2\pi i} \int_{-\infty}^{\infty} h(z) \left[(A - z - i\varepsilon)^{-1} - (A - z + i\varepsilon)^{-1} \right] dz,$$

with the limit taken in the operator topology.

Proof. Using (A.14), the result follows from the pointwise limit

$$\lim_{\varepsilon \to 0} \frac{1}{\pi} \int_{-\infty}^{\infty} \frac{\varepsilon h(z)}{(a(x) - z)^2 + \varepsilon^2} \, dz = h(a(x)). \qquad\square$$

Given an interval $I \subset \mathbb{R}$ (or more generally a Borel subset), the associated *spectral projector* is defined by

$$P_I = \chi_I(A),$$

where χ_I denotes the characteristic function.

Corollary A.12 (Stone's formula). *The spectral projectors are expressed in terms of the resolvent by*

$$\tfrac{1}{2}(P_{[\alpha,\beta]} + P_{(\alpha,\beta)}) = \lim_{\varepsilon \to 0} \frac{1}{2\pi i} \int_{\alpha}^{\beta} \left[(A - z - i\varepsilon)^{-1} - (A - z + i\varepsilon)^{-1} \right] dz.$$

This follows, as in the proof of Corollary A.11, by noting that

$$\lim_{\varepsilon \to 0} \frac{1}{\pi} \int_{\alpha}^{\beta} \frac{\varepsilon}{(z - a)^2 + \varepsilon^2} \, dz = \begin{cases} 0, & a \notin [\alpha, \beta], \\ 1/2, & a = \alpha \text{ or } \beta, \\ 1, & a \in (\alpha, \beta). \end{cases}$$

Definition. The *essential spectrum* $\sigma_{\text{ess}}(A)$ is defined to be the set of points λ for which $P_{(\lambda-\varepsilon,\lambda+\varepsilon)}$ has infinite rank for all $\varepsilon > 0$. The complement of the essential spectrum is the *discrete spectrum*, $\sigma_{\text{d}}(A)$.

The term "essential" comes from Weyl's result on the invariance of $\sigma_{\text{ess}}(A)$ under perturbation of A by compact operators. The discrete spectrum could also be defined as the set of isolated eigenvalues of A with finite multiplicity, where isolated means an isolated point of $\sigma(A)$ (not just isolated from other eigenvalues). An eigenvalue of A could be contained in $\sigma_{\text{ess}}(A)$, in which case it's called an *embedded eigenvalue*.

Theorem A.13 (Weyl criterion). *For A self-adjoint, the essential spectrum $\sigma_{\text{ess}}(A)$ is the set of $\lambda \in \mathbb{C}$ for which there exists an orthonormal sequence of vectors $\phi_n \in \mathcal{H}$ such that*

$$\|(A - \lambda)\phi_n\| \to 0. \tag{A.15}$$

Proof. If $\lambda \in \sigma_{\text{ess}}(A)$, then for each n we can choose a unit vector ϕ_n in the range of $P_{(\lambda-1/n,\lambda+1/n)}$ such that the ϕ_n's are all orthogonal, because this range is infinite-dimensional for each n. For this sequence,

$$\|(A - \lambda)\phi_n\| < \frac{1}{n}.$$

On the other hand, suppose $\lambda \notin \sigma_{\text{ess}}(A)$ and $\{\phi_n\}$ is an orthonormal sequence satisfying (A.15). For some $\varepsilon > 0$, $P_{(\lambda-\varepsilon,\lambda+\varepsilon)}$ must have finite rank, implying that

$$\lim_{n\to\infty} P_{(\lambda-\varepsilon,\lambda+\varepsilon)}\phi_n = 0. \tag{A.16}$$

But using (A.15) we also have

$$
\begin{aligned}
\|(1 - P_{(\lambda-\varepsilon,\lambda+\varepsilon)})\phi_n\| &\leq \varepsilon^{-1}\|(A - \lambda)(1 - P_{(\lambda-\varepsilon,\lambda+\varepsilon)})\phi_n\| \\
&\leq \varepsilon^{-1}\|(A - \lambda)\phi_n\| + \varepsilon^{-1}\|(A - \lambda)P_{(\lambda-\varepsilon,\lambda+\varepsilon)}\phi_n\| \\
&\leq \varepsilon^{-1}\|(A - \lambda)\phi_n\| + \|P_{(\lambda-\varepsilon,\lambda+\varepsilon)}\phi_n\| \\
&\to 0.
\end{aligned}
$$

With (A.16) this would imply $\phi_n \to 0$, contradicting the fact that $\|\phi_n\| = 1$. $\square$

The association $E \mapsto P_E := \chi_E(A)$, for a Borel subset $E \subset \mathbb{R}$, defines a projection-valued measure associated to A, and another way to decompose the spectrum is in terms of this measure.

Definition. A point $\lambda \in \sigma(A)$ is in the *continuous spectrum* $\sigma_{\text{cont}}(A)$ if $P_{\{\lambda\}} = 0$, and in the *point spectrum* otherwise. The continuous spectrum is called *absolutely continuous* if $P_E = 0$ for any subset $E \subset \sigma_{\text{cont}}(A)$ with zero Lebesgue measure.

Thus embedded eigenvalues are in the point spectrum but not the discrete spectrum. In general, one could also have a singular continuous spectrum, where the projection-valued measure is singular with respect to Lebesgue measure. Such a spectrum is problematic from the point of view of scattering theory, and in physical problems one generally hopes to rule it out (see, e.g., [181, Section XIII]).

To complete this section, we review some facts from the spectral theory of the Laplacian on a compact Riemannian manifold M, g. Let Δ denote the positive Laplacian defined by g. This is essentially self-adjoint on $C^\infty(M) \subset L^2(M, dg)$. The exact domain is the Sobolev space

$$H^2(M, dg) := \{u \in L^2(M, dg) : \Delta u \in L^2(M, dg)\}.$$

The inclusion $H^2(M, dg) \to L^2(M, dg)$ is compact (see, e.g., [207, Proposition 4.3.4]), implying that $(\Delta + 1)^{-1}$ is compact and self-adjoint as an operator on $L^2(M, dg)$. By Theorem A.9, there is an orthonormal basis $\{\varphi_j\}$ for $L^2(M, dg)$ consisting of eigenvectors of $(\Delta + 1)^{-1}$. Observing that these vectors are also eigenvectors of Δ gives the following:

Theorem A.14. *If M, g is a compact Riemannian surface, then Δ has discrete spectrum*

$$\sigma(\Delta) = \sigma_d(\Delta) = \{0 = \lambda_0 < \lambda_1 \leq \lambda_2 \leq \cdots \to \infty\},$$

with corresponding eigenfunctions $\{\varphi_j\}$ forming an orthonormal basis for $L^2(M, dg)$.

An alternative proof that avoids the use of Sobolev spaces involves applying the spectral theorem to the heat operator (see [37, Section 7.2]).

The asymptotic behavior of the sequence of eigenvalues λ_j is one of classical results of spectral theory.

Theorem A.15 (Weyl asymptotics). *The counting function for the spectrum of the Laplacian on a compact manifold of dimension n has the asymptotic behavior*

$$\#\{\lambda_j \leq x\} \sim \frac{\mathrm{Vol}(M)}{\Gamma(n/2 + 1)(4\pi)^{n/2}} x^{n/2}. \tag{A.17}$$

This is equivalent to

$$\lambda_k \sim \frac{\mathrm{Vol}(M)}{\Gamma(n/2 + 1)(4\pi)^{n/2}} k^{2/n}. \tag{A.18}$$

Proof (sketch). The proof starts from the Minakshisundaram–Pleijel expansion of the heat kernel [138]. The heat kernel $H(t; x, y)$ is the integral kernel of the heat operator $e^{-t\Delta}$. For $t > 0$ the kernel is smooth and admits an expansion as $t \to 0$,

$$H(t; x, y) \sim (4\pi t)^{-n/2} \sum_{j=0}^{\infty} t^j a_j(x, y),$$

where $a_0(x, x) = 1$. See [40, Section VI.3] or [208, Section 7.13] for derivations of the heat expansion in the general case.

On $\mathbb{H}$ we have the explicit formula for the heat kernel

$$H_\mathbb{H}(t; z, w) = \frac{\sqrt{2} e^{-t/4}}{(4\pi t)^{3/2}} \int_{d(z,w)}^{\infty} \frac{r e^{-r^2/4t}}{\sqrt{\cosh r - \cosh d(z, w)}} \, dr. \tag{A.19}$$

The derivation is nontrivial; see, e.g., [40, Section X.2] or [37, Section 7.4]. The heat expansion for a hyperbolic surface could be derived explicitly by averaging this formula over the group action.

Using the heat expansion in the trace of the heat operator gives

$$\sum_{k=1}^{\infty} e^{-\lambda_k t} \sim \frac{\text{Vol}(M)}{(4\pi t)^{n/2}},$$

as $t \to 0$. By Karamata's Tauberian theorem (see, e.g., [207, Proposition 8.3.2]), this implies (A.17). The equivalence of (A.17) and (A.18) is a reasonably straightforward exercise. See [192, Section II.13.4] for a complete proof.   □

A.4 Singular values, traces, and determinants

Let A be a compact operator acting on a Hilbert space $\mathcal{H}$. If A is self-adjoint, then by Theorem A.9, A has complete set of eigenvectors with eigenvalues accumulating only at zero. For non-self-adjoint A, we can apply this result to the absolute value $|A| := \sqrt{A^* A}$.

Definition. The *singular values* of A are the nonzero eigenvalues of $|A|$, arranged in decreasing order:

$$\mu_1(A) \geq \mu_2(A) \geq \cdots \to 0.$$

(These are sometimes called characteristic values, but the term characteristic value is also frequently used as a synonym for eigenvalue.) The operators $A^* A$ and $A A^*$ have the same nonzero eigenvalues, so that $\mu_j(A) = \mu_j(A^*)$ for all j. Since $\mu_1(A)$ is the largest eigenvalue of $|A|$, we have

$$\mu_1(A) = \|A\|$$

in particular.

We will review some standard facts in the theory of singular values. Our basic references for this material are Gohberg–Krein [72] and Simon [193]. The following characterization of singular values proves extremely useful:

Theorem A.16 (Min-max for singular values). *Let V denote an arbitrary finite-dimensional subspace of $\mathcal{H}$. Then*

$$\mu_n(A) = \min_{\substack{V \subset \mathcal{H} \\ \dim V = n-1}} \left\{ \max_{\psi \in V^\perp} \frac{\|A\psi\|}{\|\psi\|} \right\}.$$

Proof. This follows from the existence of an orthonormal basis of eigenvectors $\{\phi_n\}$ for $|A|$. An easy estimate with the triangle inequality shows that the optimal choice is $V = \text{span}(\phi_1, \ldots, \phi_{n-1})$ and $\psi = \phi_n$.   □

An immediate application is the following:

Corollary A.17. *If B is a bounded operator on $\mathcal{H}$, then*

$$\mu_j(BA) \leq \|B\| \, \mu_j(A).$$

The basic properties of singular values with respect to sums and products are expressed in the following:

Theorem A.18 (Fan inequalities). *If A and B are both compact, then*

$$\mu_{i+j-1}(A + B) \leq \mu_i(A) + \mu_j(B) \tag{A.20}$$

and

$$\mu_{i+j-1}(AB) \leq \mu_i(A)\mu_j(B). \tag{A.21}$$

Proof. By Theorem A.16, we can find an operator A_{i-1} of rank $i - 1$ such that

$$\mu_i(A) = \|A - A_{i-1}\|,$$

and an operator B_{j-1} of rank $j - 1$ such that

$$\mu_j(B) = \|B - B_{j-1}\|.$$

Since $A_{i-1} + B_{j-1}$ has rank at most $i + j - 2$,

$$\mu_{i+j-1}(A + B) \leq \|A + B - (A_{i-1} + B_{j-1})\|,$$

again by Theorem A.16. So then

$$\mu_{i+j-1}(A + B) \leq \|A - A_{i-1}\| + \|B - B_{j-1}\| = \mu_i(A) + \mu_j(B),$$

which completes the proof of (A.20).

The same method works for (A.21). Note that

$$(A - A_{i-1})(B - B_{j-1}) = AB - AB_{j-1} - A_{i-1}(B - B_{j-1}),$$

and $AB_{j-1} + A_{i-1}(B - B_{j-1})$ has rank at most $i + j - 2$, so that

$$\mu_{i+j-1}(AB) \leq \|(A - A_{i-1})(B - B_{j-1})\|. \qquad \square$$

Another basic result is an estimate relating eigenvalues to singular values.

Theorem A.19 (Weyl's inequality). *If A is compact, with eigenvalues $\lambda_j(A)$ repeated according to algebraic multiplicity and ordered so $|\lambda_j(A)|$ decreases, then*

$$\prod_{j=1}^{n} |\lambda_j(A)| \leq \prod_{j=1}^{n} \mu_j(A),$$

for any n.

Proof. Let $\wedge^n \mathcal{H}$ denote the n-fold antisymmetric tensor product of $\mathcal{H}$. Let

$$\wedge^n A := A \wedge \cdots \wedge A,$$

acting on $\wedge^n \mathcal{H}$, and note that $| \wedge^n A| = \wedge^n |A|$. If $\{\phi_j\}$ is a orthonormal basis of eigenvectors of $|A|$, with $|A|\phi_j = \mu_j(A)\phi_j$, then $\{\phi_{i_1} \wedge \cdots \wedge \phi_{i_n}\}$ forms a basis of $\wedge^n \mathcal{H}$, and

$$\wedge^n |A|(\phi_{i_1} \wedge \cdots \wedge \phi_{i_n}) = \mu_{i_1} \cdots \mu_{i_n}(\phi_{i_1} \wedge \cdots \wedge \phi_{i_n}).$$

In particular,

$$\| \wedge^n A\| = \prod_{j=1}^{n} \mu_j(A). \tag{A.22}$$

Let E_n denote the span of the first n eigenvectors of A. Applying Jordan normal form to $A|_{E_n}$ gives an orthonormal basis $\{\psi_j\}_{j=1}^{n}$ for E_n in which

$$A|_{E_n} = \begin{pmatrix} \lambda_1(A) & a_1 & \cdots & 0 \\ 0 & \lambda_2(A) & \cdots & 0 \\ \vdots & \vdots & \ddots & a_{n-1} \\ 0 & 0 & \cdots & \lambda_n(A) \end{pmatrix},$$

where $a_j = 0$ or 1 (to account for the Jordan blocks). We easily see that

$$\wedge^n A(\psi_1 \wedge \cdots \wedge \psi_n) = \lambda_1(A) \cdots \lambda_n(A)(\psi_1 \wedge \cdots \wedge \psi_n), \tag{A.23}$$

i.e., $\prod_{j=1}^{n} \lambda_j(A)$ is an eigenvalue of $\wedge^n A$. Since any eigenvalue is bounded by the operator norm, the Weyl inequality now follows from (A.22). $\qquad\square$

Definition. For a compact operator A on a Hilbert space $\mathcal{H}$, the *trace norm* is

$$\|A\|_1 := \sum_{j=1}^{\infty} \mu_j(A).$$

If $\|A\|_1 < \infty$ then A is said to be *trace class*.

For a trace class operator A, one can derive from the log of the Weyl inequality that

$$\sum_{j=1}^{\infty} |\lambda_j(A)| \le \|A\|_1. \tag{A.24}$$

This suggests a natural definition

$$\operatorname{tr} A := \sum_{j=1}^{\infty} \lambda_j(A). \tag{A.25}$$

However, another natural definition would be to take an orthonormal basis $\{\phi_j\}$ for $\mathcal{H}$, and set

$$\operatorname{tr} A := \sum_{j=1}^{\infty} \langle \phi_j, A\phi_j \rangle. \tag{A.26}$$

The equivalence of the definitions (A.25) and (A.26) is Lidskii's theorem, and actually quite challenging to prove (see [181, Section XIII.17] or [193, Chapter 3] for details).

The linearity of the trace is obvious from (A.26), but not so evident in (A.25). The trace is also cyclic, in the sense that if A is trace class and B bounded, then

$$\operatorname{tr} AB = \operatorname{tr} BA.$$

(This follows trivially from (A.26) when B is unitary, and any bounded operator can be written as a linear combination of four unitary operators. See [180, Section VI.6] for details.)

Theorem A.20. *Suppose X, g is a Riemannian manifold, and A is a trace class operator on $L^2(X, dg)$ given by a continuous, compactly supported integral kernel $A(x, y)$. Then*

$$\operatorname{tr} A = \int_X A(x, x) \, dg(x).$$

To prove this one introduces a basis of step functions such that definition (A.26) becomes a limit of Riemann sums. See [193, Theorem 3.9]. Note that one needs to know already that A is trace class to apply the theorem. This will always be the case if the kernel is smooth and compactly supported, but continuity alone is not sufficient.

Definition. If $A^* A$ is trace class, then A is said to be *Hilbert–Schmidt*. This corresponds to finiteness of the norm

$$\|A\|_2 := \left[\sum_{j=1}^{\infty} \mu_j(A)^2 \right]^{1/2} = \sqrt{\operatorname{tr} A^* A}.$$

The concept of Hilbert–Schmidt operators actually predates trace class. They are easier to handle in many ways; for example, one can actually characterize the integral kernel precisely in this case.

Theorem A.21 (Hilbert–Schmidt kernels). *For any measure space M, μ, a bounded operator A on $L^2(M, d\mu)$ is Hilbert–Schmidt if and only if it is represented by an integral kernel $A(x, y)$ such that*

$$A(\cdot, \cdot) \in L^2(M \times M, d(\mu \times \mu)).$$

In this case,

$$\|A\|_2^2 = \int |A(x, y)|^2 \, d\mu(x) \, d\mu(y).$$

Proof. Suppose we are given that A is Hilbert–Schmidt. Pick an orthonormal basis $\{\phi_j\}$ for $L^2(M, d\mu)$, and let

$$a_{jk} = \langle \phi_j, A\phi_k \rangle.$$

By assumption,

$$\|A\|_2^2 = \operatorname{tr} A^* A = \sum_{j,k} |a_{jk}|^2 < \infty.$$

This guarantees the convergence of

$$A(x, y) := \sum_{j,k} a_{j,k} \phi_j(x) \overline{\phi_k(y)}$$

in $L^2(M \times M, d(\mu \times \mu))$, with

$$\int |A(x, y)|^2 \, d\mu(x) \, d\mu(y) = \sum_{j,k} |a_{jk}|^2.$$

Starting from the kernel, we expand $A(\cdot, \cdot)$ in the basis $\phi_j(x)\overline{\phi_k(y)}$ for $L^2(M \times M, d(\mu \times \mu))$, and the argument is easily reversed.   □

The set of Hilbert–Schmidt operators is itself a Hilbert space with the inner product $\langle A, B \rangle := \operatorname{tr} A^* B$. The corresponding Schwarz inequality relates the Hilbert–Schmidt norm to the trace norm:

$$\|AB\|_1 \leq \|A\|_2 \|B\|_2.$$

Definition. For A trace class, we can define the *Fredholm determinant* $\det(I + A)$ by

$$\det(I + A) := \prod_{k=1}^n (1 + \lambda_k(A)).$$

This product is convergent by (A.24). Weyl's inequality can be adapted to give the following:

Theorem A.22 (Weyl). *For A trace class,*

$$|\det(I + A)| \leq \prod_{j=1}^{\infty} (1 + \mu_j(A)) = \det(1 + |A|) \leq e^{\operatorname{tr}|A|}. \tag{A.27}$$

Proof. Let $h(t) = \log(1 + e^t)$, which we can represent as

$$h(t) = \int_{-\infty}^{\infty} [t - x]_+ \frac{e^x}{1 + e^x} \, dx, \tag{A.28}$$

where $[\,\cdot\,]_+$ denotes the positive part. By Theorem A.19, we have

$$\sum_{j=1}^k \log |\lambda_j(A)| \leq \sum_{j=1}^k \log |\mu_j(A)|,$$

for any k. It is straightforward to verify that this implies

$$\sum_{j=1}^{k}\big[\log|\lambda_j(A)| - x\big]_+ \le \sum_{j=1}^{k}\big[\log|\mu_j(A)| - x\big]_+,$$

for each k and for all $x \in \mathbb{R}$. Then by (A.28) it's clear that

$$\sum_{j=1}^{k} h(\log|\lambda_j(A)|) \le \sum_{j=1}^{k} h(\log|\mu_j(A)|),$$

which is the same as

$$\prod_{j=1}^{k}(1 + |\lambda_j(A)|) \le \prod_{j=1}^{k}(1 + \log|\mu_j(A)|).$$

Taking $k \to \infty$ gives the first inequality in (A.27), and the second follows from $1 + x \le e^x$. $\qquad\square$

In fact, Weyl's theorem is more general. One could replace $h(t)$ with any convex function and still obtain a representation of the form (A.28), so the same proof would apply.

We have given the definition of $\det(I + A)$ used by Gohberg–Krein [72]. Fredholm's original definition (later advocated by Grothendieck) is based on the observation (A.23) that the eigenvalues of $\wedge^n A$ are products of eigenvalues of A. Thus we should be able to define

$$\det(I + A) := \sum_{k=0}^{\infty} \operatorname{tr} \wedge^k A. \tag{A.29}$$

Because $|\wedge^k A| = \wedge^k|A|$, we have

$$\|\wedge^k A\|_1 = \sum_{i_1 < \cdots < i_k} \mu_{i_1}(A) \cdots \mu_{i_k}(A), \tag{A.30}$$

from which (A.27) follows immediately. The two definitions of determinant are equivalent, but this is not obvious; see, e.g., [193, Theorem 3.7].

We conclude the section with a useful resolvent estimate from Gohberg–Krein [72, Theorem V.5.1].

Theorem A.23. *For A trace class,*

$$\|(I - A)^{-1}\| \le \frac{\det(I + |A|)}{|\det(I - A)|}.$$

Proof. To estimate $\langle u, (I - A)^{-1}v \rangle$ for two unit vectors u, v, we define a new operator

$$A_t := A + tv\,\langle u, \cdot\rangle.$$

From the observation

$$(I - A_t)(I - A)^{-1}w = w - tv\,\langle u, (I - A)^{-1}w\rangle,$$

it's easy to compute that

$$\frac{\det(I - A_t)}{\det(I - A)} = 1 - t\,\langle u, (I - A)^{-1}v\rangle.$$

This leads to the estimate

$$\langle u, (I - A)^{-1}v\rangle \leq \frac{1}{t} + \frac{1}{t}\frac{|\det(I - A_t)|}{|\det(I - A)|}. \tag{A.31}$$

Since A_t differs from A by an operator of rank one, Theorem A.16 implies

$$\mu_{j+1}(A_t) \leq \mu_j(A).$$

We can estimate $\mu_1(A_t)$ by the bound

$$\|A_t\| \leq \|A\| + t.$$

Hence, by Theorem A.22,

$$\begin{aligned}
|\det(I - A_t)| &\leq \prod_{j=1}^{\infty}(1 + \mu_j(A_t)) \\
&\leq (\|A\| + t)\prod_{j=1}^{\infty}(1 + \mu_j(A)) \\
&= (\|A\| + t)\det(I + |A|).
\end{aligned}$$

Using this in (A.31) gives

$$\langle u, (I - A)^{-1}v\rangle \leq \frac{1}{t} + \frac{(\|A\| + t)}{t}\frac{\det(I + |A|)}{|\det(I - A)|}.$$

From the limit $t \to \infty$ we obtain

$$\langle u, (I - A)^{-1}v\rangle \leq \frac{\det(I + |A|)}{|\det(I - A)|},$$

which completes the proof, since u, v were arbitrary.   □

A.5 Pseudodifferential operators

A full treatment of the calculus of pseudodifferential operators requires considerable technical detail. We present a highly abbreviated development here, as a guide to the

applications that are used in the text. For complete treatments, we refer the reader to
[75], [96], [192], [208].

The action of a differential operator $p(x, D)$ on $f \in \mathcal{S}(\mathbb{R}^n)$ could be written in
terms of the Fourier transform as

$$p(x, D_x)f(x) := \frac{1}{(2\pi)^n} \int e^{ix \cdot \xi} p(x, \xi) \widehat{f}(\xi) \, d\xi, \tag{A.32}$$

where the *symbol* $p(x, \xi)$ is a polynomial in ξ with coefficients that are smooth functions of x. Here $D_x = -i(\partial_{x_1}, \ldots, \partial_{x_n})$. We can extend the space of symbols from
polynomials to functions satisfying polynomial growth conditions. Given a multi-
index $\alpha = (\alpha_1, \ldots, \alpha_n) \in \mathbb{N}^m$, let $D^\alpha = D_1^{\alpha_1} \cdots D_n^{\alpha_n}$ and $|\alpha| = \alpha_1 + \cdots + \alpha_n$. The
basic symbol space $S_{1,0}^m$ consists of smooth functions $p(x, \xi)$ satisfying

$$|D_x^\alpha D_\xi^\beta p(x, \xi)| \le C\langle \xi \rangle^{m-|\alpha|}, \tag{A.33}$$

for any multi-indices α, β. (The subscripts 1, 0 in $S_{1,0}^m$ refer to the coefficients of
$-|\alpha|$ and $+|\beta|$ in the exponent in (A.33).)

Definition. For a smooth symbol satisfying (A.33), the operator defined on $\mathcal{S}(\mathbb{R}^n)$
by (A.32) is a *pseudodifferential operator* of order m.

The set of all such operators is denoted by $\Psi^m(\mathbb{R}^n)$. The action extends to tempered
distributions; a pseudodifferential operator maps $\mathcal{S}'(\mathbb{R}^n) \to \mathcal{S}'(\mathbb{R}^n)$.

The set of all pseudodifferential operators forms an algebra such that

$$\Psi^m(\mathbb{R}^n) \circ \Psi^{m'}(\mathbb{R}^n) \subset \Psi^{m+m'}(\mathbb{R}^n).$$

To see this, one can analyze the symbol of $q(x, D_x) = p_1(x, D_x)p_2(x, D_x)$ through
the formula

$$q(x, \xi) = \frac{1}{(2\pi)^n} \int p_1(x, \xi) p_2(y, \xi') e^{i(x-y) \cdot (\xi - \xi')} \, d\xi' \, dy.$$

Similarly, one can show that $\Psi^m(\mathbb{R}^n)$ is closed under adjoints and obtain an expression for the symbol of an adjoint.

The space of *smoothing operators* $\Psi^{-\infty}(\mathbb{R}^n)$ is defined by requiring that the
symbol satisfy (A.33) for all $m \in \mathbb{Z}$. In other words, the symbol looks like a Schwartz
function in the ξ variable. Such operators map $\mathcal{S}' \to C^\infty$, hence the terminology.

Pseudodifferential operators were developed to describe inverses and resolvents
and other operators associated to differential operators. For example, $\Delta_{\mathbb{R}} + 1$ is
invertible on $\mathcal{S}(\mathbb{R})$, and its inverse is the pseudodifferential operator with symbol
$p(x, \xi) = (\xi^2 + 1)^{-1}$. Pseudodifferential operators that arise in this way typically
have symbols polyhomogeneous in ξ. This means that there are smooth functions
$p_j(x, \xi)$, homogeneous in ξ of degree $m - j$ for $|\xi| \ge 1$, such that

$$p(x, \xi) - \sum_{j=0}^{N} p_j(x, \xi) \in S_{1,0}^{m-N},$$

for any N. Operators with such symbol expansions are called *classical*, and they form a subalgebra. The leading term $p_0(x, \xi)$ in the expansion of the symbol is called the *principal symbol*. (One can define the principal symbol for a nonclassical pseudodifferential operator, but not so directly.)

The kernel of $p(x, D)$ is defined as a distribution by the oscillatory integral

$$K(x, y) = \frac{1}{(2\pi)^n} \int e^{i(x-y)\cdot\xi} p(x, \xi)\, d\xi.$$

Multiplying by powers of $(x - y)$, we can write

$$(x - y)^\alpha K(x, y) = \frac{1}{(2\pi)^n} \int e^{i(x-y)\cdot\xi} D_\xi^\alpha p(x, \xi)\, d\xi. \qquad (A.34)$$

The symbol estimates (A.33) then give us the following lemma.

Lemma A.24. *An operator $p(x, D) \in \Psi^m(\mathbb{R}^n)$ has kernel that is smooth away from the diagonal. If $m < -n$, then the kernel is continuous across the diagonal.*

For $m > -n$ the kernel will be singular on the diagonal, and this diagonal singularity can be described precisely. Setting $z = x - y$, we have

$$|D_x^\alpha D_z^\beta K(x, x - z)| \leq C|z|^{-n-m-|\beta|}.$$

This type of behavior of the kernel on the diagonal, called a conormal singularity, essentially characterizes pseudodifferential operators (see for example [208, Proposition 7.2.7] for a precise statement of this). Classical pseudodifferential operators can be similarly characterized by polyhomogeneity in the variable z.

The space $\Psi^m(M)$ of pseudodifferential operators of order m on a d-dimensional manifold M is defined by identifying coordinate patches of M with $\mathbb{R}^d$. One requires that the operator have smooth kernel away from the diagonal and that its restrictions to coordinate patches defined by a partition of unity be identified with pseudodifferential operators on $\mathbb{R}^n$. For M compact this definition is sufficient. For M noncompact the lack of growth restriction at infinity means that pseudodifferential operators cannot necessarily be composed. In this case the standard fix for this problem is to assume that the operators are properly supported, meaning that the projections from the support of the kernel to either factor of M are proper maps. For our purposes it will suffice to consider compactly supported operators whose kernels are supported in a compact subset of $M \times M$.

Proposition A.25. *Given a Riemannian manifold M, g, a compactly supported pseudodifferential operator of order m is bounded on $L^2(M, dg)$ for $m \leq 0$.*

Proof. The assumption of compact support allows us to reduce the problem, via a partition of unity, to $\mathbb{R}^n$. Assume first that $A \in \Psi^m(\mathbb{R}^n)$ for $m < -1$. Then from (A.34) we can deduce that the integral kernel $K(x, y)$ is integrable with respect to either variable (i.e., $\int |K(x, y)|\, dx$ and $\int |K(x, y)|\, dy$ are finite). From this it is easy to deduce that A is bounded.

Now assume $m < 0$. Boundedness of A is equivalent to boundedness of $(A^*A)^k \in \Psi^{-2km}$, so by taking k sufficiently large we reduce to the previous case.

For $m = 0$, we must work a little harder. Take $q(x, D_x) = A^*A \in \Psi^0(\mathbb{R}^n)$. Use the boundedness of $q(x, \xi)$ to choose $M, \varepsilon > 0$ such that $M - |q(x, \xi)| \geq \varepsilon$. Then form $B = b(x, D_x)$, where

$$b(x, \xi) = \sqrt{M - \operatorname{Re} q(x, \xi)} \in \mathcal{S}^0_{1,0}.$$

The leading part of the symbol of B^*B (the principal symbol in the classical case) matches that of $M - A^*A$. From this we can derive that

$$A^*A + B^*B - M = R \in \Psi^{-1}(\mathbb{R}^n).$$

Since R is bounded by the arguments above and B^*B is positive, it follows that $\|A\|^2 \leq M + \|R\|$. □

On a compact Riemannian manifold M, g the operator $\Delta + 1$ is invertible and powers $(\Delta + 1)^a$ are well-defined for $a \in \mathbb{R}$ by the spectral theorem. Indeed, taking the orthonormal basis of eigenfuctions $\{\phi_j\}$ for Δ, the operator $(\Delta + 1)^a$ acts by $(\lambda_j + 1)^a$. The asymptotics of Theorem A.15 then show immediately that $(\Delta + 1)^a$ is compact for $a < 0$. These powers of the Laplacian are also pseudodifferential operators of order $2a$, a fact that we will exploit in the following:

Proposition A.26. *Suppose A is a compactly supported pseudodifferential operator of order $-m$ on a manifold M, g of dimension d. Then for $m > 0$, A is compact on $L^2(M, dg)$ and its singular values satisfy*

$$\mu_k(A) \leq Ck^{-m/d}. \tag{A.35}$$

In particular, a compactly supported pseudodifferential operator of order $-m$ is trace class for $m > d$.

Proof. It suffices to assume that M is compact. If it is not, we can replace M by a compact manifold $\overline{M}, \bar{g}$ that agrees with M, g within the support of the kernel of A. Extending A to act on $L^2(\overline{M}, dg)$ gives an operator with the same singular values.

Under the assumption that M is compact, we use powers of $\Delta + 1$ to write

$$A = (\Delta + 1)^{-m/2}(\Delta + 1)^{m/2}A.$$

The operator $(\Delta + 1)^{m/2}A$ is pseudodifferential of order zero, hence bounded on $L^2(M, dg)$. The compactness of A then follows from the compactness of $(\Delta + 1)^{-m/2}$. By Corollary A.17,

$$\mu_k(A) \leq \left\| (\Delta + 1)^{m/2}A \right\| \mu_k\big((\Delta + 1)^{-m/2}\big).$$

By Theorem A.15,

$$\mu_k\big((\Delta + 1)^{-m/2}\big) \sim k^{-m/d}. \qquad \square$$

References

1. M. Abramowitz and I. Stegun (eds.), *Handbook of Mathematical Functions*, Dover Publications Inc., New York, 1966.
2. S. Agmon, On the spectral theory of the Laplacian on noncompact hyperbolic manifolds, *Journées "Équations aux dérivées partielles" (Saint Jean de Monts, 1987)*, École Polytech., Palaiseau, 1987, Exp. No. XVII, 16 pp.
3. S. Agmon, A perturbation theory of resonances, *Comm. Pure Appl. Math.* **51** (1998), 1255–1309.
4. J. W. Anderson, *Hyperbolic Geometry*, second ed., Springer-Verlag, London, 2005.
5. J. W. Anderson and A. C. Rocha, Analyticity of Hausdorff dimension of limit sets of Kleinian groups, *Ann. Acad. Sci. Fenn. Math.* **22** (1997), 349–364.
6. T. M. Apostol, *Introduction to Analytic Number Theory*, Springer-Verlag, New York, 1976.
7. N. Aronszajn, A unique continuation theorem for solutions of elliptic partial differential equations or inequalities of second order, *J. Math. Pures Appl. (9)* **36** (1957), 235–249.
8. M. Babillot and M. Peigné, Closed geodesics in homology classes on hyperbolic manifolds with cusps, *C. R. Acad. Sci. Paris Sér. I Math.* **324** (1997), 901–906.
9. M. Babillot and M. Peigné, Homologie des géodésiques fermées sur des variétés hyperboliques avec bouts cuspidaux, *Ann. Sci. École Norm. Sup. (4)* **33** (2000), 81–120.
10. M. Babillot and M. Peigné, Asymptotic laws for geodesic homology on hyperbolic manifolds with cusps, *Bull. Soc. Math. France* **134** (2006), 119–163.
11. N. L. Balazs and A. Voros, Chaos on the pseudosphere, *Phys. Rep.* **143** (1986), 109–240.
12. C. Bardos, J.-C. Guillot, and J. Ralston, La relation de Poisson pour l'équation des ondes dans un ouvert non borné. Application à la théorie de la diffusion, *Comm. Partial Differential Equations* **7** (1982), 905–958.
13. A. Sá Barreto, Radiation fields, scattering, and inverse scattering on asymptotically hyperbolic manifolds, *Duke Math. J.* **129** (2005), 407–480.
14. A. F. Beardon, The exponent of convergence of Poincaré series, *Proc. London Math. Soc.* **18** (1968), 461–483.
15. A. F. Beardon, Inequalities for certain Fuchsian groups, *Acta Math.* **127** (1971), 221–258.
16. A. F. Beardon, *The Geometry of Discrete Groups*, Springer-Verlag, New York, 1995.
17. T. Bedford, M. Keane, and C. Series (eds.), *Ergodic Theory, Symbolic Dynamics, and Hyperbolic Spaces*, Oxford University Press, New York, 1991.
18. P. Bérard, Transplantaion et isospectralité. I., *Math. Ann.* **292** (1992), 547–560.

19. L. Bers, A remark on Mumford's compactness theorem, *Israel. J. Math.* **12** (1972), 400–407.

20. L. Bers, An inequality for Riemann surfaces, *Differential Geometry and Complex Analysis* (I. Chavel and H. M. Farkas, eds.), Springer, Berlin, 1985, pp. 87–93.

21. R. P. Boas, *Entire Functions*, Academic Press Inc., New York, 1954.

22. D. Borthwick, Scattering theory for conformally compact metrics with variable curvature at infinity, *J. Funct. Anal.* **184** (2001), 313–376.

23. D. Borthwick, C. Judge, and P. A. Perry, Determinants of Laplacians and isopolar metrics on surfaces of infinite area, *Duke Math. J.* **118** (2003), 61–102.

24. D. Borthwick, C. Judge, and P. A. Perry, Selberg's zeta function and the spectral geometry of geometrically finite hyperbolic surfaces, *Comment. Math. Helv.* **80** (2005), 483–515.

25. D. Borthwick, A. McRae, and E. C. Taylor, Quasirigidity of hyperbolic 3-manifolds and scattering theory, *Duke Math. J.* **89** (1997), 225–236.

26. D. Borthwick and P. A. Perry, Scattering poles for asymptotically hyperbolic manifolds, *Trans. Amer. Math. Soc.* **354** (2002), 1215–1231.

27. B. H. Bowditch, Geometrical finiteness for hyperbolic groups, *J. Funct. Anal.* **113** (1993), 245–317.

28. R. Bowen and C. Series, Markov maps associated with Fuchsian groups, *Inst. Hautes Études Sci. Publ. Math.* **50** (1979), 153–170.

29. R. Brooks and O. Davidovich, Isoscattering on surfaces, *J. Geom. Anal.* **13** (2003), 39–53.

30. R. Brooks, R. Gornet, and P. A. Perry, Isoscattering Schottky manifolds, *Geom. Funct. Anal.* **10** (2000), 307–326.

31. R. Brooks and P. A. Perry, Isophasal scattering manifolds in two dimensions, *Comm. Math. Phys.* **223** (2001), 465–474.

32. U. Bunke and M. Olbrich, Gamma-cohomology and the Selberg zeta function, *J. Reine Angew. Math.* **467** (1995), 199–219.

33. U. Bunke and M. Olbrich, *Selberg Zeta and Theta Functions: A Differential Operator Approach*, Akademie-Verlag, Berlin, 1995.

34. U. Bunke and M. Olbrich, Fuchsian groups of the second kind and representations carried by the limit set, *Invent. Math.* **127** (1997), 127–154.

35. U. Bunke and M. Olbrich, Group cohomology and the singularities of the Selberg zeta function associated to a Kleinian group, *Ann. Math.* **149** (1999), 627–689.

36. U. Bunke and M. Olbrich, Scattering theory for geometrically finite groups, Arxiv preprint, 1999.

37. P. Buser, *Geometry and Spectra of Compact Riemann Surfaces*, Birkhäuser Boston Inc., Boston, MA, 1992.

38. J. Button, All Fuchsian Schottky groups are classical Schottky groups, *The Epstein birthday schrift*, Geom. Topol. Publ., Coventry, 1998, pp. 117–125.

39. G. Carron, Déterminant relatif et la fonction Xi, *Amer. J. Math.* **124** (2002), 307–352.

40. I. Chavel, *Eigenvalues in Riemannian Geometry*, Academic Press Inc., Orlando, FL, 1984, Including a chapter by B. Randol, with an appendix by J. Dodziuk.

41. J. Chazarain, Formule de Poisson pour les variétés riemanniennes, *Invent. Math.* **24** (1974), 65–82.

42. T. Christiansen, Weyl asymptotics for the Laplacian on asymptotically Euclidean spaces, *Amer. J. Math.* **121** (1999), 1–22.

43. T. Christiansen, Weyl asymptotics for the Laplacian on manifolds with asymptotically cusp ends, *J. Funct. Anal.* **187** (2001), 211–226.

44. T. Christiansen and M. Zworski, Spectral asymptotics for manifolds with cylindrical ends, *Ann. Inst. Fourier (Grenoble)* **45** (1995), 251–263.

45. T. Christiansen and M. Zworski, Resonance wave expansions: two hyperbolic examples, *Comm. Math. Phys.* **212** (2000), 323–336.

46. J. B. Conway, *Functions of One Complex Variable*, second ed., Graduate Texts in Mathematics, vol. 11, Springer-Verlag, New York, 1978.

47. C. Cuevas and G. Vodev, Sharp bounds on the number of resonacnes for conformally compact manifolds with constant negative curvature near infinity, *Comm. PDE* **28** (2003), 1685–1704.

48. Y. Colin de Verdière, Spectre du laplacien et longueurs des géodésiques périodiques, *C. R. Acad. Sci. Paris Sér. A–B* **275** (1972), A805–A808.

49. Y. Colin de Verdière, Spectre du laplacien et longueurs des géodésiques périodiques. I, II, *Compositio Math.* **27** (1973), 83–106; 159–184.

50. Y. Colin de Verdière, Théorie spectrale des surfaces de Riemann d'aire infinie, *Astérisque* (1985), no. 132, 259–275, Colloquium in honor of Laurent Schwartz, Vol. 2 (Palaiseau, 1983).

51. M. P. do Carmo, *Differential Geometry of Curves and Surfaces*, Prentice-Hall Inc., Englewood Cliffs, NJ, 1976.

52. H. Donnelly and C. Fefferman, Fixed point formula for the Bergman kernel, *Amer. J. Math.* **108** (1986), 1241–1258.

53. J. J. Duistermaat and V. W. Guillemin, The spectrum of positive elliptic operators and periodic bicharacteristics, *Invent. Math.* **29** (1975), 39–79.

54. I. Efrat, Determinants of Laplacians on surfaces of finite volume, *Comm. Math. Phys.* **119** (1988), 443–451, Erratum, *Comm. Math. Phys.*, **138**, (1991), 607.

55. J. Elstrodt, Die Resolvente zum Eigenwertproblem der automorphen Formen in der hyperbolischen Ebene. I, *Math. Ann.* **203** (1973), 295–300; II, *Math. Z.* **132** (1973), 99–134; III, *Math. Ann.* **208** (1974), 99–132.

56. C. L. Epstein, Asymptotics for closed geodesics in a homology class, the finite volume case, *Duke Math. J.* **55** (1987), 717–757.

57. A. Erdélyi, W. Magnus, F. Oberhettinger, and F. G. Tricomi, *Higher Transcendental Functions. Vol. I*, McGraw-Hill, New York–Toronto–London, 1953, Based, in part, on notes left by Harry Bateman.

58. L. D. Faddeev, The eigenfunction expansion of Laplace's operator on the fundamental domain of a discrete group on the Lobačevskiĭ plane, *Trudy Moskov. Mat. Obšč.* **17** (1967), 323–350.

59. H. M. Farkas and I. Kra, *Riemann Surfaces*, second ed., Springer-Verlag, New York, 1992.

60. J. D. Fay, Fourier coefficients of the resolvent for a Fuchsian group, *J. Reine Angew. Math.* **293/294** (1977), 143–203.

61. W. Fenchel and J. Nielsen, *Discontinuous Groups of Isometries in the Hyperbolic Plane*, Walter de Gruyter & Co., Berlin, 2003, edited and with a preface by Asmus L. Schmidt.

62. J. Fischer, *An Approach to the Selberg Trace Formula via the Selberg Zeta-Function*, Springer-Verlag, Berlin, 1987.

63. G. B. Folland, *Real Analysis: Modern Techniques and Their Applications*, Wiley & Sons Inc., New York, 1984.

64. R. Fricke and F. Klein, *Vorlesungen über die Theorie der elliptischen Modulfunktionen/Automorphenfunktionen*, G. Teubner, Leipzig, 1896.

65. D. Fried, The zeta functions of Ruelle and Selberg. I, *Ann. Sci. École Norm. Sup. (4)* **19** (1986), 491–517.

66. F. G. Friedlander, *Introduction to the Theory of Distributions*, second ed., Cambridge University Press, Cambridge, 1998, With additional material by M. Joshi.

67. R. Froese and P. Hislop, On the distribution of resonances for some asymptotically hyperbolic manifolds, *Journées "Équations aux Dérivées Partielles" (La Chapelle sur Erdre, 2000)*, Univ. Nantes, Nantes, 2000, p. Exp. No. VII.

68. R. Froese, P. Hislop, and P. Perry, The Laplace operator on hyperbolic three manifolds with cusps of nonmaximal rank, *Invent. Math.* **106** (1991), 295–333.

69. R. Froese, P. Hislop, and P. Perry, A Mourre estimate and related bounds for hyperbolic manifolds with cusps of nonmaximal rank, *J. Funct. Anal.* **98** (1991), 292–310.

70. I. M. Gel'fand and G. E. Shilov, *Generalized Functions. Vol. I: Properties and Operations*, Translated by Eugene Saletan, Academic Press, New York, 1964.

71. C. Gérard, Asymptotique des pôles de la matrice de scattering pour deux obstacles strictement convexes, *Mém. Soc. Math. France (N.S.)* **31** (1988), 1–146.

72. I. C. Gohberg and M. Krein, *Introduction to the Theory of Linear Nonselfadjoint Operators*, Translations of Mathematical Monographs, vol. 18, American Mathematical Society, Providence, RI, 1969.

73. I. C. Gohberg and E. I. Sigal, An operator generalization of the logarithmic residue theorem and the theorem of Rouché, *Math. U. S. S. R. Sbornik* **13** (1971), 603–625.

74. R. C. Graham and M. Zworski, Scattering matrix in conformal geometry, *Invent. Math.* **152** (2003), 89–118.

75. A. Grigis and J. Sjöstrand, *Microlocal Analysis for Differential Operators*, Cambridge University Press, Cambridge, 1994.

76. C. Guillarmou, Absence of resonance near the critical line on asymptotically hyperbolic spaces, *Asymptot. Anal.* **42** (2005), 105–121.

77. C. Guillarmou, Generalized Krein formula and determinants for Poincaré-Einstein manifolds, preprint, 2005.

78. C. Guillarmou, Meromorphic properties of the resolvent on asymptotically hyperbolic manifolds, *Duke Math. J.* **129** (2005), 1–37.

79. C. Guillarmou, Resonances and scattering poles on asymptotically hyperbolic manifolds, *Math. Res. Lett.* **12** (2005), 103–119.

80. C. Guillarmou, Resonances on some geometrically finite hyperbolic manifolds, *Comm. Partial Differential Equations* **31** (2006), 445–467.

81. C. Guillarmou and F. Naud, Wave 0-trace and length spectrum on convex co-compact hyperbolic manifolds, *Comm. Anal. Geom.* **14** (2006), 945–967.

82. L. Guillopé, Sur la distribution des longeurs des géodésiques fermées d'une surface compacte à bord totalement géodésique, *Duke Math. J.* **53** (1986), 827–848.

83. L. Guillopé, Fonctions zeta de Selberg et surfaces de géométrie finie, *Zeta Functions in Geometry (Tokyo, 1990)*, Adv. Stud. Pure Math., vol. 21, Kinokuniya, Tokyo, 1992, pp. 33–70.

84. L. Guillopé, K. Lin, and M. Zworski, The Selberg zeta function for convex co-compact Schottky groups, *Comm. Math. Phys.* **245** (2004), 149–176.

85. L. Guillopé and M. Zworski, Polynomial bounds on the number of resonances for some complete spaces of constant negative curvature near infinity, *Asymptotic Anal.* **11** (1995), 1–22.

86. L. Guillopé and M. Zworski, Upper bounds on the number of resonances for non-compact Riemann surfaces, *J. Funct. Anal.* **129** (1995), 364–389.

87. L. Guillopé and M. Zworski, Scattering asymptotics for Riemann surfaces, *Ann. Math.* **145** (1997), 597–660.

88. L. Guillopé and M. Zworski, The wave trace for Riemann surfaces, *Geom. Funct. Anal.* **9** (1999), 1156–1168.

89. A. Hassell and S. Zelditch, Determinants of Laplacians in exterior domains, *Internat. Math. Res. Notices* (1999), 971–1004.

90. D. A. Hejhal, *The Selberg Trace Formula for* PSL(2, $\mathbb{R}$), *Vol. II*, Springer-Verlag, Berlin, 1983.

91. Dennis A. Hejhal, *The Selberg Trace Formula for* PSL(2, R). *Vol. I*, Springer-Verlag, Berlin, 1976.

92. P. D. Hislop, The geometry and spectra of hyperbolic manifolds. spectral and inverse spectral theory, *Proc. Indian Acad. Sci. Math. Sci.* **104** (1994), 715–776.

93. P. D. Hislop and I. M. Sigal, *Introduction to Spectral Theory*, Springer-Verlag, New York, 1996.

94. E. Hopf, Statistik der geodätischen Linien in Mannigfaltigkeiten negativer Krümmung, *Ber. Verh. Sächs. Akad. Wiss. Leipzig* **91** (1939), 261–304.

95. L. Hörmander, Uniqueness theorems for second order elliptic differential equations, *Comm. Partial Differential Equations* **8** (1983), 21–64.

96. L. Hörmander, *The Analysis of Linear Partial Differential Operators. III*, Springer-Verlag, Berlin, 1994, Corrected reprint of the 1985 original.

97. H. Huber, Zur analytischen Theorie hyperbolischen Raumformen und Bewegungsgruppen, *Math. Ann.* **138** (1959), 1–26; II, *Math. Ann.* **142** (1960/1961), 385–398; Nachtrag zu II, *Math. Ann.* **143** (1961), 463–464.

98. M. Ikawa, On the poles of the scattering matrix for two strictly convex obstacles, *J. Math. Kyoto Univ.* **23** (1983), 127–194.

99. A. Intissar, A polynomial bound on the number of the scattering poles for a potential in even-dimensional spaces $\mathbf{R}^n$, *Comm. Partial Differential Equations* **11** (1986), 367–396.

100. V. Ja. Ivrii, On the second term of the spectral asymptotics for the Laplace–Beltrami operator on a manifold with boundary, *Funct. Anal. Appl.* **14** (1980), 98–106.

101. N. Jacobson, *Basic Algebra. I*, second ed., W. H. Freeman, New York, 1985.

102. M. S. Joshi and A. Sá Barreto, Inverse scattering on asymptotically hyperbolic manifolds, *Acta Math.* **184** (2000), 41–86.

103. M. S. Joshi and A. Sá Barreto, The wave group on asymptotically hyperbolic manifolds, *J. Funct. Anal.* **184** (2001), 291–312.

104. J. Jost, *Compact Riemann Surfaces*, second ed., Springer-Verlag, Berlin, 2002.

105. A. Juhl, *Cohomological Theory of Dynamical Zeta Functions*, Birkhäuser Verlag, Basel, 2001.

106. R. P. Kanwal, *Generalized Functions: Theory and Applications*, third ed., Birkhäuser Boston Inc., Boston, MA, 2004.

107. T. Kato, *Perturbation Theory for Linear Operators*, Springer-Verlag, Berlin, 1995, Reprint of the 1980 edition.

108. S. Katok, *Fuchsian Groups*, University of Chicago Press, Chicago, IL, 1992.

109. A. Katsuda and T. Sunada, Homology of closed geodesics in certain Riemannian manifolds, *Proc. Amer. Math. Soc.* **96** (1986), 657–660.

110. I. Kra, *Automorphic Forms and Kleinian Groups*, W. A. Benjamin, Reading, MA, 1972.

111. S. P. Lalley, Renewal theorems in symbolic dynamics, with applications to geodesic flows, non-Euclidean tessellations, and their fractal limits, *Acta Math.* **139** (1976), 241–273.

112. P. Lancaster, *Theory of Matrices*, Academic Press, New York, 1969.

113. P. Lax and R. S. Phillips, The asymptotic distribution of lattice points in Euclidean and non-Euclidean spaces, *J. Funct. Anal.* **46** (1982), 280–350.

114. P. Lax and R. S. Phillips, Translation representation for automorphic solutions of the wave equation in non-Euclidean spaces. I, *Comm. Pure Appl. Math.* **37** (1984), 303–328; II, *Comm. Pure Appl. Math.* **37** (1984), 779–813; III, *Comm. Pure Appl. Math.* **38** (1985), 179–207.

115. P. Lax and R. S. Phillips, *Scattering Theory*, second ed., Academic Press Inc., Boston, MA, 1989.

116. P. D. Lax and R. S. Phillips, Decaying modes for the wave equation in the exterior of an obstacle., *Comm. Pure Appl. Math.* **22** (1969), 737–787.

117. P. D. Lax and R. S. Phillips, *Scattering Theory for Automorphic Functions*, Princeton Univ. Press, Princeton, N.J., 1976, Annals of Mathematics Studies, No. 87.

118. J. M. Lee, *Riemannian Manifolds*, Springer-Verlag, New York, 1997, An introduction to curvature.

119. O. Lehto, *Univalent Functions and Teichmüller Spaces*, Springer-Verlag, New York, 1987.

120. H. Maass, Über eine neue Art von nichtanalytischen automorphen Funktionen und die Bestimmung Dirichletscher Reihen durch Funktionalgleichungen, *Math. Ann.* **121** (1949), 141–183.

121. B. Maskit, A characterization of Schottky groups, *J. Analyse Math.* **19** (1967), 227–230.

122. B. Maskit, *Kleinian Groups*, Springer-Verlag, Berlin, 1988.

123. W. S. Massey, *A Basic Course in Algebraic Topology*, Springer-Verlag, New York, 1991.

124. J. N. Mather, Characterization of Anosov diffeomorphisms, *Nederl. Akad. Wetensch. Proc. Ser. A* **71** (1968), 479–483.

125. R. Mazzeo, The Hodge cohomology of a conformally compact metric, *J. Differential Geom.* **28** (1988), 309–339.

126. R. Mazzeo, Elliptic theory of differential edge operators I, *Comm. PDE* **16** (1991), 1615–1664.

127. R. Mazzeo, Unique continuation at infinity and embedded eigenvalues for asymptotically hyperbolic manifolds, *Amer. J. Math.* **113** (1991), 25–45.

128. R. Mazzeo and R. B. Melrose, Meromorphic extension of the resolvent on complete spaces with asymptotically constant negative curvature, *J. Funct. Anal.* **75** (1987), 260–310.

129. J. McGowan and P. Perry, Closed geodesics in homology classes for convex co-compact hyperbolic manifolds, *Proceedings of the Euroconference on Partial Differential Equations and Their Applications to Geometry and Physics (Castelvecchio Pascoli, 2000)*, vol. 91, 2002, pp. 197–209.

130. H. P. McKean, Selberg's trace formula as applied to a compact Riemann surface, *Comm. Pure Appl. Math.* **25** (1972), 225–246.

131. R. B. Melrose, Scattering theory and the trace of the wave group, *J. Funct. Anal.* **45** (1982), 29–40.

132. R. B. Melrose, Polynomial bound on the number of scattering poles, *J. Funct. Anal.* **53** (1983), 287–303.

133. R. B. Melrose, Polynomial bounds on the distribution of poles in scattering by an obstacle, *Journées "Équations aux dérivées partielles" (Saint Jean de Monts)*, 1984.

134. R. B. Melrose, Weyl asymptotics for the phase in obstacle scattering, *Comm. PDE* **13** (1988), 1431–1439.

135. R. B. Melrose, Spectral and scattering theory for the Laplacian on asymptotically Euclidian spaces, *Spectral and Scattering Theory (Sanda, 1992)*, Lecture Notes in Pure and Appl. Math., vol. 161, Dekker, New York, 1994, pp. 85–130.

136. R. B. Melrose, *Geometric Scattering Theory*, Cambridge University Press, Cambridge, 1995.

137. J. Milnor, Hyperbolic geometry: the first 150 years, *Bull. Amer. Math. Soc. (N.S.)* **6** (1982), 9–24.

138. S. Minakshisundaram and A. Pleijel, Some properties of the eigenfunctions of the Laplace-operator on Riemannian manifolds, *Canadian J. Math.* **1** (1949), 242–256.

139. W. Müller, Spectral geometry and scattering theory for certain complete surfaces of finite volume, *Invent. Math.* **109** (1992), 265–305.

140. D. Mumford, A remark on Mahler's compactness theorem, *Proc. Amer. Math. Soc.* **28** (1971), 289–294.

141. D. Mumford, C. Series, and D. Wright, *Indra's Pearls*, Cambridge University Press, New York, 2002, The Vision of Felix Klein.

142. J. R. Munkres, *Topology*, second ed., Prentice-Hall, Englewood Cliffs, NJ, 1999.

143. F. Naud, Expanding maps on Cantor sets and analytic continuation of zeta functions, *Ann. Sci. École Norm. Sup.* **38** (2005), 116–153.

144. F. Naud, Precise asymptotics of the length spectrum for finite-geometry Riemann surfaces, *Int. Math. Res. Not.* (2005), 299–210.

145. L. Nedelec, Asymptotique du nombre de résonances de l'opérateur de Schrödinger avec potentiel linéaire et matriciel, *Math. Res. Lett.* **4** (1997), 309–320.

146. L. Nedelec, Multiplicity of resonances in black box scattering, *Canad. Math. Bull.* **47** (2004), 407–416.

147. D. J. Newman, Simple analytic proof of the prime number theorem, *Amer. Math. Monthly* **87** (1980), 693–696.

148. P. J. Nicholls, A lattice point problem in hyperbolic space, *Michigan Math. J.* **30** (1983), 273–287.

149. P. J. Nicholls, *The Ergodic Theory of Discrete Groups*, Cambridge University Press, Cambridge, 1989.

150. M. Olbrich, Cohomology of convex cocompact groups and invariant distributions on limit sets, preprint, 2002.

151. B. Osgood, R. Phillips, and P. Sarnak, Compact isospectral sets of surfaces, *J. Funct. Anal.* **80** (1988), 212–234.

152. L. B. Parnovski, Spectral asymptotics of Laplace operators on surfaces with cusps, *Math. Ann.* **303** (1995), 281–296.

153. L. B. Parnovski, Spectral asymptotics of the Laplace operator on manifolds with cylindrical ends, *Internat. J. Math.* **6** (1995), 911–920.

154. S. J. Patterson, The Laplace operator on a Riemann surface, *Compositio Math.* **31** (1975), 83–107.

155. S. J. Patterson, A lattice-point problem in hyperbolic space, *Mathematika* **22** (1975), 81–88; II, *Compositio Math.* **32** (1976), 71–112; III, *Compositio Math.* **33** (1976), 227–259.

156. S. J. Patterson, The limit set of a Fuchsian group, *Acta Math.* **136** (1976), 241–273.

157. S. J. Patterson, Lectures on measures on limit sets of Kleinian groups, *Analytical and geometric aspects of hyperbolic space (Coventry/Durham, 1984)*, London Math. Soc. Lecture Note Ser., vol. 111, Cambridge Univ. Press, Cambridge, 1987, pp. 281–323.

158. S. J. Patterson, On a lattice-point problem in hyperbolic space and related questions in spectral theory, *Ark. Mat.* **26** (1988), 167–172.

159. S. J. Patterson, The Selberg zeta-function of a Kleinian group, *Number Theory, Trace Formulas, and Discrete Groups: Symposium in Honor of Atle Selberg, Oslo, Norway, July 14–21, 1987*, Academic Press, New York, 1989.

160. S. J. Patterson and P. A. Perry, The divisor of Selberg's zeta function for Kleinian groups, *Duke Math. J.* **106** (2001), 321–390, Appendix A by Charles Epstein.

161. P. A. Perry, The Laplace operator on a hyperbolic manifold. I. Spectral and scattering theory, *J. Funct. Anal.* **75** (1987), 161–187.

162. P. A. Perry, The Laplace operator on a hyperbolic manifold. II. Eisenstein series and the scattering matrix, *J. Reine Angew. Math.* **398** (1989), 67–91.

163. P. A. Perry, The Selberg zeta function and a local trace formula for Kleinian groups, *J. Reine Angew. Math.* **410** (1990), 116–152.

164. P. A. Perry, The Selberg zeta function and scattering poles for Kleinian groups, *Bull. Amer. Math. Soc.* **24** (1991), 327–333.

165. P. A. Perry, Inverse spectral problems in Riemannian geometry, *Inverse Problems in Mathematical Physics (Saariselkä, 1992)*, Lecture Notes in Phys., vol. 422, Springer, Berlin, 1993, pp. 174–182.

166. P. A. Perry, A trace-class rigidity theorem for Kleinian groups, *Ann. Acad. Sci. Fenn. Ser. A I Math.* **20** (1995), 251–257.

167. P. A. Perry, Asymptotics of the length spectrum for hyperbolic manifolds of infinite volume, *Geom. Funct. Anal.* **11** (2001), 132–141.

168. P. A. Perry, Spectral theory, dynamics, and Selberg's zeta function for Kleinian groups, *Dynamical, Spectral, and Arithmetic Zeta Functions (San Antonio, TX, 1999)*, Contemp. Math., vol. 290, Amer. Math. Soc., Providence, RI, 2001, pp. 145–165.

169. P. A. Perry, A Poisson summation formula and lower bounds for resonances in hyperbolic manifolds, *Int. Math. Res. Not.* (2003), 1837–1851.

170. P. A. Perry, The spectral geometry of geometrically finite hyperbolic manifolds, *Spectral Theory and Mathematical Physics: A Festschrift in Honor of Barry Simon's 60th Birthday*, Proc. Sympos. Pure Math., vol. 76, Amer. Math. Soc., Providence, RI, 2007, pp. 289–328.

171. P. Petersen, *Riemannian Geometry*, Springer-Verlag, New York, 1998.

172. V. Petkov and M. Zworski, Semi-classical estimates on the scattering determinant, *Ann. Henri Poincaré* **2** (2001), 675–711.

173. R. Phillips and P. Sarnak, Geodesics in homology classes, *Duke Math. J.* **55** (1987), 287–297.

174. R. S. Phillips and P. Sarnak, On cusp forms for co-finite subgroups of PSL(2, r), *Invent. Math.* **80** (1985), 339–364.

175. M. Pollicott, Some applications of thermodynamic formalism to manifolds with constant negative curvature, *Adv. Math.* **85** (1991), no. 2, 161–192.

176. M. Pollicott and A. C. Rocha, A remarkable formula for the determinant of the Laplacian, *Invent. Math.* **130** (1997), 399–414.

177. A. Pressley, *Elementary Differential Geometry*, Springer-Verlag London Ltd., London, 2001.

178. B. Randol, On the asymptotic distribution of closed geodesics on compact Riemann surfaces, *Trans. Amer. Math. Soc.* **233** (1977), 241–247.

179. J. G. Ratcliffe, *Foundations of Hyperbolic Manifolds*, Springer-Verlag, New York, 1994.

180. M. Reed and B. Simon, *Methods of Modern Mathematical Physics. I. Functional Analysis*, Academic Press, New York, 1972.

181. M. Reed and B. Simon, *Methods of Modern Mathematical Physics. IV. Analysis of Operators*, Academic Press, New York, 1978.

182. D. Robert, Sur la formule de Weyl pour des ouverts non bornés, *C. R. Acad. Sci. Paris Sér. I Math.* **319** (1994), 29–34.

183. D. Ruelle, Zeta-functions for expanding maps and Anosov flows, *Invent. Math.* **34** (1976), 231–242.

184. P. Sarnak, Prime geodesic theorems, Ph.D. thesis, Stanford University, 1980.

185. P. Sarnak, Determinants of Laplacians, *Comm. Math. Phys.* **110** (1987), 113–120.

186. P. Sarnak, Arithmetic quantum chaos, *The Schur Lectures (1992) (Tel Aviv)*, Israel Math. Conf. Proc., vol. 8, Bar-Ilan Univ., Ramat Gan, 1995, pp. 183–236.

187. P. Sarnak, Quantum chaos, symmetry and zeta functions. Lectures I and II, *Current Developments in Mathematics, 1997 (Cambridge, MA)*, Int. Press, Boston, MA, 1999, pp. 127–159.

188. A. Selberg, Harmonic analysis and discontinuous groups in weakly symmetric Riemannian spaces with applications to Dirichlet series, *J. Indian Math. Soc. (N.S.)* **20** (1956), 47–87.

189. A. Selberg, Göttingen lectures, *Collected Works, Vol. I*, Springer-Verlag, Berlin, 1989, pp. 626–674.

190. M. Seppälä and T. Sorvali, *Geometry of Riemann Surfaces and Teichmüller Spaces*, North-Holland Publishing Co., Amsterdam, 1992.

191. R. Sharp, Uniform estimates for closed geodesics and homology on finite area hyperbolic surfaces, *Math. Proc. Cambridge Philos. Soc.* **137** (2004), 245–254.

192. M. A. Shubin, *Pseudodifferential Operators and Spectral Theory*, Springer-Verlag, Berlin, 1987.

193. B. Simon, *Trace Ideals and Their Applications*, second ed., American Mathematical Society, Providence, RI, 2005.

194. J. Sjöstrand, Geometric bounds on the density of resonances for semiclassical problems, *Duke Math. J.* **60** (1990), 1–57.

195. J. Sjöstrand, A trace formula and review of some estimates for resonances, *Microlocal Analysis and Spectral Theory (Lucca, 1996)*, NATO Adv. Sci. Inst. Ser. C Math. Phys. Sci., vol. 490, Kluwer, Dordrecht, 1997, pp. 377–437.

196. J. Sjöstrand, A trace formula for resonances and application to semi-classical Schrödinger operators, *Séminaire sur les Équations aux Dérivées Partielles, 1996–1997*, École Polytech., Palaiseau, 1997, Exp. No. II, 13 pp.

197. J. Sjöstrand and M. Zworski, Complex scaling and the distribution of scattering poles, *J. Amer. Math. Soc.* **4** (1991), 729–769.

198. J. Sjöstrand and M. Zworski, Distribution of scattering poles near the real axis, *Comm. Partial Differential Equations* **17** (1992), 1021–1035.

199. J. Sjöstrand and M. Zworski, Lower bounds on the number of scattering poles, *Comm. P. D. E.* **18** (1993), 847–857.

200. J. Sjöstrand and M. Zworski, Lower bounds on the number of scattering poles. II, *J. Funct. Anal.* **123** (1994), 336–367.

201. P. Stefanov, Sharp upper bounds on the number of resonances near the real axis for trapping systems, *Amer. J. Math.* **125** (2003), 183–224.

202. D. Sullivan, The density at infinity of a discrete group of hyperbolic motions, *Publ. Math. IHES* **50** (1979), 171–202.

203. D. Sullivan, Discrete conformal groups and measurable dynamics, *Bull. Amer. Math. Soc. (N.S.)* **6** (1982), 57–73.

204. D. Sullivan, Entropy, Hausdorff measures old and new, and limit sets of geometrically finite Kleinian groups, *Acta Math.* **153** (1984), 259–277.

205. N. Tarkhanov, Fixed point formula for holomoprhic functions, *Proc. Amer. Math. Soc.* **132** (2004), 2411–2419.

206. M. E. Taylor, *Noncommutative Harmonic Analysis*, American Mathematical Society, Providence, RI, 1986.

207. M. E. Taylor, *Partial Differential Equations. I. Basic Theory*, Springer-Verlag, New York, 1996.

208. M. E. Taylor, *Partial Differential Equations. II. Qualitative Studies of Linear Equations*, Springer-Verlag, New York, 1996.

209. A. J. Tromba, *Teichmüller Theory in Riemannian Geometry*, Birkhäuser Verlag, Basel, 1992.

210. A. B. Venkov, *Spectral Theory of Automorphic Functions and Its Applications*, Kluwer Academic Publishers, Dordrecht, 1990.

211. G. Vodev, Sharp polynomial bounds on the number of scattering poles for metric perturbations of the Laplacian in $\mathbf{R}^n$, *Math. Ann.* **291** (1991), 39–49.
212. G. Vodev, Sharp bounds on the number of scattering poles for perturbations of the Laplacian, *Comm. Math. Phys.* **146** (1992), 205–216.
213. G. Vodev, Sharp bounds on the number of scattering poles in even-dimensional spaces,
Duke Math. J. **74** (1994), 1–17.
214. G. Vodev, Sharp bounds on the number of scattering poles in the two-dimensional case,
Math. Nachr. **170** (1994), 287–297.
215. A. Voros, Spectral functions, special functions and the Selberg zeta function, *Comm.
Math. Phys.* **110** (1987), 439–465.
216. P. Walters, *An Introduction to Ergodic Theory*, Springer-Verlag, New York, 1982.
217. E. T. Whittaker and G. N. Watson, *A Course of Modern Analysis*, Cambridge Mathematical Library, Cambridge University Press, 1996, Reprint of the fourth edition.
218. D. Zagier, Newman's short proof of the prime number theorem, *Amer. Math. Monthly*
104 (1997), 705–708.
219. S. Zelditch, The inverse spectral problem, *Surveys in Differential Geometry. Vol. IX*, Int.
Press, Somerville, MA, 2004, pp. 401–467.
220. M. Zerzeri, Majoration du nombre de résonances près de l'axe réel pour une perturbation abstraite à support compact, du laplacien, *Comm. Partial Differential Equations* **26**
(2001), 2121–2188.
221. M. Zworski, Distribution of poles for scattering on the real line, *J. Funct. Anal.* **73**
(1987), 277–296.
222. M. Zworski, Sharp polynomial bounds on the number of scattering poles, *Duke Math. J.*
59 (1989), 311–323.
223. M. Zworski, Sharp polynomial bounds on the number of scattering poles of radial potentials, *J. Funct. Anal.* **82** (1989), 370–403.
224. M. Zworski, Counting scattering poles, *Spectral and Scattering Theory*, Lecture Notes
in Pure and Appl. Math., vol. 161, Dekker, New York, 1994, pp. 301–331.
225. M. Zworski, Dimension of the limit set and density of resonances for convex co-compact
hyperbolic quotients, *Invent. Math.* **136** (1999), 353–409.
226. M. Zworski, Resonances in physics and geometry, *Notices Amer. Math. Soc.* **46** (1999),
319–328.
227. M. Zworski, Density of resonances for Schottky groups, talk, 2002.
228. M. Zworski, Quantum resonances and partial differential equations, *Proceedings of the
International Congress of Mathematicians, Vol. III (Beijing, 2002)*, Higher Ed. Press,
Beijing, 2002, pp. 243–252.

Notation Guide

$:=$	definition
$\sim$	asymptotic to (ratio approaches 1)
$\asymp$	comparable to (ratio bounded above and below)
$\alpha(T)$	axis of the hyperbolic transformation T, see Section 2.1
$\mathbb{B}$	hyperbolic unit disk, see Section 2.1
$B(w; r)$	ball of hyperbolic radius r, centered at w
C_j	cusp component of X, see Section 6.1
C_ℓ	hyperbolic cylinder of diameter ℓ, $\Gamma_\ell \backslash \mathbb{H}$, see Section 2.4
C_∞	parabolic cylinder, $\Gamma_\infty \backslash \mathbb{H}$, see Section 2.4
$\mathcal{D}_w$	Dirichlet fundamental domain with center $w \in \mathbb{H}$, see (2.12)
dA	hyperbolic area form on $\mathbb{H}$ or X, see (2.8)
δ	exponent of convergence of Γ, see (2.19)
Δ_X	positive Laplacian on X.
$\partial_0 \overline{F}_\ell$	infinite ($\rho = 0$) boundary of compactified funnel, see Section 7.4.
dh	measure induced on $\partial \overline{X}$, see Section 7.4.
$E_X(s)$	Poisson kernel of X, see Section 7.4
$\mathcal{F}$	fundamental domain for Γ, see Section 2.2
F_j	funnel component of X, see Section 6.1
F_ℓ	funnel of diameter ℓ, see Section 2.4
$G_\infty(s)$	the entire function $\Gamma(s)G(s)^2$, where $G(s)$ is Barnes G-function
Γ	geometrically finite Fuchsian group, see Section 2.2
Γ_ℓ	cyclic hyperbolic group, see Section 2.4
Γ_∞	cyclic parabolic subgroup, see Section 2.4
$\mathbb{H}$	hyperbolic upper half-plane, see Section 2.1

H^s Hausdorff measure of dimension s, see (14.21)

$I(w;r)$ shadow of $B(w;r)$ on $\partial\mathbb{B}$, see (14.23)

$\ell(T)$ displacement length of the hyperbolic transformation T, see Section 2.1

$\Lambda(\Gamma)$ limit set of Γ, see Section 2.2

$\mathcal{L}_X$ primitive length spectrum of X, see Section 2.6

$L(s)$ parametric error term, see (6.9)

$M(s)$ parametrix for $(\Delta_X - s(1-s))$, see (6.12)

m_ζ multiplicity of a resonance at ζ, see (8.3)

μ Patterson–Sullivan measure on $\Lambda(\Gamma)$, see Section 14.1

$\mu_j(A)$ jth singular value of A, see Section A.4

$\mathbb{N}_0$ nonnegative integers $\mathbb{N} \cup \{0\}$

n_c number of cusps, see Section 6.1

n_f number of funnels, see Section 6.1

$N_X(r)$ resonance counting function, see (9.1)

ν_ζ multiplicity of a scattering pole at ζ, see (8.22)

ord_ζ order of a meromorphic function at ζ, positive for zeros

$P_X(s)$ Hadamard product over the resonance set, see (10.2)

$\Phi_X(s)$ regularized trace of $R_X(s) - R_{\mathbb{H}}(s)$, see (10.13)

$\pi_X(t)$ counting function for length spectrum, see (2.18)

$\Psi(s)$ digamma function $\Gamma'/\Gamma(s)$

ρ boundary-defining function for $\overline{X}$, see Section 6.1

ρ_c cusp boundary-defining function, see Section 6.4

ρ_f funnel boundary-defining function, see Section 6.4

$\mathcal{R}_X$ resonance set, repeated according to multiplicity, see Chapter 8

$R_X(s)$ resolvent $(\Delta_X - s(1-s))^{-1}$, see Chapter 6

$S_X(s)$ scattering matrix, see Section 7.4

$\sigma(z,z')$ $\cosh^2(d(z,z')/2) = \left[(x-x')^2 + (y+y')^2\right]/(4yy')$, see Section 4.1

$\tau_X(s)$ relative scattering determinant, see (10.4)

$\Theta_X(t)$ wave 0-trace, see (11.2)

$\Upsilon_X(s)$ regularized trace of $R_X(s) - R_X(1-s)$, see (10.23)

X hyperbolic surface

$\overline{X}$ compactification of X, see Section 6.1

χ Euler characteristic

$Z_X(s)$ Selberg zeta function, see (2.22)

Index